Building Knowledge-Based Systems for Natural Resource Management

Building Knowledge-Based Systems for Natural Resource Management

Daniel L. Schmoldt
H. Michael Rauscher

CHAPMAN & HALL

I(T)P An International Thomson Publishing Company

New York • Albany • Bonn • Boston • Cincinnati • Detroit • London • Madrid • Melbourne
Mexico City • Pacific Grove • Paris • San Francisco • Singapore • Tokyo • Toronto • Washington

Cover design: Andrea Meyer, emDASH inc.

Printed in the United States of America

Chapman & Hall
115 Fifth Avenue
New York, NY 10003

Chapman & Hall
2-6 Boundary Row
London SE1 8HN
England

Thomas Nelson Australia
102 Dodds Street
South Melbourne, 3205
Victoria, Australia

Chapman & Hall GmbH
Postfach 100 263
D-69442 Weinheim
Germany

Nelson Canada
1120 Birchmount Road
Scarborough, Ontario
Canada M1K 5G4

International Thomson Publishing Asia
221 Henderson Road #05-10
Henderson Building
Singapore 0315

International Thomson Editores
Campos Eliseos 385, Piso 7
Col. Polanco
11560 Mexico D.F
Mexico

Internation Thomson Publishing–Japan
Hirakawacho-cho Kyowa Building, 3F
1-2-1 Hirakawacho-cho
Chiyoda-ku, 102 Tokyo
Japan

1 2 3 4 5 6 7 8 9 10 XXX 01 00 99 98 97 96

Library of Congress Cataloging-in-Publication Data

Schmoldt, Daniel L. (Daniel Lee).
Building knowledge-based systems for natural resource management/
Daniel L. Schmoldt, H. Michael Rauscher.
p. cm.
Includes bibliographical references and index.
ISBN 0-412-01921-3
1. Expert systems (Computer science) 2. Conservation of natural resources--Data processing. I. Rauscher, Harold M. II. Title.
QA76.76.E95S34 1995
006.3'3--dc20 95-5099
CIP

To order this or any other Chapman & Hall book, please contact **International Thomson Publishing, 7625 Empire Drive, Florence, KY 41042.** Phone: (606) 525-6600 or 1-800-842-3636. Fax: (606) 525-7778. e-mail: order@chaphall.com.

For a complete listing of Chapman & Hall titles, send your request to **Chapman & Hall, Dept. BC, 115 Fifth Avenue, New York, NY 10003.**

CONTENTS

PREFACE

If one were forced to use a single key word to describe the decade of the 1980's, a very prominent one would be "technology." Leading the forefront of technology advancement were breakthroughs in electronics. Devices that were uncommon or unknown in 1980 became commonplace, and almost indispensable, by 1989. This trend has continued into the 1990's and it does not seem to be abating in any way. Microwave ovens, video recorders, telephone answering machines, compact disc players, computers, and a host of smaller or less sophisticated devices now appear in most households. The development of small and inexpensive computers, i.e., personal computers, has placed computing resources within reach of many more people. In addition, many traditional, and largely mechanical devices, have been enhanced by electronics. For example, specialized microprocessors are combined with arrays of electronic sensors to control and monitor sophisticated engineering components in most new automobiles. In this and many other ways, we are touched by the new electronics in almost every aspect of our daily lives.

Initially, personal computers were little more than toys. They contained only a small fraction of the computing power of their immediate ancestors, the minicomputers and mainframe computers. However, rapid improvements in integrated circuit design and chip manufacture produced regular reductions in size and cost of computer components. During the same time, processor speed and sophistication increased. By the end of the decade, desktop machines possessed the power of mainframes from a few years earlier. However, maybe more important than computer hardware advances, were the tremendous increases in the numbers and sophistication of software tools available for personal computers. As widely available personal computers have become more powerful, their usefulness has been extended to individuals that previously had little or no access to computing resources.

Computers have the potential to increase productivity in many sectors of society and, in particular, for natural resource professionals. Hardware advances, however, have only made computers more readily available and faster. *Powerful and functional software is necessary to make computers truly useful.* From the viewpoint of this book, the field of artificial intelligence has begun to provide software methodologies that can substantially improve productivity. This productivity manifests itself as an ability to do more things, to do them faster, and also to do things that could not previously be accomplished at all with the old technology. The way we view and the way we use computers is experiencing an ideological shift. Computers have become more like advisors and assistants, and less like calculators.

Knowledge is an important commodity. By producing more knowledge and by applying it more wisely, we are able to work smarter. Knowledge-based systems (KBS) have been shown to contribute to the application of knowledge. By using knowledge-based systems to organize and deliver knowledge in a useful format scientists and managers are able to deal more effectively with natural resource problems. The knowledge and expertise of a few individuals can be distributed to support and advise a larger number of less-skilled individuals.

OUR TREATMENT OF KNOWLEDGE-BASED SYSTEMS

Because the authors are convinced of the value of knowledge-based systems, one of our aims for this book is to help expand the number of people in the natural resource community with the skill and interest to build these systems. We hope that this will occur in two different ways. First, academic training might include some exposure to knowledge-based systems as part of existing or new natural resource management classes. And second, scientists and managers may develop and distribute their own systems based on knowledge that they have acquired in their particular area of specialty. In these ways, both existing practitioners and natural resource professionals in training can benefit from some understanding of this useful technology.

Despite the importance that this book places on KBS technology, it is only one aspect of the overall information technology that is critical for resource managers. Knowledge-based systems fit into a larger category of software and hardware tools that is often referred to as decision support systems (DSS). There are many other useful knowledge and information management tools contained in DSSs, for example, geographic information systems, data base management systems, operations research methods, simulation models, operations management tools, and experimental design methods, to mention a few. By focusing on KBSs in this text, we are neither ignoring nor minimizing the importance of broad-scope DSSs and their other components.

There are a number of different AI techniques that can be applied to building KBSs. One of the most common is the use of IF-THEN rules to represent heuristic human knowledge. There are other methods of knowledge representation, and there are also a variety of KBS control strategies and KBS architectures. While we briefly discuss many of these different techniques, most of our examples in the book use IF-THEN rules and a popular control method (backward chaining). We do this for three reasons: (1) the application example that we use throughout the text (RP-FMAS) employs these methods, (2) despite the demonstrated usefulness of the many other methods, it remains a fact that the majority of developed system use rules and backward chaining, and (3) our choice of Prolog as the KBS implementation environment further reinforces the use of rules and backward chaining, because they are included as part of the language. At least initially, most people find those two techniques easy to understand and implement.

THIS TEXT IN RELATION TO OTHER TEXTS

One of the reviewers for an early draft of this text noted that there are three different needs that might be fulfilled by a book of this sort. These are: (1) a description of knowledge-based system technology, (2) a tutorial on knowledge-based programming environments, e.g., Prolog, and (3) a case study of expert system development. Most texts only address one or two of these uses. We believe that each one of these is important and works in synergy with the other two.

What almost all AI texts fail to do, however, is to describe in detail all the steps that one must take in order to develop and implement a KBS. As an analogy, it is one thing to know about various building materials and about various construction tools and how they work, but it is an entirely different matter to know how to build a house. So, it is with texts that only contain case studies, or tutorials on programming environments, or descriptions of knowledge-based system technology. This book provides a brief review of many AI techniques in the earlier chapters, with the intent being to organize many of the techniques that are scattered throughout the literature. This review portion of the book is deliberately brief and does not cover any particular topic in great detail. The bulk of the text focuses, instead, on what steps should be followed to plan, design, implement, test, and deliver a completed system.

A large number of new ideas and unfamiliar techniques have blossomed from the work in AI and KBSs. Many literature sources contain very fine descriptions of these new methods. Consequently, we do not feel that it is necessary to reiterate what someone else has already done quite satisfactorily. There is not sufficient room in such a broad text to cover all methods in detail. Nor are we comfortable in declaring that one method is good and another method has no merit. For us to say that things should always be done "this way" or "that way" is an oversimplification of the state of the science, and of the world at large. Instead, *we have chosen to place as many of these new AI concepts as possible into a cohesive framework that describes how to build knowledge-based systems.* We have chosen to organize many of the ideas that other individuals have developed separately by creating a strategic KBS development paradigm within which those techniques can be placed. In doing so, we have intentionally abbreviated the presentation of some material and have provided references to better and more comprehensive descriptions in other sources.

USE OF THE TEXT

The book contains three logical sections. These are: (1) AI and knowledge-based systems—descriptions and components (Chapters 1–3), (2) system planning, design, and development—issues and techniques (Chapters 4–6), and (3) system implementation—programming in Prolog, prototype development, and system testing and delivery (Chapters 7–10). Sections 2 and 3 include a forest management example that we use to illustrate various ideas. Also, many other small examples are included from our experiences in other subject areas.

The chapters are arranged in order of difficulty. They also exhibit the logical progression of a tutorial on KBS development, from review material on KBSs to KBS planning and design to implementation, testing, and delivery. This organization is intended to target a wide and diverse audience, from forest managers to computer programmers, from domain experts to knowledge engineers. Forest managers and students may only desire a broad review of the concepts of KBSs, this is provided in chapters 1–3. Other texts are devoted to this preliminary material in their entirety, we have selected to highlight design, development, and implementation between these covers. Chapters 4–6 and 10 would be useful to those who are most likely to develop a KBS, advanced students and scientists. Those readers who actually implement a KBS may desire to use Prolog, chapters 7–9, as we have found many developers do.

We asked Don J. Latham, a colleague of ours, to present, as an appendix, a version of a talk that we have heard him present on several occasions. Appendix A documents that talk and it serves as a good introduction to the AI material present in the remainder of the text. While the tone of this appendix is light-hearted—almost whimsical—we feel that it is entertaining, and establishes several important points without being needlessly technical or obtuse. In his overview, Don briefly covers some of the historical, cultural, and philosophical backgrounds of artificial intelligence. Because he covers many of the AI fundamentals that appear in various components of knowledge-based applications, at least a cursory reading is recommended. A major theme of the appendix suggests that *artificial* intelligence is not a recent invention by scientists, but rather interest in AI has been a part of our culture and literature for many centuries.

Chapter 1 sets the stage for subsequent discussions of AI and knowledge-based systems. We present some of the unique features of natural resource problems and resource management that motivate the application of AI methods. A key aspect of our motivating discussion is the need to manage knowledge, i.e., to create, organize, and deliver it. Both AI and knowledge-based systems are defined in general terms. Benefits and limitations of this technology are also enumerated.

Chapters 2 and 3 investigate in more depth many of the knowledge-based system topics from Chapter 1. This thorough review of KBS components sets the stage for the system development methods presented in the remainder of the book. Included are the topics of knowledge representation, inferencing, control strategies, working memory, explanation facilities, user interfaces, learning capabilities, and interfacing to conventional programs. Chapter 3 also surveys some of the most commonly use methods for inexact reasoning (i.e. reasoning under uncertainty).

Any successful software development project requires careful planning. Some characteristics of a minimal system, personnel involvement, and construction techniques are items that need to be discussed by the development team prior to investing substantial resources. These topics appear in Chapter 4. There are several different development methods that can be used. Several of the common ones are introduced and a hybrid methodology is suggested by the authors. Regardless of which method is chosen by the developer, each requires a certain amount of parallelism among development steps. These parallel aspects of the development process cannot easily be realized in practice, however, they must be simulated by cycling repeatedly through the development steps. We feel that one of the best means for describing and explaining new techniques such as KBSs is through examples. To avoid concocting different examples for each aspect of KBS development, a red pine stand management example is introduced in Chapter 4 and is used extensively throughout the remainder of the text. This provides a measure of continuity from one KBS topic to another.

Chapter 5 covers the role of knowledge acquisition in the system development process. Depending on where the knowledge, that is applicable to a particular problem, resides, different approaches may be needed to capture and organize that knowledge for imbedding into a KBS. Surface (heuristic) and deep (fundamental) knowledge are the two categories into which subject area knowledge is often classified. Because most knowledge acquisition methods deal with elicitation from one or more experts, many acquisition methods are tailored specifically for eliciting knowledge from human sources of expertise. Elicitation methods selected may depend on how many people are involved in the project and how many experts are used. The chapter also examines knowledge acquisition from the standpoint of capturing different "levels" of knowledge, that is, factor, structure, or strategic knowledge. Certain acquisition methods are better suited to extracting knowledge from different levels.

Part of our system development methodology (Chapter 6) includes construction of two conceptual models. The first, a knowledge model, describes in detail the subject matter knowledge required to solve problems in the application. The second, a human factors model lies between the user and the problem-solving ability of the KBS. A number of pre-implementation issues are also discussed in this chapter, e.g. selecting a knowledge representation scheme and search methods and choosing development software.

Knowledge systems differ substantially from traditional computer programs. Several new programming paradigms have been generated by work in AI. These have resulted in several new computer programming languages and some sophisticated programming environments. Chapter 7 introduces the programming language Prolog. Some of the topics in this section may, however, be too advanced for readers completely unfamiliar with Prolog. Nevertheless, because Prolog will be used for many of the examples in Chapters 8 and 9, a cursory introduction is provided.

Two basic methods exist for implementing a knowledge system in Prolog. Chapter 8 presents the first of these—native Prolog. System development using native Prolog utilizes the backward chaining control structure and the quasi-predicate logic knowledge representation features of the Prolog language. An extensive example of this approach is presented using one of the modules from our red pine management advisory system. Implementation of a knowledge system must include three separate levels of knowledge. These are identified as domain, tactical, and strategic. Within the tactical level of knowledge, inferencing, user-interface, explanation, and uncertainty methods are implemented in this chapter.

In Chapter 9, the second approach to implementing a knowledge system in Prolog is used—meta Prolog. Much greater flexibility and additional system features can be added to an implementation by including an extra layer of Prolog. This extra layer of code allows the developer to include features that are not directly supported by Prolog itself. One powerful idea that follows from the use of meta Prolog is the creation of a toolkit of modules to implement various aspects of a knowledge system, e.g., facts, rules, inference, control, explanation, and inexact reasoning. The result is a collection of interchangeable and shareable pieces of code that can be used for many different applications by many different developers. Consequently, as a development environment, a toolkit lies somewhere between a programming language and an expert system shell.

Chapter 10 covers field testing and system delivery. Tests at this stage include distributing the system to users and recording their responses, and/or conducting experiments that compare system performance with that of an expert. Revisions suggested by the tests should then be incorporated into the final delivery system. Before delivery can be completed, however, documentation must be written and included with the system. This stage requires repeated interactions with the users to ensure that the software manual is informative and comprehensive. Also at this time, maintenance and updating responsibilities should be assigned, so that state-of-the-art knowledge can be added regularly.

We have organized the text to be flexible to many different readers' needs. Chapter 1 is optional due to its introductory nature. It does not contain any information that is essential to understand other topics in the book. The reader must grasp, however, the ideas cover in Chapter 2, whether they are learned from this text or some other. The more advanced KBS components presented in Chapter 3 are seen again in Chapters 8 and 9, but need not be understood to read most of Chapters 4–6 and 10. There is a logical flow through Chapters 4–6, however, none of these three chapters depends on any of the other two. Because chapters 7–9 deal with increasingly complex Prolog programming, Chapters 8 and 9 each require an understanding of material in the previous chapter and also require an understanding of the knowledge model (chapter 6) that describes our forest management example. Chapter 9 treats very advanced Prolog topics; only the most dedicated Prolog programmers may be interested in reading this chapter. For readers that have little or no interest in Prolog as an implementation language, Chapters 7–9 can be skipped entirely. Chapter 10 can be read immediately following Chapter 2 or need not be read at all depending on the reader's interests. In general, except for Chapters 8 and 9, the reader can skip to any chapter once the material in Chapter 2 has been digested. Therefore, each of the remaining chapters is independent.

The independent nature of the chapters provides extreme flexibility in teaching situations. An instructor can skip from chapter to chapter as particular topics fit into the workplan of the course. Undergraduate classes may only need to cover the material in the first two chapters, possibly supplemented by other readings that are more introductory in nature; this would give students a good feel for what the technology is without engaging in development. Graduate courses might focus on Chapters 2 and 3 along with several chapters on design, development, and implementation to provide a more hands-on treatment of knowledge-based systems.

TERMINOLOGY USED IN THE TEXT

Jargon permeates the literature on artificial intelligence. Neither terminology nor concepts are immune to the large volume of new perspectives on knowledge and intelligence that are emerging from AI. Many ideas are entirely new and innovative. However, even the ones that have a familiar ring often sport a new term to label them. This trend is not unusual, it occurs in most new fields. The rapid growth of AI, however, has accelerated that tendency to a confusing level.

Active researchers and developers of AI theory and applications struggle to keep abreast of the latest developments. Moreover, newcomers to the field are often overwhelmed by AI's excessive jargon. Many people express justified frustration at this state of affairs. As authors of a book in AI, we would dearly like to institute a wonderful and innovative solution to this dilemma. Unfortunately, the fact remains—serious students in AI *do* need to master a certain amount of the jargon. This situation is similar to visiting a foreign county, e.g., if you go to France unprepared to converse in French, you will be strongly disappointed by your experiences there, and will probably avoid France and French in the future. On the other hand, if you immerse yourself in the language, making some mistakes along the way, you will also experience some successes. You will come away from the experience with an appreciation for novel customs, culture, and ideas, and will have acquired new tools that are applicable elsewhere. The same can be said for learning some of the jargon of AI.

In this text we have adopted the following approach for dealing with the profusion of unfamiliar terminology. Many techniques and ideas are scattered throughout the AI literature. One of our aims in this book has been to bring many of those topics together in a meaningful and structured way. As new terminology is introduced we try to provide a reasonably precise, yet intuitive, definition for each term and also indicate how that concept or term relates to similar or competing ideas. Where space does not permit this much detail, we refer the reader to references that produce a more in-depth treatment of the topic. In either case, we position the jargon in relation to more general topics. By doing so, we hope that the reader can understand how everything fits, without necessarily understanding everything in detail.

Most of the terminology specific to AI has been highlighted in bold type face. Each highlighted term has a definition in the Glossary at the end of the book. While a brief glossary description cannot truly do justice to an important idea, we hope that this is satisfactory for the intents and purposes of this book.

ACKNOWLEDGMENTS

When giving due credit to individuals that have made a contribution to a project like this, there is always the danger that someone will be left out. Nevertheless, in full knowledge of this danger, we still feel that it is important to specifically recognize friends and colleagues who have donated their time and assistance.

Many people reviewed earlier drafts of chapters. They helped keep us on the right track and pointed out things that we left out of the book or left out of our thinking processes entirely. We thank Susan C. Barro, William G. Bradshaw, Richard L. Olson, Constance Rauscher, James M. Saveland, and Alan J. Thompson We also thank the anonymous reviewers that worked with our publisher to suggest many valuable changes to later drafts. There are also those people (some know who they are and some do not) that provided moral support and encouragement in one way or another. These contributions were no less important than technical and editorial assistance. Lastly, but certainly not least, we owe Gregory Payne, our editor from Chapman & Hall, a great deal of gratitude for his patience during, what must have seemed, an interminable authoring period.

There are always more topics that can be included in a text of this sort and there are always better ways to present the material. However, at some point we must stop revising and re-revising, and feel confident that we have done our best to produce something that can be useful to someone else. We hope that our intentions have been properly placed.

Example Programs

The example programs used for illustrative purposes in Chapter 8 and 9 of this book can be found at the Chapman & Hall FTP site (ftp://ftp.thomson.com/chapman&hall/schmoldt/). Chapter 8's files are rp_read.me, rp_file1.zip, and rp_file2.zip; Chapter 9's files are fm_read.me, forest.zip, plfile.zip, aspen.exe, and aspfiles.lzh. There is one set of files for the Red Pine Forest Management Advisory System using the native PROLOG programming approach discussed in Chapter 8. Please read the "RP_READ.ME" file for a description of each file and how to install these files on an IBM-compatible personal computer for execution. There is a second set of files for the Aspen & Red Pine (combined) Forest Management Advisory using the DSSTOOLS toolkit described in Chapter 9. Please read the "FM_READ.ME" file for a description of each file and how to install these files on an IBM-compatible personal computer. Both sets of files include the compiled executable version of the Advisory System software as well as the source code modules in ASCII text format so the reader may study the techniques discussed in the book in detail.

CHAPTER 1

AI and Natural Resource Management

In some sense, all subjects are really AI. All fields discuss the nature of man. AI tries to do something about it.

R. C. Schank

It has become increasingly common to hear or read news stories that involve the use or misuse of natural resources. Some examples are, for instance, the deforestation of tropical rain forests, timber harvesting and other activities on national forest (i.e. public) lands, or the loss of habitat for animal species. In most cases these stories include some controversy. Such disagreements arise for two primary reasons: (1) the complex character of natural resource phenomena and (2) the diversity of a resource's user group. This book addresses the first of these, while the latter deals with social, political, and economic questions of extreme complexity that are only beginning to be discussed in academic as well as legal circles.

The complexity of natural resources makes it difficult for stewards of the land to know everything about any particular resource. Even so, if we did have very accurate knowledge about the state of a resource and how it functions, we still may not have an understanding of how to put that resource knowledge into practice. Having knowledge about some resource and understanding how to use that knowledge are two very different matters.

As scientific progress helps us learn more and more about our natural surroundings, we are gradually becoming overwhelmed with greater amounts of information—information that can be applied to natural resource problems. The scientific community has been extremely productive in providing volumes of new information. Scientific inquiry has been very effective in generating information and knowledge by reducing complex problems into smaller and smaller pieces. These efforts follow naturally from the *reductionist* paradigm of scientific research. By *reducing* the scope of each scientific question to a narrower and narrower focus, the issue becomes easier to understand. While this strategy has allowed science to accumulate vast amounts of knowledge about many things, it does not address the difficulty of organizing and synthesizing such a large body of knowledge and applying it to real-world problems.

Traditionally there has been a smaller effort on the part of the scientific community to synthesize this information into knowledge that can be readily applied to problem solving. The ecological and environmental sciences are somewhat of an exception to the lack of holism in scientific investigation. But even in these disciplines there has not been a substantial effort to translate knowledge into application, i.e. solving management problems. Large collections of static information do little to help us answer questions about a dynamic world that contains many complex systems, each with many interacting components. We need to impose some structure on our accumulated information to make it work for us.

The unproductive use of what we know is becoming increasingly critical as more and more of our decisions about natural resources are interdisciplinary in nature. For example, to manage forests for the "good" of many publics requires the application of knowledge from many different fields, such as soil science, entomology, pathology, tree physiology, silviculture, economics, wildlife ecology, etc. This means that we must find some way to integrate the information about particular land resources together with our knowledge concerning various scientific disciplines.

To divide a resource into smaller components and manage each component independently is no longer a viable approach to providing products and services demanded by a growing and diverse population. More users necessarily create more and varied uses for our natural resources. Many of these uses may conflict to some degree. It then becomes much more difficult to manage any resource in an equitable way. A decision that affects one aspect of a resource, that is important to one interest group, frequently affects several other aspects of the resource. Hence, many additional interest groups become involved in the decisions that affect how our natural resources are managed.

In the remainder of this chapter we motivate and introduce the topic of knowledge-based systems as it may be applied to natural resource management. Although most readers know what natural resources are, we include an operational definition for natural resources and illustrate by example why their management is often very difficult. Knowledge is critical to the management of natural resources, therefore we expend considerable effort to present the origins, characteristics, and uses of knowledge. The last half of the chapter provides a brief overview of knowledge-based systems—their origin, structure, benefits, limitations, and application.

1.1 NATURAL RESOURCES

Before we discuss the management of natural resources it seems appropriate to provide some sense for what natural resources are. By **natural resources** we include *any physical or biological asset that occurs in nature and may possess some measure of value, tangible or perceived.* This definition of natural resources may also include resources that are not so *natural,* i.e. resources that are created or manipulated by people, e.g., a pine plantation managed for timber or a dammed river that produces a lake recreation area. Less human-influenced resources, such as wilderness areas, are also included. So, the term "natural" does not necessarily mean that these resources are untouched or unaffected by humans. There are very few places on earth that remain in this day and age untrammeled by people.

The ways in which we measure and valuate resources has a tremendous impact on how we manage them. Resources such as trees can readily be measured and can be related to timber commodities with substantial monetary value. Therefore, impacts of various management alternatives can be compared objectively using economic and silvicultural knowledge—their management is a straightforward task. Less tangible benefits, such as aesthetics and wilderness experience, however, are more difficult to quantify and measure, and their values are more spiritual and hence resist objective valuation. Our limited knowledge of how to inventory and monitor these resources further compounds our ignorance about what value to place on them. Consequently, our management of these intangible resources can only be haphazard and inconsistent, at best, and destructive and wasteful, at worst. Still, other resources, such as water quality, can readily be measured, but we don't always have good agreement on how to value them. In this case, our management activities are still hampered by ignorance about resource value, especially in relation to other, possibly, competing resources. Resource management efforts are always going to be controversial because we value resources differently, whereas better measurement skills can mitigate disagreements over what resources exist and to what extent they are present.

A particularly valuable feature of many natural resources is their renewability. The processes that govern the genesis, morphology, and functioning of renewable resources ensure continued production of those resources as existing stores are consumed. This seemingly limitless abundance of a resource does not come free because natural processes do not always preserve any particular resource to perpetuity. That is, resource A at time T_1 may eventually transform into resource B at time T_2. People, acting as stewards of these resources, must often intercede to guarantee production of those resources that are especially beneficial to humanity. Wisely managed renewable resources can yield a continual supply of resource values and human benefits.

In a larger context, chemical and physical natural resources such as oil, coal, natural gas, and geothermal energy may also be considered natural resources. However, most of the authors' experiences lie within the managerial aspects of the natural resources of a *forest* landscape. Our emphasis and examples throughout the book will reflect this bias, therefore. Nevertheless, the techniques discussed in this book are completely applicable to those other, non-forest resources.

1.2 RESOURCE MANAGEMENT

The essence of management is *synthesis*. To make an informed decision, a manager must have full knowledge, that is, all pertinent information about a subject area must be collected, analyzed, and integrated. These activities are governed by the aim to achieve some desired goal(s). When several goals conflict or compete, then management becomes much more difficult and demands that the manager know more about each resource to make "good" management decisions.

We do not define "good" here, but recognize that most people would agree that it includes principles such as, long-term sustainability, ecologically sound, and socially and economically palatable. *Good* management will be defined within the context of the goals and objectives that are established, and upon which management decisions are based. Any further discussion of this topic is beyond the scope of this book.

Management requires a holistic approach with a focus on problem solving. In contrast, scientific investigation often reduces a subject to smaller and smaller pieces for the purpose of understanding its inner workings. However, once the pieces are understood to some degree, they must be glued back together because that is how a system functions, as a whole and not a pile of disjoint pieces. Scientists attempt to reconstruct these pieces into a model that reflects the workings of the real-world system. For the manager, understanding how the real-world system works is only part of the task. The manager must go further and understand how to use this knowledge to achieve certain goals and objectives.

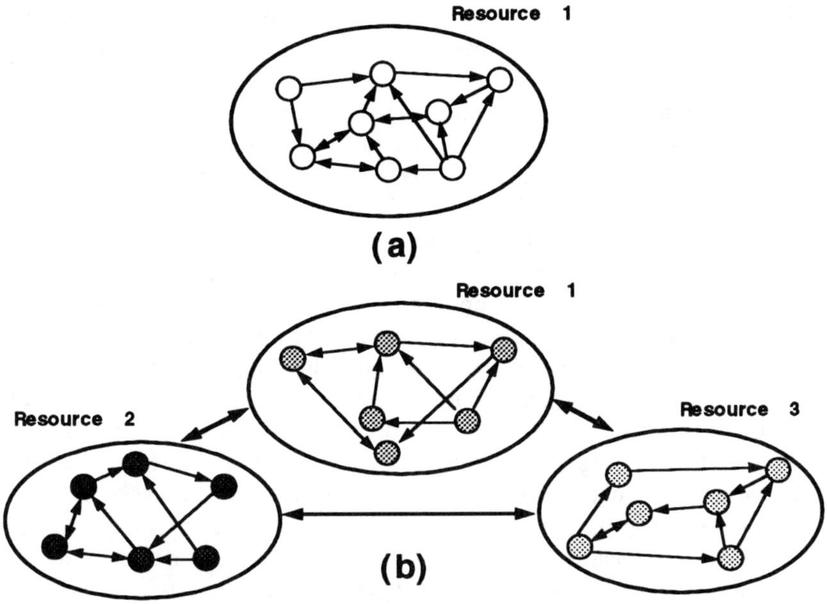

Figure 1–1. Two approaches may be applied to modeling natural resources. The first method (a) treats a resource as relatively autonomous to other resources; management decisions are based upon factors directly related to the resource of interest without concern for their impact on or by exogenous resources. Type (b) focuses more on the interaction between diverse resources for a holistic approach to management modeling.

Regardless of the resource with which we deal, each may be managed in two ways: (1) with a narrow-focused, isolated-system approach (Figure 1–1 a), or preferably, (2) with a broader concern for and integration of inter-related resources (Figure 1–1 b). In the first approach, a resource is treated as an independent system. Most factors pertinent to management exist internally; only a few exogenous variables, possibly representing management objectives, affect the system from without. This type of representation is not necessarily simplistic—it may in fact be quite complex. There may be very many important factors related to the resource which interact in very complex ways. A red pine plantation managed for wood pulp is a good example of how an isolated-system approach usually operates. Soils, topography, and climate local to the plantation are integrated with red pine silvicultural knowledge to produce the highest economic return for pulp production.

Each parcel of land contains many different resources. An alternative management approach (Figure 1–1 b) considers two or more resources simultaneously and addresses their cumulative effects and interactions. The effect of managing Resource 2, for example, has some impact on Resource 3 management. These considerations result in a management scheme of increased complexity, but it is also more realistic and complete. In the simplest case, if the separate components associated with each resource are somewhat independent, then designing and prototyping an abstract model for managing all the resources can proceed in a relatively modular fashion. As interactions with other resources become more numerous, management models necessarily increase in complexity.

An example of an integrated-system approach might be the management of an aspen stand for timber and wildlife values simultaneously. Under this scenario, silvicultural activities that improve the timber interests for the stand might have no, some, or severe implications for wildlife habitat. By the same token, the complementary situation exists for wildlife management actions that potentially might affect the vigor or merchantability of the aspen timber. Even something as indirect as building a road for timber harvesting access might fragment the habitat for some wildlife species, while at the same time permitting access for recreational hunting. Just considering two, relatively simple resources, the interactions can become both numerous and complex quite rapidly.

Overseeing the management of any resource involves manipulating some underlying processes to achieve stated objectives. The underlying process may not necessarily be one that explains everything about the resource. Such a process should provide, however, some explanatory mechanism which allows a manager to exert control on the resource to realize specific objectives. For example, for many purposes it can be sufficient to consider a stand of trees as a silvicultural system which reacts in certain ways to specific forestry practices. This particular process view allows control over the resource's outputs, but does not explain many things that some other process viewpoint, e.g. nutrient cycling, might. When only one process is operating and the objectives are few, it becomes a relatively simple matter to control the mechanism of interest to fulfill the stated objective.

In a natural resource situation, we are able to model the growth processes of trees, composition and growth of herbaceous vegetation, responses of animal populations to predators and food supplies, nutrient availability in the soil, above ground water movement and storage, and many other systems. Difficulties arise when we attempt to merge the macro with the micro, and the biological with the physical. The limitations of our scientific understanding about these interface areas become readily apparent. Resource management entails synthesizing assorted layers of interdependent, interdisciplinary mechanisms (Figure 1–2). As broader management objectives are defined and implemented, additional processes must be added and greater detail of existing processes must be understood and manipulated. Political scientists would argue that natural resource management decisions are social allocation issues, i.e. political decisions. Nevertheless, a thorough understanding of physical and

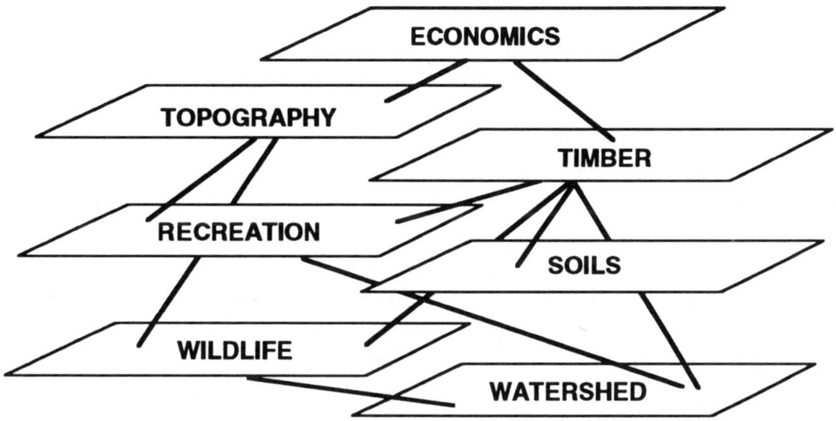

Figure 1–2. Management of a land base for multiple resource values involves consideration of complex, highly-interrelated components. Possible complexity has not been fully represented here, and often the individual components are not as modular as this illustration depicts.

biological systems becomes essential when making informed political decisions that impact millions of people.

1.2.1 A Common Management Scenario

To illustrate the difficulties associated with managing natural resources, we would like to introduce a brief, and entirely hypothetical, example. Despite the fabricated nature of our story, it addresses some very important difficulties associated with making intelligent decisions regarding a complex, highly interrelated set of natural processes and possible resource outputs.

Our case deals with the concerns of a moderately large manufacturing company, Outdoor Products Inc., which owns some large acreages of land. This land contains, among other things, forests, lakes, and streams spread out geographically over a large area. Outdoor Products Inc. employs three land managers to perform necessary stewardship duties; each has a designated geographical area of responsibility. Several political jurisdictions are represented and there is some adjacency to public lands and population centers.

The remainder of this example deals with one of the geographical areas and its land manager. Because this business enterprise comprises a monetary investment by many individuals (i.e. stockholders), there are strong incentives for Outdoor Products Inc. (OP) to produce profits from their extensive land holdings. Therefore, the economic concerns of OP's stockholders lead to an corporate emphasis on timber products which can be marketed to several local mills.

Stockholders' desires, however, represent only one aspect of OP's overall objectives for the land. Their economic concerns must be combined with the need to maintain an unobtrusive timber harvesting profile in areas near population centers. For tax reasons, it may also be desirable to provide for hunting activities—therefore, wildlife considerations are also of concern. Several of the lakes present on, and rivers flowing through, their land are also used by the public, whose interests in clean water and good fish populations must be addressed. Suppose further that adjacent national forests are managed by the USDA Forest Service and, not unrealistically, the agency has intense concern for periodic insect and disease outbreaks. Pest outbreaks do not stop at ownerships boundaries, however, they can, in many cases, be mitigated by proper silvicultural practices. It's apparent that simple financial interests can quickly become entangled with a host of other management goals. In our example, wildlife, esthetics, pest management, water quality, and fisheries are all exogenous to the organizational goal of supplying wood to mills, but it appears that these influences may be difficult to ignore or to treat as entirely separate.

The fate of OP's land base in this area depends on increasingly difficult management decisions for the designated land manager, as opposed to relatively easy single-objective timber management. Ultimate land management decisions in this multiple resource situation will need to include input from hydrologists, fisheries biologists, silviculturalists, entomologists, pathologists, wildlife biologists, and economists. Under the most desirable circumstances all these professionals will be in the employ of OP and also be available to help with subject matter questions. OP does maintain some of the required expertise in these specialty areas to solve these sorts of managerial problems. However, as we'll see, it's not always easy to apply this expert knowledge in sufficient quantity and in a timely manner.

As you might expect, OP scientists and professionals have responsibilities throughout the compartments of all three land managers. Consequently, these specialists may not be available in a specific location when their expertise is needed. Also, OP does not employ a hydrologist or a fisheries biologist, so these skills are lacking entirely; and their wildlife biologist retired several months prior. Selecting acreages of land to be cut for their timber harvesting operations is performed by a complex, mathematical optimization model. To provide the input data and to interpret the model output requires understanding that only one of the economists at OP possesses. Because each of these specialists has time constraints, the land manager must schedule time with the available specialists as their busy calendars permit. Absent expertise (hydrology and fisheries biology) may be obtained through consultant companies or may be ignored entirely. More often, some of the other in-house professionals must substitute for deficiencies in hydrology, ecology, and fisheries biology. Knowledgeable individuals are an extremely valuable resource. Their time constraints, scarcity, and transient nature, however, cause problems for organizations and, in a larger sense, for society as a whole.

1.2.2 An Alternative Management Scenario

In light of the topics to be covered in this book, we can imagine another possible set of options which might ease the predicament of this land manager. The management situation remains as before, but this time the manager has alternative avenues for procuring expertise. Now, the knowledge of consultants and company specialists resides in computer programs as well as in professional people. These software advisors are readily available, don't have busy schedules, and don't retire.

In this alternative reality of the OP circumstance, much of the private knowledge of OP's specialists has been recorded previously. Before the retirement of their wildlife biologist, a large portion of her lifetime of experience was transcribed and represented on computer. It remains as a functional record of her long-acquired knowledge. Similarly, the knowledge of OP's silviculturalists, pest specialists, and economists have also been recorded. In anticipation of the need for specialists not employed by the company (hydrologist and biologist), OP contracted for acquisition of other advisory support software. Now when land management problems arise a preliminary or alternative source of know-how may be exploited in the absence of, or in concert with, human experts.

We don't wish to imply that these programs make decisions for the manager, instead they provide technical expertise on complex and difficult problems. Final decisions must still be made by someone with organizational responsibility. Those determinations, however, can be more timely and effective when essential expertise is at hand. An ancillary benefit is that the human experts have greater time to devote to challenging problems rather than day-to-day decisions which oftentimes may seem, from their highly-trained perspective, to be mundane and repetitive. They enjoy additional freedom to pursue creative solutions to new problems without constant interruptions.

An advantage of this approach, that may not be immediately obvious, is that *knowledge has been multiplied* (more people have access to and "possess" it), making it more available and, hence, more useful. The following section creates a more complete picture of knowledge and why it is important.

1.3 KNOWLEDGE: A HISTORICAL FRAMEWORK

To place the importance of knowledge in perspective, we must realize that it is one of the three basic human resources: food, energy, and knowledge (Branscomb 1979). The greatest breakthrough in food production occurred about 12,000 BC. when, after 2.5 million years of biological and social evolution, people learned how to farm. Farming brought relative food security to many, leisure to a few, and led to specialization allowing increases in social complexity. The defining breakthrough in energy production occurred sometime between 1704 and 1705 AD. when an English iron monger named Thomas Newcomen developed the first steam engine. After major improvements by James Watt, it resulted in the first entirely man-made energy producing system the world had ever known (Burke 1978). The Industrial Revolution had begun.

We have not yet experienced an equivalent breakthrough capable of providing cheap, instant, useful knowledge to all individuals on earth. As pointed out in our example above, available and useful knowledge can be very beneficial for natural resource management. It is claimed that the electronic computer, invented in the United States during 1937-1942 (Mackintosh 1988), is the key knowledge management device (Branscomb 1979). Most likely it is only part of the answer. The breakthrough we are looking for will not occur until we have the "right" computer software. This software should evolve into a suite of personal knowledge management tools (providing support for the intellect) as a consequence of research and applications in many fields such as computer science, telecommunications, artificial intelligence, etc.

1.3.1 A Formal Definition of Knowledge

Defining knowledge is difficult. We often use the terms *data*, *information*, and *knowledge* relatively interchangeably. Nevertheless, we'd like to take a slightly more formal approach to these concepts here. From a practical viewpoint, knowledge is a set of assertions and relationships which can be used to reason about the world and to solve problems. In cognitive psychology, however, **knowledge** is formally defined as *justified true belief* (Giere 1984). Below is a rule for determining what it means to know something.

FORMAL RULE OF KNOWLEDGE CERTAINTY
  IF A person BELIEVES a statement is true
  AND A person is JUSTIFIED in so believing
  AND The statement is actually TRUE
THEN A person KNOWS the statement is TRUE
  (Certainty Factor = 100%)

However, our observational abilities are not always faithful to us; occasionally we perceive something as true when, in actuality, it is not. Consequently, a statement can be false, and knowledge does not exist. If a person believes a true statement without proper justification, knowledge also does not exist. We need to be aware of what beliefs we have and ask ourselves whether they are

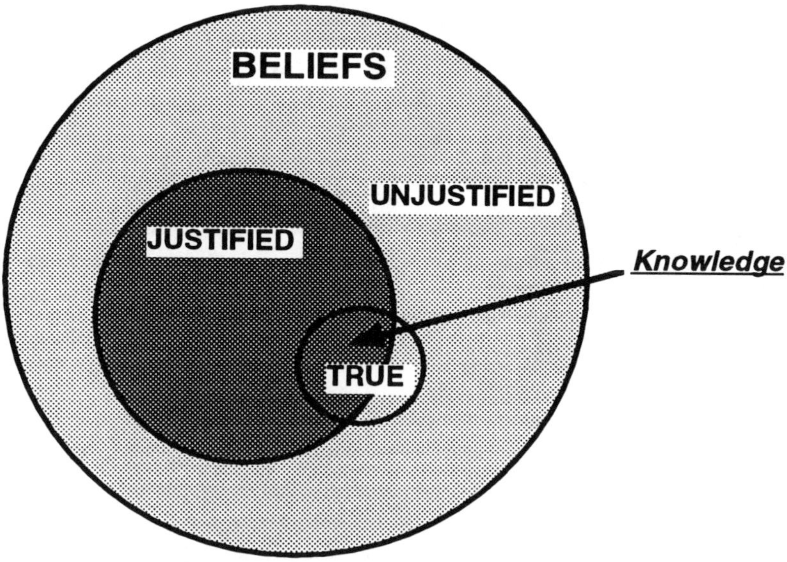

Figure 1–3. Human knowledge contains both justified and unjustified beliefs. Justified beliefs that are also TRUE constitute knowledge (dark area within TRUE circle). Unjustified TRUE beliefs result in uncertain knowledge (lightly shaded area within TRUE circle).

justified. In Figure 1–3, beliefs are partitioned into justified and unjustified sets. Of those beliefs, both justified and unjustified, some are actually true. Those BELIEFS that are JUSTIFIED *and* TRUE constitute knowledge.

It's possible to believe a TRUE statement, but not have any justification for it—for example, being told by someone that a particular statement is true. You may consider that person's testimony to be justification, but generally, direct observation or rationalization is the only valid justification. The degree of certainty we have in our knowledge may legitimately be associated with a strong belief or a strong justification or both, as long as we claim only a strongly "justified belief" (Giere 1984). It is furthermore true that most knowledge in natural resource management and science is conditional, not absolute. We know few things to be absolutely true.

1.3.2 Knowledge Versus Data and Information

Data may be defined as symbols representing undigested observations and relationships. Data are unorganized and unevaluated. For example, in Figure 1–4 numerical data are tabulated for two quantities, the height, H, and the diameter, D, of some trees. These height and diameter measurements are

Height	Diameter
20.4	5.0
25.1	5.5
26.1	6.5
23.5	5.4
27.1	6.2

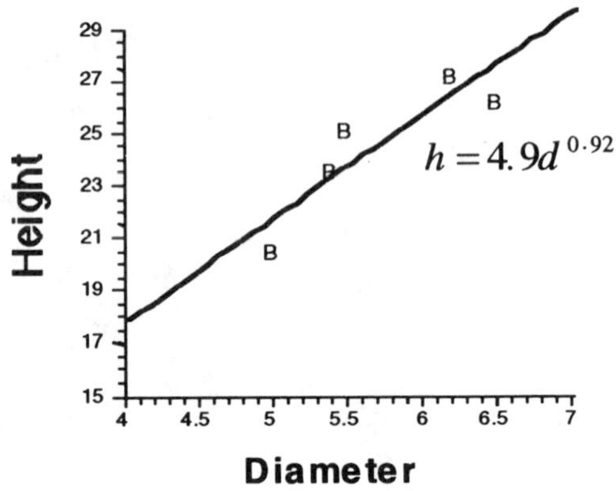

Figure 1–4. A table containing tree measurements shows that data items alone possess little usefulness. A graphic display of those values and the mathematical relationship imposed on them give the data regularity, i.e. organization.

evaluated only in the sense that, given careful measurement techniques, we may be relatively sure that they are reliable estimates of the quantities we intended to measure. That is, we are reasonably sure that the value 6.5 is not 6.4 or 6.6. Also, the values are unorganized, except that they represent corresponding measurements from members of a population, and that our measurement scales and instruments have imposed some order on them. But, beyond measurement reliability and pairwise association, these values have little meaning for us.

In contrast, **information** may be defined as a collection of data that has at least some level of organization. The second portion of Figure 1-4 organizes the two columns of data by providing a mathematical relationship describing the data, i.e., $H = 4.9D^{0.92}$. The graphical display of these data on a Cartesian coordinate system is also a way of organizing data, and it aids our understanding due to its visual nature. We often impose some logical construct on natural phenomena, in this case it is mathematical. In doing so, we are implying that this mathematical pattern mirrors some pattern in the physical reality of trees. Information remains unevaluated however, in the sense that we don't know whether to believe in its validity (TRUTH value) or not. In this example we merely have an empirical relationship with no support for believing it.

Information is converted to **knowledge** when we develop a justified belief in its truth value. In terms of this height-diameter relationship, justification would consist of some causal or biological support of our beliefs—some justification that says trees grow in height and girth simultaneously, and in the regular, mathematical way of Figure 1–4. Corroborating empirical evidence from other populations, using a similar mathematical relationship, might also be used to justify our belief. Once we have justified true belief in this height-diameter relationship, then we can use this knowledge to imply similar relationships for other tree populations and possibly for other organisms.

Data organized into information, which goes into the formation of knowledge, may be likened to clay formed into bricks used in the construction of a building (Forscher 1963). Information, as meaningful data, is like a brick of a particular kind. Data are like the clay used to make a brick. Bricks can be used to form many different structures leading to different functions. The critical difference, however, is one of structure (Bronowski 1973). Consequently, how we structure our knowledge greatly affects the things that we are able to do with it.

One of the most important aspects of the structure of knowledge is its *applicative* functionality. That is, knowledge not only contains relationships between facts and information, but it also contains *specifications* for their use (Luger and Stubblefield 1993). For example, conditional "if-then" rules "tell" us to verify specific antecedent conditions if we want to draw a particular conclusion. The following if-then rule tells us how to use two pieces of data to determine that any particular object *A* is a tree seedling.

IF *A* is a tree,
AND *A* is less than 4.5 feet in height,
THEN *A* is a seedling.

Other forms of knowledge representation (Chapter 2) have different applicative mechanisms.

1.3.3 Knowledge and Thinking

The ability to consolidate and organize mental impressions (experiences) as part of goal-oriented planning is a very sophisticated capability, one that few other organisms besides people possess to any degree. Certain collections (or groups) of mental impressions can by experiential association assume a collective identity, that may be represented abstractly in our minds as a single concept with a specific name or **symbol**. For example, after several exposures to a hairy animal with four legs that barks, we are able to form a concept for that object with the label "dog". In fact, the concept "hairy," itself, is a labeled concept that is based on experience with many different objects that have hair—similarly, for "bark" and "legs." It's then possible to link these symbols together into a *structured* network of symbols, i.e. we are doing symbolic processing. By linking the labeled concepts, "bark," "hairy," and "four legs," together we can form a new concept that we label "dog." **Learning** is the process of establishing and revising the *content* and *structure* of these symbols, which have been called "chunks" (Harmon and King 1985, Hofstadter 1979, Newell 1972).

The term **chunk** is often used to mean a unit of data or information or knowledge, but at the lowest neuro-biological level, a chunk may really be a group of neurons in our brain which recognizes some object or stores some memory fact. Hofstadter (1979) proposes a hierarchical organization of the brain in which the lowest-level chunks are interconnected through various higher-level chunks. These higher-level chunks represent more abstract ideas or concepts. In turn, these higher-level chunks are also connected to other chunks at even higher levels of cognition. The result is a *tangled hierarchy* (Hofstadter 1979) of symbols, each with some underlying neuro-physiological structure and a behavioral association to the outside world.

In another sense we can envision a knowledge chunk as analogous to a paragraph. It is composed of a primary concept, analogous to a topic sentence, and the necessary secondary concepts to support and develop the primary concept. A good chunk has unity, coherence, and adequate development. Unity means singleness of purpose. Coherence refers to the joining of the constituent concepts into a logical, understandable pattern. Chunks may be composed of data, information, or knowledge and may themselves be the data, information, or knowledge for other chunks. Learning may be described as the process of creating new chunks and linking them with appropriate other chunks into an organized knowledge structure.

1.3.4 Knowledgeable People

Knowledgeable people are called **experts**. Experts perform better than non-experts because they have a large amount of "compiled", domain-specific knowledge stored in long-term memory and they have become skilled at using that knowledge to solve specific problems (Harmon and King 1985). A world-class expert is said to have between 50,000 and 100,000 chunks of knowledge stored in long-term memory. When young, we begin to organize our experiences into chunks centered around objects, either concrete or abstract. As we learn more, we cluster more and more knowledge around successively more abstract chunks. But no matter how complex our chunks become, our ability to manipulate them seems to stay about the same. Experts ordinarily emerge after about 10 years or more of study and work in their fields because it takes that long to acquire, chunk, and organize the necessary knowledge and experience.

One of the distinguishing characteristics of knowledgeable people (experts) is the *inaccessible* nature of their knowledge. Their knowledge is inaccessible (Hofstadter 1979) in the sense that an expert encounters difficulty when relating the how's and why's of his or her knowledge to others. Much of an individual's knowledge is *implicitly* (Barr and Feigenbaum 1981) attached to and integrated with prior experiences and ways of doing things. It is situational knowledge that is very difficult to extract when it is taken out of context.

This phenomenon has also been referred to as the *compiled* nature of expertise. The characterization of human expertise as *compiled* arises from an analogy between knowledge and computer programming languages. A compiled program is useful because it can be executed on some machine to perform some computation, i.e. to solve some problem. If one looks at the compiled code, however, it's just machine code (0's and 1's), and it's impossible to understand how the program's logic works. If we can "reverse compile" it back into the source code of some programming language which we understand, then the rationale becomes much more apparent. It is similar with human expertise. A human expert's knowledge is extremely useful for problem solving, that is, for applying ("executing" in computer science terminology) to some tangible problem. However, it is generally unavailable to the expert for communicating to others; the expert experiences difficulty introspecting his or her knowledge. They are generally unable to reverse compile their knowledge in a way that others can comprehend. This represents the crux of what is referred to as the knowledge acquisition *bottleneck* discussed in Chapter 5.

1.4 ARTIFICIAL INTELLIGENCE

A small, but active, branch of computer science called Artificial Intelligence (AI) has been working for nearly 40 years on problems associated with knowledge—its acquisition, representation, and utilization. We have proposed that the results of those AI efforts can be applied to Outdoor Products' need for advisory software in the scenario above. AI has created a new philosophy of information processing and has produced some useful techniques to help make the computer more like a consultant rather than solely as a calculator. These ideas portend major changes in how we'll eventually view our own intelligence and how we'll perform useful work in the future.

There are always problems associated with the introduction of a new technology. One of these problems is the tendency to view a new technology within the perspective of an old technology (Hill 1989), hence, terms like "horse-less carriage" for the automobile, "mechanical scribe" for the printing press, and now "artificial intelligence" for smart machines. Naming a new technology with the verbiage of an old one encourages people to interpret the new as very similar to the old, where often it is not. A new technology often originates from a new philosophy that cannot be adequately grasped using the terminology of the past and its associated culture. A disadvantage of this mal-interpretation of a technology is that it clouds our understanding of it, and then limits our utilization of its full potential (Hill 1989).

Commentators on AI converge upon two goals they see as defining the field:

1. to better understand the human mind through specifying and testing computational models, and
2. to construct computer systems that do things traditionally regarded as requiring some intelligence.

Whereas these two goals generally cover the bulk of AI work to date, there are new areas of AI investigation that aren't easily categorized as either. These new endeavors align generally under the topic of *intellect-support tools*, that is, the establishment and construction of new computation-based representational media in which human intellect may come to express itself with different clarity and force (Hill 1989). If we can use the computer to help us express our ideas more effectively and extend them more easily, then the computer becomes an extension of our own intellect and, hence, non-threatening. In some sense, this new role can be viewed as functionally analogous to human language: it gives us the capability to represent, communicate, and record our mental states.

Given the context of this book (i.e. knowledge-based systems) and our discussion of knowledge, above, we can perhaps offer another goal for AI. Knowledge can be treated as a tangible product separate from its storage in the minds of humans. Therefore, *AI can better help us understand how to create, organize, manage, and deliver knowledge* (McRoberts *et al.* 1991).

It is beyond the scope of this book to delve very deeply into the question, "What is intelligence?" There are many philosophical issues that can leave the inquirer between a rock and a hard place. Nor do we care to entertain questions like, "Is artificial intelligence possible?" The applicability and validity of AI technologies speak for themselves by touching our lives almost every day. For the purposes of this book, we accept AI methods as useful tools, and our aim then becomes to demonstrate how to use them effectively for building knowledge-based systems.

1.4.1 Work in AI

Three important themes continually reoccur in AI (Barr and Feigenbaum 1981; Stillings *et al.* 1987); they are knowledge representation, search, and learning. **Knowledge representation** includes an investigation of the types of knowledge required by a task, their acquisition, and their storage in the computer. **Search**, which is composed of inference and control methods, explores the selection of alternative moves (or states) toward a problem solution; it also selects strategies and resources necessary for solving a problem. **Learning** concerns the ability of a system to improve itself, i.e., perform a task better with experience. An even more difficult aspect of learning is the ability to accomplish a task which has not been previously seen—this requires abstracting essential aspects of prior, related experiences.

The field of AI encompasses a broad range of topics. Many of them can be described as engineering oriented, such as, robotics, vision systems, speech recognition and synthesis, and knowledge-based systems. In engineering these systems, an attempt is made to mimic human interaction with, or cognition of, the outside world. These engineering endeavors are in contrast to scientific AI research (pure research) to investigate psychological, linguistic, or philosophical aspects of human thought processing. Some of these basic research issues are knowledge representation, inexact reasoning, learning, natural language understanding, and cognitive modelling. Even though a science versus engineering distinction is made here, all these topics are closely tied together. Science provides a solid foundation and a set of techniques that system engineers can apply to specific problems. On the other hand, while building applications, engineers must often extend old techniques and develop new ones. Also, engineering's final products provide the economic incentive to continue science-related studies.

In the most general sense, themes in this book are centered around engineering applications of AI (i.e., construction of working systems) with emphasis on KBSs. Reasoning, problem solving, understanding, and knowledge structure represent some of the cognitive issues to be discussed and will serve as the basis for the application methods presented.

1.4.2 Motivation for Knowledge-Based Systems

Very generally, the impetus behind building knowledge-based systems is to help yourself or someone else solve a problem. That objective should not imply that this is the exclusive reason. Starfield and Bleloch (1986) suggest that like all models, the primary purpose of KBSs should be to help the user think about his or her problem. Tanimoto (1987) feels that AI instills new vigor into all fields it touches because of the essential questions it asks about a subject's knowledge.

Much of science's ability to understand and manipulate the world around us has been understood and stored by certain individuals (domain experts) by means of their experiences. The loss of an individual with substantial private expertise forces society to re-learn knowledge that has not been transmitted to others. By recording and storing the extensive knowledge of a specialist (through the use of KBSs), we can have more continuity between the loss of one expert and the development of another, who must fill that void.

Many authors have suggested that knowledge-based systems can provide a practical way to help organize and transmit problem-solving expertise based on research knowledge (Buech et al. 1989, Coulson and Saunders 1987, Davis et al. 1987, Rauscher 1987, Schmoldt and Martin 1986). Knowledge-based systems seem to offer a noteworthy benefit that print technology cannot deliver. Feigenbaum and others (1988) estimate that the average enterprise in the United States spends only one third of its operating budget on "touch" type of work, such as cutting down trees and hauling them to market. Roughly two thirds is spent on "non touch" activities such as deciding when and where to cut trees. Therefore, major productivity gains can be achieved by increasing the ability of the user to make decisions—planning, problem solving, teaching, learning, and advising. The essential improvement that knowledge-based systems are expected to offer natural resource managers, scientists, and students is a way to apply the power of automation to management decision-making processes (Feigenbaum and others 1988).

Within any group, organization, or society there exists a distribution of task performance (knowledge) across individuals. Figure 1–5 depicts three possible scenarios for this distribution. In the first case, all or most individuals within the population exhibit some acceptable level of ability for a task. Knowledge-based systems are not useful here for increasing the average level of knowledge; they may, however, be useful in other ways as noted above. When no one in the population possess a great deal of skill (Figure 1–5b), then it will not be possible for a KBS to increase overall skill level because there exists no model of good performance. Figure 1–5c differs from 1–5b by the existence of a skilled group within the population. Consequently, a model of good performance is present and a KBS may be built to capture the knowledge of this group for the benefit of the less-skilled performers in the population. Even if worker performance cannot be improved, there may still be other good reasons for creating a KBS.

The above paragraphs describe several possible motivations for developing a knowledge system. These are:

1. provide a focal point for thinking about the subject area and asking questions,
2. provide a mechanism for technology transfer, and.
3. enhance the productivity of an organization.

There may be other reasons that are specific to a particular circumstance, but these three reasons motivate most KBS development efforts. The following paragraph provides one final, natural resource specific justification for building KBSs.

Increasingly, a more interested and involved public must be satisfied with the process and management of natural resources. To meet this need, decision support systems for ecosystem management have been elevated in importance (Rauscher and others 1995). Decision support systems (DSS) are interactive, flexible and adaptable computer programs, specially developed for supporting

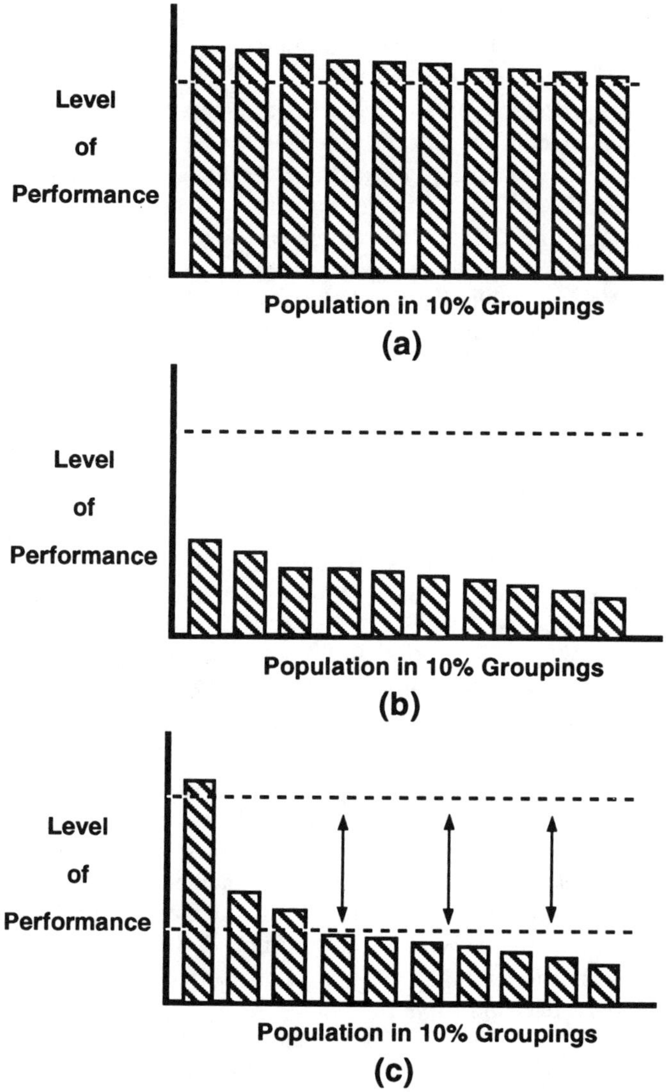

Figure 1–5. Each of the three graphs contains a hypothetical population separated into 10% performance groupings. Each groups performance can be compared along the vertical axis. A dashed line across the top of the graph indicates some acceptable level of skill. All population groups function at or above the acceptable level in (a), all groups function well below acceptable in (b), and a single group functions at the acceptable level in (c). In terms of increasing overall population performance, KBSs would be valuable in (c), where there exists a skilled group and also the potential to raise the mean skill level of other groups (lower dashed line).

the solution of a particular management problem. They facilitate improved decision making. They utilize spatial and non-spatial data, provide a friendly user interface, and integrate the decision maker's own insights as part of the decision making process (Turban 1993). The most sophisticated DSSs also utilize simulation models, include knowledge-based systems, and support all phases of the decision making process (Sage 1991). In contrast, to date most knowledge-based systems for natural resource management have been developed primarily to perform as stand-alone computer programs addressing important but very small parts of the entire ecosystem management decision support field (Rauscher and Hacker 1989, Davis and Clark 1989, Durkin 1993). The use of KBSs as an important component of natural resource management DSS provides another powerful motivation for learning how to build and apply knowledge-based systems successfully.

1.5 KNOWLEDGE-BASED SYSTEMS

A variety of terms have been used to reference AI applications in decision making and problem solving. Some of these are knowledge-based systems, expert systems, inference systems, and rule-based systems. **Expert systems** seem to be the most common for historical reasons. This label, however, carries with it many undesirable connotations (e.g. human-like, surrogate expert, human replacement, or know-all) which we would like to avoid if possible. We prefer the terms **knowledge-based systems** (KBS) or **knowledge systems** (KS). This terminology is more consistent with the construction, "knowledge engineering", and use, "knowledge management", of these programs.

This book focuses primarily on *knowledge-based systems*, their application, construction, and use. The Outdoor Products example, created above, was intended to provide a quasi-realistic situation in which computer-aided decision making could be beneficial. We will now further define and describe knowledge-based systems and how they differ from more traditional ways of providing computer-based decision support.

1.5.1 KBSs and Expert Systems Contrasted

Expert systems have been defined as *computer programs which solve problems that would require expertise if solved by a human* (Harmon and King 1985). The knowledge of an expert system consists of **facts** and **heuristics**. Facts constitute a body of information that is widely shared, publicly available, and generally agreed upon by experts in a field (*public knowledge*). Heuristics are mostly private, little-discussed rules of good judgment (rules of plausible reasoning, rules of good guessing) that characterize expert level decision making in a field (*private knowledge*). Heuristic knowledge distinguishes expert systems as a specialization of knowledge-based systems, wherein some of the system knowledge has been acquired from the experiences of one or more human experts.

In contrast, knowledge-based systems may or may not contain expertise. Non-expert, knowledge-based systems may perform tasks which do not require capturing and representing human expertise directly and explicitly (Mills, 1987). Instead, they may assist with the execution of some complex process—a system to guide the use of a complex simulation program would be an example, or the collection and organization of some body of knowledge from the scientific literature, e.g., an encyclopedia of flora fire effects (Fischer, 1987). Such systems contain knowledge of a particular domain, but may not necessarily include human expertise explicitly or capture heuristic, problem-solving methods.

The concept of a knowledge-based system emphasizes the *structural* nature of computer programs. That is, knowledge about a particular domain is represented and stored separately from the remainder of computer program code. Of course, knowledge is an inherent component of every computer program—or else the program won't function very well. *But, unless the knowledge in the program has been extracted into a separate entity (often called a knowledge base), the program is not a knowledge-based system.* A strong distinction is made between *knowledge* about a particular problem (referred to as domain knowledge) and the *algorithms* used to manipulate that knowledge for problem solving (Nau 1983).

1.5.2 A Structural Definition of KBSs

Figure 1–6 illustrates the basic components of a knowledge-based system and their interactions. This is a rather sketchy diagram that will be expanded in later chapters. An expert system's structure is also captured by this diagram, but for the remainder of this text we will adopt the generic term *knowledge-based system*. A KBS operates in two ways: first, to acquire knowledge, it interacts with a human expert and/or a knowledge engineer to create a knowledge base. A knowledge engineer and a domain specialist together provide a representation for the application-specific knowledge and strategies for solving problems using that knowledge. And second, in production mode, the system provides advisory consultation to end-users who have particular problems to solve.

During knowledge acquisition, knowledge must be acquired and deposited in the knowledge base. This if often accomplished by reviewing the literature, interviewing human experts, or incorporating existing management models, process models, problem-solving methods, simulation models or statistical methods. Some appropriate knowledge representation scheme must be selected

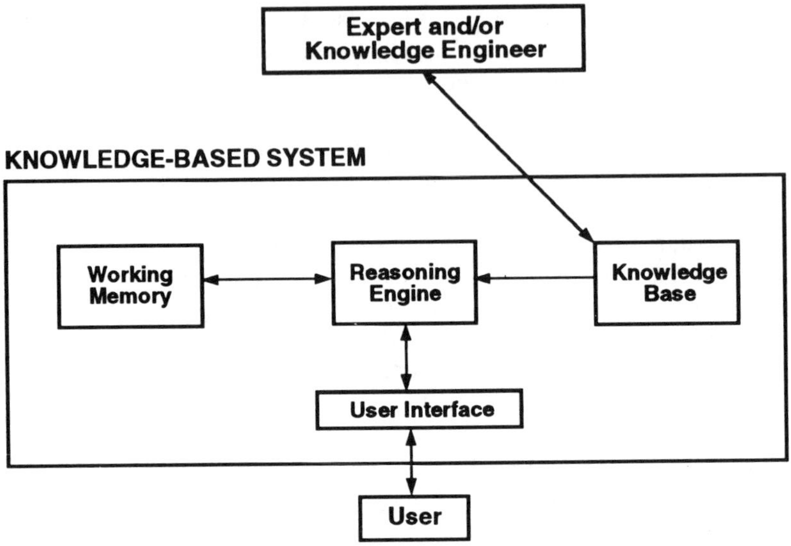

Figure 1–6. The basic components of a knowledge-based system are depicted here, arrows indicate information flows. The two modes of operation, knowledge acquisition and consultation, are shown at the top and bottom, respectively.

depending on the type of knowledge and how it will be used in decision making. Chapter 2 describes various knowledge representation methods and chapter 6 provides some guidelines for how one might select which one(s) to use for any particular application. Once a knowledge-based system contains some knowledge about a subject area, users may interact with the system to solve particular problems.

Consultation with the system usually occurs through some user-interface program, which may be menu driven or may incorporate some sophisticated natural language capability or, possibly, a graphical user-interface. Facts about a current consultation problem are provided by the user and entered into the working memory. The reasoning engine then attempts to validate other facts from working memory facts and the knowledge base "understanding" of the application area. By seeking to validate new facts, the KBS is generating and corroborating hypotheses. Flexibility and user-friendliness of the system depend on how each of these components are implemented. The *power* of knowledge-based systems exists, however, in the knowledge and its representation (Schmoldt and Martin 1986).

From a theoretical perspective, this program organization has some very nice advantages. Because program code (knowledge-manipulation algorithms) and knowledge are separated, operation of the program becomes more easily visible and comprehensible. Subject-matter knowledge is no longer embedded in code, therefore examination and modification of the knowledge can be accomplished more easily. Secondly, for a wide variety of applications, there may be no dependency between particular knowledge bases and the other system components. Consequently, a general purpose KBS (or **shell**) can be useful for many completely different types of problems. Different systems can be created by substituting one knowledge base for another and re-using the remainder of the system (reasoning engine, user-interface, and working memory structure). This *potential* modularity has been promoted as one of the major advantages for adopting a knowledge-based approach and provides the basis for producing and using expert system development tools (see Chapter 6 for a discussion of this software development approach).

1.5.3 Qualitative (Symbolic) Modelling

In many circumstances, managers make decisions which are non-quantitative, i.e., decisions whose primary output or result is not a numeric quantity. For example, selecting a tree species to plant on a particular site or deciding which insects may eventually cause damage in a stand of trees are qualitative decisions. These decisions are characterized by their ability to answer questions like "What?", "When?", or "Where?", rather than "How much?". Stocking levels and timber harvest levels are *how-much* type questions. *How-much* questions are very important in natural resources, but they are usually less important than "what" type questions. Consequently, qualitative models, that answer questions qualitatively, can help a manager solve many problems. Some of the basic differences between quantitative and qualitative modelling are summarized in Table 1.

Because the questions asked of qualitative models differ from those of their quantitative counterparts, methods by which answers are formulated differ also. Quantitative questions can usually be answered in some algorithmic manner, provided that all the pertinent variables are measured and the modeled process can be described mathematically. "What" type questions often require an experienced practitioner to interpret scant and loosely structured information and arrive at some solution through inexplicable and intuitive judgment. Starfield and Bleloch (1986) remark about knowledge-based systems in wildlife and conservation management, "... modelling can be qualitative instead of quantitative and might lend itself particularly well to the kind of information usually available to managers and to the types of decisions they often have to make."

As opposed to the use of numerical values and continuous-valued functions characteristic of mathematical models, qualitative models deal with symbolic representations of phenomena. Often these phenomena are understood in discrete quanta only. Although many biological phenomena could be described mathematically if understood in sufficient detail, our current level of scientific

Table 1–1. Several attributes of quantitative and qualitative models are compared.

Attribute	Quantitative Model	Qualitative Model
Basic computational unit	Numbers	Symbols
Solution process	Mathematical	Heuristic
Form of output	Numerical relationship	Symbolic description
Output range	Continuous	Discrete
Interpretation of output accuracy	Precise	Uncertain
Explanation of output	Hidden in the mathematical models	Trace of reasoning available
Applicability	Well-understood problems, wealth of empirical data	Poorly-understood problems, little data available
Problem-solving representation	Idealized mathematical relationships	Intuitive, subjective notions

knowledge may only permit understanding at a more macro level—at a larger grain size where relationships appear more discrete. From a pragmatic standpoint, it may be questionable whether a highly detailed mathematical description is superior to a useful and more intuitively understandable model which operates at a grain size that corresponds to available information and ultimate decisions.

When we use a mathematical model and receive a single numerical answer as output, there's a strong sense of power associated with that result. A mathematical result seems very true and correct because we've been taught throughout our education that mathematics does not lie. But, in fact, the answer is not dependent on the mathematics. It is generally more sensitive to incomplete models or limited data, pitfalls which mathematics cannot ameliorate. The false sense of accuracy which analytical models convey belies incomplete understanding of "true" (but unknown) mathematical relationships. A number implies accuracy which oftentimes may be no valuable improvement over an educated guess. On the other hand, qualitative models don't obscure the fact that much of their capability relies on subjectivity, uncertainty, and heuristics. Instead, they emulate our *modus operandi* in the real world.

When an answer is produced by a model, it's more reliable, from the user's standpoint, if some explanation accompanies that answer. Understanding the result produced by a quantitative model requires that the user has some familiarity with the underlying mathematics and understands the model assumptions. On the other hand, qualitative models, being symbolic, have an organization that allows the construction of explanations from the knowledge they store and from a record of the deductive steps taken. A user with access to this type of output can infuse his or her own understanding into the decision process by agreeing or disagreeing or learning.

Lest our intent be misunderstood, we are not adamantly opposed to numerical models—they are most certainly useful and we do not discourage their proper usage. Oftentimes, however, they require large amounts of data from statistically sound studies. These are not available in many cases or may require many years to obtain. In the mean time, however, it may be possible to use qualitative models as a precursor to quantitative ones. If we have some general understanding of particular relationships, that describe trends between different components of a resource problem, we could model those trends qualitatively (D'Ambrosio 1989). Then, by changing some variable in the qualitative model, we could observe whether other variables in the system increase or decrease (Puccia and Levins 1985). In many situations, these results alone can be very informative and may even suffice for most practical uses. As more hard data become available from research studies, the model can be extended to include quantitative relationships. In this way, our utilization of what we know does not suffer while basic research catches up with our resource management needs.

Knowledge-based systems, as a particular case of qualitative models, help make subjective judgment and intuitive rationale more explicit and less enigmatic. In the process, imperfect expertise and faulty rationale become apparent and are subjected to revision. It is an undeniable fact that human models of real-world events are potentially biased and error prone (Cleaves *et al.* 1985). Nevertheless, if we can identify their systematic nature, there's no reason to treat them as inferior to analytical models and mathematical relationships.

1.6 SOME POTENTIAL BENEFITS OF KNOWLEDGE-BASED SYSTEMS

Many *potential* benefits have been proposed for the application of knowledge systems in science, and to society as a whole. Many of these benefits are relate to the management of our scientific and technical knowledge (Rauscher 1987, Schmoldt 1989, Schmoldt and Rauscher 1994). Three key **knowledge management** ideas are:

1. knowledge should take an explicit and tangible form, much like the consumer product that it is,
2. knowledge should be organized and interrelated to all other pieces of associated knowledge so that we are aware of what we know and what we don't know, and
3. knowledge should be delivered in a usable form so that all interested persons—rather than a few "inner circle" disciples—have ready access to it.

The remainder of this section delineates some of the benefits of building and using KBSs. Some of these potential benefits are speculative in nature, but others follow directly from the construction of, or from the use of, knowledge-based systems.

Most knowledge found in textbooks, handbooks, research papers, and other "public" sources address general principles and make recommendations for the average case. Knowledge-based systems are designed to translate this general understanding into a solution process for specific problems. Differences between routine situations and those requiring more analysis or data collection or attention by a human expert can be identified and treated appropriately (Starfield and Bleloch 1986). The centuries old gap between the producers and the consumers of knowledge may be substantially reduced.

KBSs can be used to bring together much of the heuristic knowledge in a particular domain in a formulated, testable, and maintainable way (Cooney 1986). KBSs provide an explicit record of expertise to ensure some protection from catastrophic loss and to provide a concrete record of current decision strategies. This can be especially desirable when decisions may be interpreted as controversial in nature. Also, strategies used in solving problems and making decisions become less mysterious. *Artificial* expertise provides a focus for upgrading and improving management strategies over time (Starfield and Bleloch, 1983). As new information is produced, it can be incorporated into our understanding (knowledge) of that field, and lead to new strategies for the application of that knowledge in the real world.

Not every day is a good day and not everyone works in top form all the time. The day-to-day variation in proficiency, that most people exhibit, is undesirable in critical decision-making situations. Knowledge systems are consistent from one day to the next, unaffected by day-to-day biases. Human experts or other decision makers can utilize this reliability to perform checks on their own skills. The explicit record provided by a KBS affords the decision maker a measure of accountability—a means for, at least partially, justifying decisions. Regular consultation with an advisor program might help decision makers avoid omissions of crucial decision elements and also, over time, validate the performance of the system. Many KBS users have commented on the usefulness of these systems for reminding them of critical decision elements.

It is a continuing problem that human experts retire from their occupations after a lifetime of accumulating expertise. If human experts are allowed to depart an organization without proper care to nurture their replacements, the power to solve problems can erode, adversely affecting productivity and competitiveness. It's not always possible for specialists to transfer their expertise to younger colleagues, largely because experts' time is highly demanded from others during their tenure. KBSs provide one way to slow the institutional erosion of knowledge and to retain some of this hard-earned expertise for future use.

KBSs are currently available that enable employees to perform tasks that previously required close supervision or access to knowledgeable senior specialists (Harmon and King 1985). By capturing and distributing knowledge, knowledge systems can be used as assistants by people with less than expert skills in some area and as training vehicles for novices (Bowerman and Glover 1988). New employee transition periods may be shortened and the concomitant loss of productivity, that traditionally accompanies employee turnover, could be greatly reduced. Continuity can be maintained despite changes in staff.

Knowledge-based systems can be valuable in domains where our understanding of cause and effect is incomplete or totally absent. Knowledge in these domains tends to be based on experience and observation primarily. Experience-based knowledge is often difficult and costly to acquire and organize. Often it cannot be verified for accuracy, it is difficult to know which cases it applies to, and in what circumstances it does apply to a case. It also provides only probabilistic results, not certainty (Sell 1985). These vague and poorly-understood applications were discussed above and qualitative modelling was suggested as a means of utilizing the limited knowledge that exists in those fields.

One of the primary advantages of the knowledge system approach is the ability to create systems which fuse the knowledge of several separate disciplines (e.g. Schmoldt and Martin 1989, Rauscher 1989). Breadth of expertise endows KBSs with much greater potential to solve complex problems than human experts who are often knowledgeable only in a single domain. In a sense we can theoretically create "super" experts that have a greater breadth of knowledge and can address a wider range of problems.

Once a knowledge-based system has been built, it becomes relatively inexpensive to expose large segments of the society to this expert knowledge, making expertise much more universally accessible and usable (Cooney 1986). The most current knowledge becomes available for many users rather than the few individuals acquainted with the human expert. A previously centralized resource becomes distributed for democratic application by a wider audience.

While no single system built will do all these things, as a class, KBSs have the potential for some of these benefits. For any system to be useful it should provide at least one of these uses. The following list summarizes the benefits we have identified above:

1. a compendium of applicative knowledge may be stored,
2. catastrophic loss of expertise may be prevented,
3. some accountability for decision makers can be provided,
4. decision become less enigmatic and more open to analysis,
5. decisions are more consistent and less biased,
6. KBSs can be used as a checklist to avoid omissions,

7. the institutional erosion of knowledge may be lessened,
8. KBSs can handle poorly understood domains,
9. the knowledge from several disciplines can be combined,
10. KBSs can provide for more equal distribution of knowledge,
11. decision making can be more democratic, everyone may have access to the important knowledge, and
12. KBSs can reduce training time for inexperienced personnel.

Overall there is little doubt that the world's demand for experts will continue to far outreach the supply. We're all in need of all the knowledge sources we can get (Michie and Johnston 1985).

1.7 LIMITATIONS OF KNOWLEDGE-BASED SYSTEMS

The previous section painted KBSs in a very favorable light, indeed. Nevertheless, they are not without their limitations. KBSs, being part of and reliant on AI, have all the same limitations associated with the inability to field truly intelligent AI systems. As AI develops and continues to create more sophisticated and intelligent systems, KBSs will improve to the point where many limitations may eventually disappear. Bowerman and Glover (1988) note several categories of KBS weaknesses:

1. Systems function in a subdomain of the full human potential, i.e. cognitive and logical. They are not adept in managing sensory input or output, or dealing with less analytical aspects of human reasoning abilities, e.g. creativity, intuition.
2. The total knowledge available to a human expert appears in many forms which KBSs have no understanding of, e.g. social knowledge, analogical reasoning, random memories, feelings, emotions, common sense, and irrational information. Their problem-solving skills have a very narrow focus.
3. KBSs degrade almost immediately when confronted with situations outside their domain. The ability to recognize and recover from errors is non-existent.
4. Human intellectual development relies extensively on learning abilities; few current systems have any capacity for learning.
5. KBSs exhibit no sense of self; therefore, they possess no understanding of what they know—or what they don't know.
6. The current state of scientific understanding limits KBS capabilities; they can acquire no new knowledge. Some systems are able to solve problems more effectively than humans because of their exhaustive search methods, but they have not increased their knowledge in the domain.

Limitations do not imply any theoretical defect, currently apparent, that would prevent KBS developers from eventually overcoming these problems. This technology remains in its infancy. Vigorous research in AI has as its goal the solution of several of these problems; in particular, learning methods, alternative logics, and common sense reasoning are receiving substantial effort. The critical breakthrough for solving many of these problems may likely involve some capacity for learning (Schank 1987). Learning will enable systems to recognize familiar situations and react quickly, to adapt to new and unfamiliar circumstances, and to expand their narrow focus to larger, more general problems.

A knowledge-based system can only reasonably be expected to incorporate in its reasoning that part of the domain that human experts would regard as routine (Sell 1985). Obviously, what is routine to a medical doctor or a physicist or a beekeeper is probably far from routine for the rest of us. Many of the current knowledge systems—while very competent in their specific subject area using their narrow problem-solving skills—are not as good as the best human experts or even the "average" human expert, when one considers the full range of capabilities that a human brings to bear on a problem. Jackson (1986) observes, however, that like their human parents, KBSs need not be right all the time in order to be useful. While such inconsistency may create difficulties for system users, the unavailability of expertise, that will likely persist without the use of KBSs, may be a more widespread and troublesome problem than periodically incorrect KBS decisions.

1.8 SUITABLE APPLICATION AREAS

Many authors have attempted to identify application problems that seem appropriate for a KBS approach (e.g. Casey 1989, Prerau 1985, Slagle and Wick 1988). This has been done to help developers avoid applications that may be doomed to failure from the start. The following characteristics are often used to help identify appropriate applications for KBSs:

1. largely symbolic processing demanded,
2. combinatorial problem—put together a large number of primitive elements in many ways,
3. the scope of the problem is limited,
4. it can be solved in a 1/2 to 1 hour consultation with an expert,
5. general agreement must be available by experts,
6. the problem must be important,
7. a library of test cases is available,
8. incremental progress/development is possible.
9. solution approaches do not contradict established social/political norms.

These criteria are not completely determinant or excluding. There are very few applications which satisfy all these conditions. One of the challenges of knowledge engineering is working around the limitations imposed by a particular application problem.

To avoid the pitfall of working on a KBS application that is doomed to failure for one of the above reasons, the development team can produce a feasibility study that carefully analyses many questions regarding the viability of a KBS approach for a particular problem. Chapter 4 covers the planning phase of a KBS development project and suggests that a feasibility study can be a very important pre-development activity. Feasibility assessment might include an evaluation questionnaire which can be used to survey parties regarding both technical and non technical aspects of a KBS project (Laufmann and others 1990). Chapter 4 discusses this approach in more detail.

Probably the most common application of KBSs will be as expert systems, where human expertise will be captured in a computer program and distributed to others. They will also be useful as "intelligent" front ends or back ends to complex analytical tools (Hahn 1985, Martin 1980, ÒKeefe 1985, Olson and Hanson 1987)—for example, statistical software, large data bases, geographic information systems, analysis of data, or the operations research model used by the economist at Outdoor Products. KBSs may also be employed as large-system subcomponents which might perform discrete, qualitative modelling as part of larger quantitative models (Rauscher 1989). KBSs can be used effectively to replace conventional manuals of all kinds from repair manuals to reference manuals to manuals of rules and regulations. They can feature user-friendly, mixed-initiative interaction more useful than hard copy formats for new-situation learning (Schmoldt and Bradshaw 1991). The broad applicability of the knowledge base paradigm ensures that there will always be problems which are amenable to this approach.

Allan Turing pleaded for the creation of programs that can learn and must in fact be taught by humans much the same way humans teach each other. It would be very desirable to write such a program with an empty knowledge base and then start teaching it in a very small domain so it could gain expertise quickly (Michie and Johnston 1985). Such an accomplishment would open the door to powerful learning systems and, in a short time, to systems with general intelligence as a consequence of their learning abilities.

1.9 SUMMARY

Understanding and managing natural resources can easily become a very complex problem. However, just knowing the various factors involved and how they are interrelated does not guarantee that someone will be able to make rational management decisions. There must also be some interpretation of these factors, some knowledge of how they associate with desired states of the world, and some capability to manipulate controlling variables for desiderata. Certain individuals possess this knowledge in terms of mental models, they are called "experts." Knowledge can also reside in other more public forms, e.g., textbooks and mathematical or computer models. Regardless of where knowledge is found, it remains a valuable commodity, it is scarce, tenuous, and often mysterious to the untrained.

Throughout history, philosophers have engaged in the study of human intelligence. Advances in computer hardware and software now permit a more experimental investigation of this field. They are now joined by computer scientists, anthropologists, linguists, and psychologists within the scientific study known as artificial intelligence.

Artificial intelligence has as one of its goals the creation of intelligent agents (computer programs). These programs can supplement and, in some instances, replace expertise that might otherwise be unavailable. Expert systems and, more generally, knowledge-based systems constitute one area of applied AI where the ideas of knowledge representation, search, and learning take form in advisory computer systems. These systems can be viewed as models, but in a much different sense than traditional mathematical models. By applying heuristic knowledge to a symbolic representation of a problem, KBSs are able to answer *what* types of questions. They can avoid some of the undesirable aspects of numerical answers and they possess the ability to explain their reasoning steps.

KBSs are applicable to a wide variety of real-world problems and can provide many social and organizational benefits. They don't represent a panacea because they do exhibit some limitations in application and functionality. The fact that KBSs and AI methods are firmly infused into current software ensures that these ideas do not represent a passing fad.

In addition to the works cited in the above text, there are numerous other books and articles covering the philosophical bases of AI. Seminal contributions by Herbert Simon (Simon 1969, Simon 1979) and Marvin Minsky (Minsky 1969) are highly recommended. More advanced essays by Dennett (1985, 1986) and Pylyshyn (1986) delve into tough philosophy, psychology, and cognitive science issues. On the other hand there are a large number of less technical books that provide some good background (e.g., Johnson 1986, Moravec 1988).

CHAPTER

2

Knowledge-Based Systems: Representation and Search

Knowledge must be represented before it can be manipulated.

Gregory Lazarev

In Chapter 1, we briefly introduced the idea of knowledge-based systems and included a simple structural representation for such a system (Figure 1–6). This chapter and the next will focus more closely on those ideas and introduce new ones using the more comprehensive knowledge-based system architecture in Figure 2–1. Examples of specific implementations of these concepts (using the Prolog language) can be found in chapters 7, 8 and 9.

As noted in the Preface, it is not our intent to provide detailed coverage on all the topics presented in this chapter and the next one. Rather, our aim is to organize and synthesize what others have done, to create a comprehensive framework within which most KBS topics can be placed, and to suggest how these techniques can be applied to real system development projects. This is particularly evident by our emphasis (in terms of numbers of pages) in chapters 4–10 on planning, design, implementation, and testing. For elaborate detail on particular topics in chapters 2–3, there are many other reference books and papers available. We have tried to reference the appropriate sources as much as possible, and have not attempted to duplicate the explanatory detail of other authors. Readers are referred to those sources for explicit technical details.

The following sections explain the major components of a KBS in more detail. Topics included in this chapter are knowledge representation schemes, knowledge-base systems architectures, working memory, and inference and control mechanisms. The next chapter presents the remaining components of a knowledge-based system: explanation, the user and program interfaces, learning, and extending inference methods to include uncertain knowledge.

2.1 AN OVERVIEW OF KBS ORGANIZATION

The structure of a knowledge-based system may contain many component parts (Figure 2–1). Not all of these elements are necessarily included in any one system. We've illustrated, however, a relatively comprehensive collection of components. The essential components were introduced in Chapter 1: working memory, reasoning engine, knowledge base, and some user interface facility.

It should be noted that Figure 2–1 and other similar figures in this chapter and the next chapter portray a KBS system as the central component, with other components optionally connected to it. KBS primacy in these figures reflects the focus of this text and is not intended to suggest any relative importance for KBSs versus these other technologies or imply any preferential organization for decisionmaking software in general. In fact, a knowledge-based system could instead be one of the appended components with, for example, a geographic information system as the unifying central component (q.v., Reynolds et al. 1995). This latter arrangement may actually be a more appropriate organization for decision problems that rely greatly on spatially referenced data.

The process of running a knowledge-based system is commonly referred to as a "consultation", alluding to the analogous scenario of consulting a human expert. The **reasoning engine** is a computer program (consisting of algorithms) that navigates through the knowledge base using inference methods and control strategies to solve specific problems posed to it. The **knowledge base** contains the domain-pertinent knowledge. A **working memory** serves as the work area (scratch pad) of the inference engine where all current assertions are stored. **User interface** software enables the system to elicit input from the user—often in terms of answers to questions—and to display results, images, and screen reports. The **explanation system** along with the user interface constitutes the **human-machine interface**. Explanations may answer "why" a question is asked of the user, "how" an answer has been derived, the meaning of terms, the intent and source of references for the knowledge in the knowledge base, and other knowledge-specific informa-

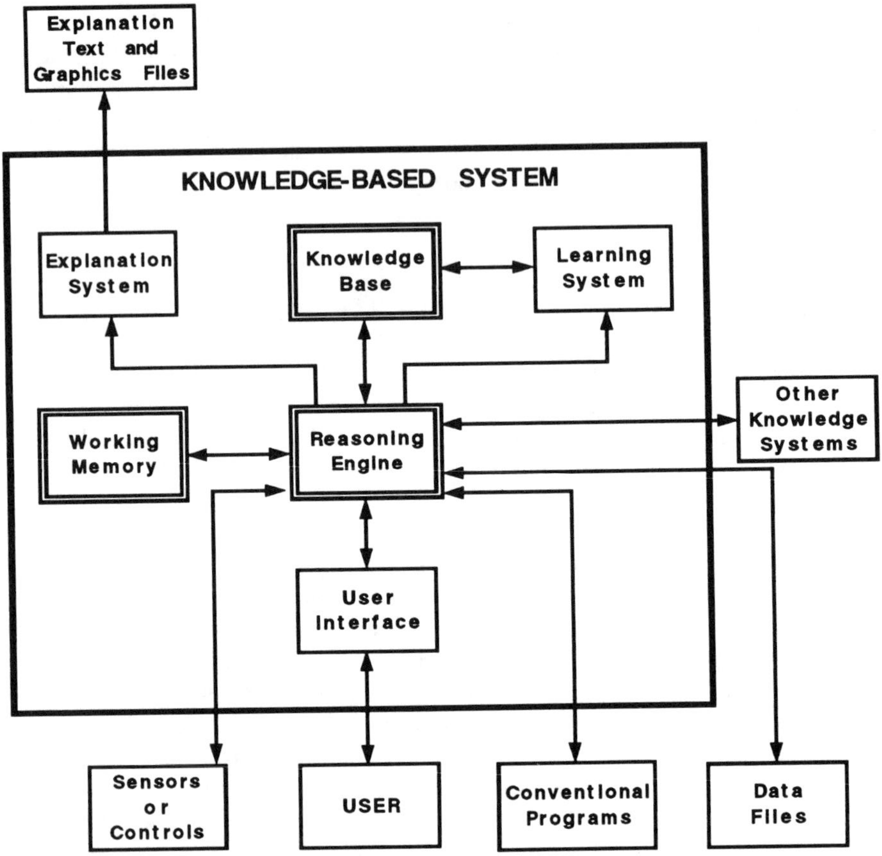

Figure 2–1. A typical knowledge-based system used for consultation might include some of these structural components. However, components with double borders are essential for any such system. Aspects of the system that are related to knowledge acquisition are not included in this rendition.

tion. A KBS may *input* data from simulation models (external programs) or database/spreadsheet files as well as *output* data to simulation models and change database/spreadsheet files. A KBS may also include connections to external devices, in which case input comes from sensors or output goes to electromechanical controls. Finally, when a KBS finds a dead-end in its chain of logic (i.e. a point in the line of reasoning from which no solution can be reached), a **learning** component might report this finding to the user and help the user remedy the situation by making changes to the knowledge base during the running of the consultation.

2.2 THE KNOWLEDGE BASE AND KNOWLEDGE REPRESENTATION

Consider for a moment how we humans think about things in the real world. Because the real world exhibits a high degree of complexity and a tremendous amount of interrelatedness, in order for us to reason about phenomena around us we must necessarily internally represent those things in some way. For any portion of the real world on which we may focus our attention, there often exists many different perspectives, that is, different ways of looking at things, different, but equally valid, models of things and events. Each unique perspective may lead to a different mechanism for understanding and reasoning about the world. Consequently, a single representation or model cannot usually capture all aspects of real-world objects and events that we would like to reason about. This use of multiple perspectives applies to knowledge representation in KBSs, also. There are different ways to represent knowledge, and under different circumstances one knowledge representation scheme may be preferred over others.

A knowledge base (Figure 2-2) contains the scientific knowledge and experience for a particular area of expertise. In addition, it may also contain some **common sense** or world knowledge, knowledge which is not specific to the subject domain. For example, to diagnose automobile engine malfunctions it is important to know about valves and spark plugs and batteries and carburetors, etc. (domain knowledge). On the other hand, it is fairly common knowledge that there must also be gas in the fuel tank for the engine to function properly. Most current KBSs only contain small amounts of this "common sense" knowledge.

One of the fundamental problems of KBS research and development is how to represent large amounts of knowledge in a fashion that permits its effective use (Waldrop 1984). This is known as the knowledge representation problem, one of the hardest problems so far identified in AI (Jackson 1986). **Knowledge representation** is a systematic way of codifying what knowledge exists about some subject area (or domain). It deals with the structuring, encoding, and storing of information so that its attributed meaning is clear. This representation of knowledge must be such that it can be stored and manipulated by a computer. Codification formalizes knowledge, contributes to the repeatability and reliability of judgments, and helps establish consistency among a group of specialists (Bowerman and Glover 1988).

The form of knowledge representation varies from one system to another (Waterman and Hayes-Roth, 1978; Barr and Feigenbaum, 1981; Buchanan and Duda, 1983; Nau, 1983). The major knowledge representation methods developed from research in AI are (Bowerman and Glover 1988):

1. rules
2. frames & scripts

3. objects
4. networks
5. logic.

In addition, there are many other methods for encoding knowledge, some of which are relatively conventional and others are new (Rauscher and Host 1990, Jeffers 1991, Salton 1991):

6. databases (e.g., historical cases)
7. spreadsheets
8. computer algorithms
9. hypertext
10. mathematical equations
11. non-language sources (i.e., pictures, sound, smells, and other sentient means)

The AI research literature displays little agreement as to the strengths and weaknesses of the various methods of representing knowledge (Jackson 1986). Neither is it clear which type of representation is best used under what circumstances (Bowerman and Glover 1988). There is some indication that rules are the most frequently used knowledge representation system in the development of KBS. For example, one survey determined that 62% of knowledge-based system developers used rules to represent knowledge (Doukidis and Paul 1992). Some successful systems use a mixture of knowledge representation schemes while others rely on only one. We provide some guidance for selecting knowledge representations in section 6.2.1.

To deal adequately with the world, an intelligent agent often uses many different representations to display the necessary understanding of objects and to organize their relationships to other objects (Fischler and Firschein 1987). An ability to switch between different models and perspectives when problem solving illustrates one of the critical aspects of human creativity. While choosing a uniform knowledge representation method makes a knowledge-based system easier to develop and understand, it does so at the cost of inefficiency in execution and difficulties in making all knowledge conform to one representation (Jackson 1986). Like truly intelligent agents, KBSs should ideally contain multiple representations for entities that are part of their domain of understanding. It should be recognized, however, that using multiple knowledge representations in a knowledge-based system requires proportionate increases in development and run-time complexity. There is no free lunch!

The following sections briefly explain several methods of representing knowledge and discuss the known strengths and weaknesses of each.

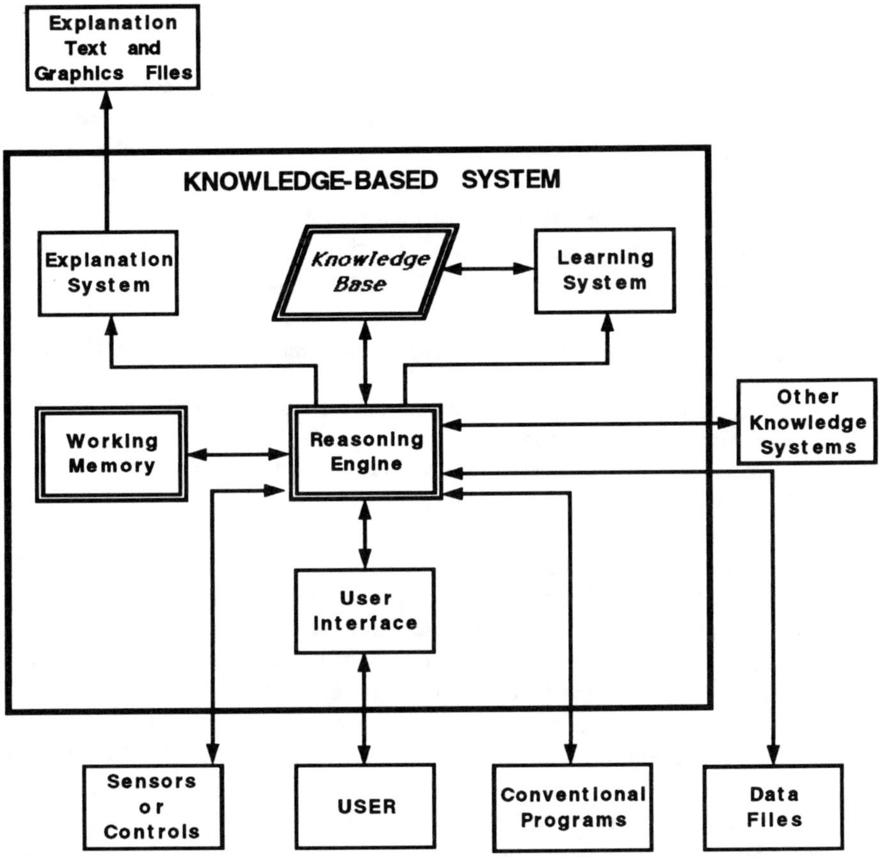

Figure 2–2. The knowledge base contains the domain specific knowledge required to solve problems in a particular area of expertise. Because knowledge representation methods are seminal to the understanding of KBSs and are critical to their effectiveness, we introduce them first in this chapter.

2.2.1 Rule-Based Knowledge Representation

The most widely applied type of knowledge representation is the **rule-based** approach (Buchanan and Duda 1983; Doukidis and Paul 1992). At their inception rule-based systems were often referred to as **production systems** (Newell and Simon 1972), and often still are today. Historically, production systems originated as a model of computing developed by Post (1943). Most rule-based representations today use a variant of the general production system model in which *productions* are represented as conditional if-then rules. A rule base encodes knowledge about an area of expertise (domain) as a set of condition/action statements of the form:

IF **condition(s)** THEN **action(s)**,

where the condition(s) are assertions in working memory. The specified action(s) is (are) performed if the antecedent condition(s) is (are) present (asserted as true) in working memory. Often the action performed will be a modification of the facts or hypotheses in the current working memory context. It's also possible to have the action evaluate another set of rules or execute some external program to perform some function. For example, the following rule is from the PREDICT (Schmoldt 1987) knowledge base:

IF The stand location is a presence county for *Gremmeniella abietina* (scleroderris canker),
AND Dead buds are present lower than six feet in the tree OR Branch dieback is present lower than six feet in the tree,
AND Green discoloration is present beneath the bark on affected stems,
THEN There is suggestive evidence (0.7) for *G. abietina* as the causative agent.

If all three antecedent conditions in this rule are true (i.e., they are found or matched in working memory), then *G. abietina* is hypothesized as the damaging agent. When this happens, we often say that the rule has "fired." The value, 0.7, indicates the strength of the rule, and its interpretation and use varies between systems (see 3.5 Inexact Reasoning). This strength value assumes that the user is certain that the premise statements are true. A reduction in the strength of the conclusion would result from less than certain premises.

Often rules are grouped into sets which perform some specific function. Figure 2–3 depicts the organization of rules for the PREDICT knowledge base (Schmoldt and Martin 1989). For example, the rule above might be one rule in a set that is used to infer scleroderris canker injury. There may be other rules, as in Figure 2–3, to infer tree injury from other causes. Other rule sets may provide disfavoring evidence for certain causes of tree injury, they might be called *elimination* rules or *disconfirming* rules or *negative evidence* rules. Still other rule sets might determine the truth value of hypotheses in the antecedent clauses of pest inference rules. These types of hypotheses would then be referred to as **intermediate hypotheses** because they lie between **atomic** or basic statements and the final conclusions, and hence, they may require substantiation prior to being used to deduce any final conclusions. Intermediate hypothesis rules help ensure completeness of a rule set's logic by providing for automatic deduction of some hypotheses, rather than asking the user for every piece of information. Common sense knowledge is often represented as intermediate-hypothesis rules. For example, the rule below deduces that a tree is dead based on evidence concerning other aspects of the tree. Intermediate hypotheses are

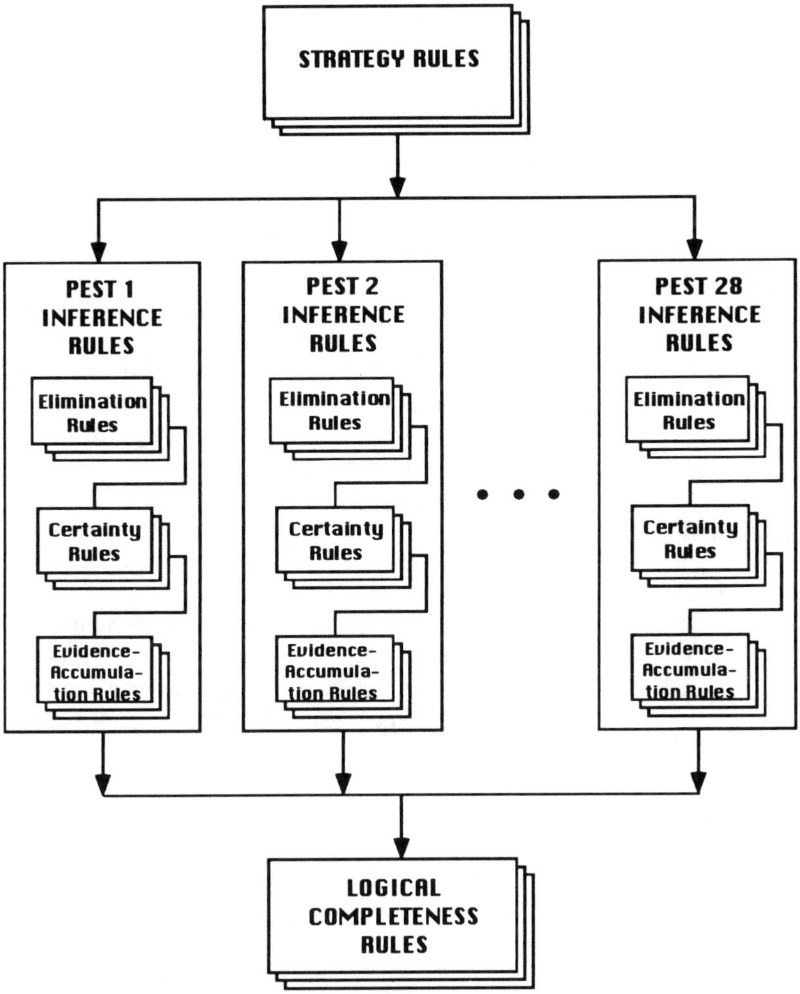

Figure 2–3. A rule-based knowledge system may contain many different types of rules. The knowledge base of PREDICT contains: strategy rules, logical completeness rules (intermediate hypotheses), and a separate block of rules for inferring damage by each of twenty-eight pests of red pine. Each block of pest inference rules also contains sub-blocks of rules to: eliminate a pest from consideration, infer damage by a particular pest with certainty, and accumulate favoring evidence for a pest.

discussed further later in this chapter as a way to organize a knowledge base to facilitate the search for solutions.

IF the condition of the buds is dead,
AND the condition of the shoots is dead,
AND the entire tree is discolored brown,
THEN the tree is dead.

Meta-rules (rules about rules) may be used to orchestrate the use of the other rule sets mentioned above (Schmoldt and Martin 1989). These higher-level rules can be viewed as strategy rules (Figure 2–3) for finding solutions in a more effective manner. Likewise, it's not hard to imagine meta-meta-rules which might select the appropriate strategy to use in a particular situation.

In theory, knowledge expressed in rules provides uniformity and modularity. Each rule represents a "chunk" of domain-specific knowledge (Feigenbaum 1979, Duda and Gaschnig 1981). Modularity permits incremental development, and uniformity makes modifications easy. In addition, rules can be simple to develop. Many experts utilize and describe what they know in terms of "rules of thumb" and programmers find it easy to encode knowledge in this form (Jackson 1986).

While these aspects of rule-based systems are their greatest strengths, rule-based representations sacrifice richness of expression and rigor. This restricts their usefulness when a subject area requires more detailed reasoning (deep knowledge) or more complex relationships. Some of the weaknesses are:

1. loops and recursion are very tricky to distinguish and handle;
2. it is difficult to predict which rules will fire in what sequence—this may be critical when adding new rules to the system;
3. rule bases must be carefully ordered and structured, otherwise it becomes difficult to understand the major knowledge patterns once a few hundred rules have been written;
4. rules cannot take advantage of the natural hierarchical structure found in many knowledge domains—rules are independent pieces of knowledge, so there is no connective (e.g., hierarchical) capability to exploit;
5. rule-based systems differ substantially in structure from traditional procedural programs and, hence, may be initially awkward for developers;
6. condition-action knowledge (surface-level) is handled well but deeper knowledge is difficult to represent; and
7. rule independence is difficult if not impossible to maintain in real systems making the knowledge base sensitive to rule order.

A knowledge base of even a few hundred rules becomes complex and unwieldy, and it is difficult to understand the impact of adding a few more rules. For this reason, the use of rules with intermediate hypotheses, the use of meta-rules, and a well-thought out rule organization incorporating rule sets become imperative. "The limitations of unstructured rule sets running under uniform control regimes... are now widely acknowledged" (Jackson 1986). Nevertheless, many successful KBSs have been built, and continue to be built, using nothing but rules to represent relevant knowledge.

2.2.2 Frame-Based Knowledge Representation

Frames were originally proposed by Minsky (1975) as a basis for understanding visual perception, natural language dialogues, and other complex behaviors (Barr and Feigenbaum 1981). In **frame-based** representations, a frame can be viewed as a collection of related knowledge about a topic (Walters and Nielsen 1988). This knowledge can include related facts, relationships, rules, events, or any entity in the domain about which it is important to reason, e.g., forest stand, tree, needle injury, elk herd, fire danger, or management unit. As you can see from these examples, frames can be used for entities that are detailed or general, singular or aggregates.

A reasonable analogy for understanding the structure of a frame can be seen by considering a box with many separate compartments. The box is representative of our frame. Each compartment represents a property of the topic being represented and it holds some relevant information, a property value. Each box may be of a specific type and contain compartments typical for boxes of that type; it may also contain special compartments particular to that box. One or several items may be contained in a compartment, and different *types* of boxes (classes) and *specific* boxes (instances) may be created.

Thus, a frame contains properties of an entity, **slots**; each slot has some value. These values can be specific values of that entity-property combination, pointers to other frames, pointers to procedures for calculating values, and pointers to inference rules. Pointers to other, higher-concept frames can provide for the inheritance of properties from a more generic frame, i.e. child frames of a parent frame (Buchanan and Duda 1983).

The example in Figure 2–4 depicts what some frames might look like for a continuous forest inventory (CFI) plot. CFI plots are typically scattered throughout the forest land of an ownership. By periodically measuring trees on these small plots, a landowner is able to track various characteristics of the forest as a whole, e.g. growth, volume, and mortality of the trees. Figure 2–4 contains two *class* frames and two *instance* frames. Class frames contain generic information that is typical of a class of objects. Instance frames contain specific information relating to a specific instantiation of that class, that is an individual within the class. A class frame, such as *continuous-forest-inventory-plot*, often contains default values for various properties. These are inherited by

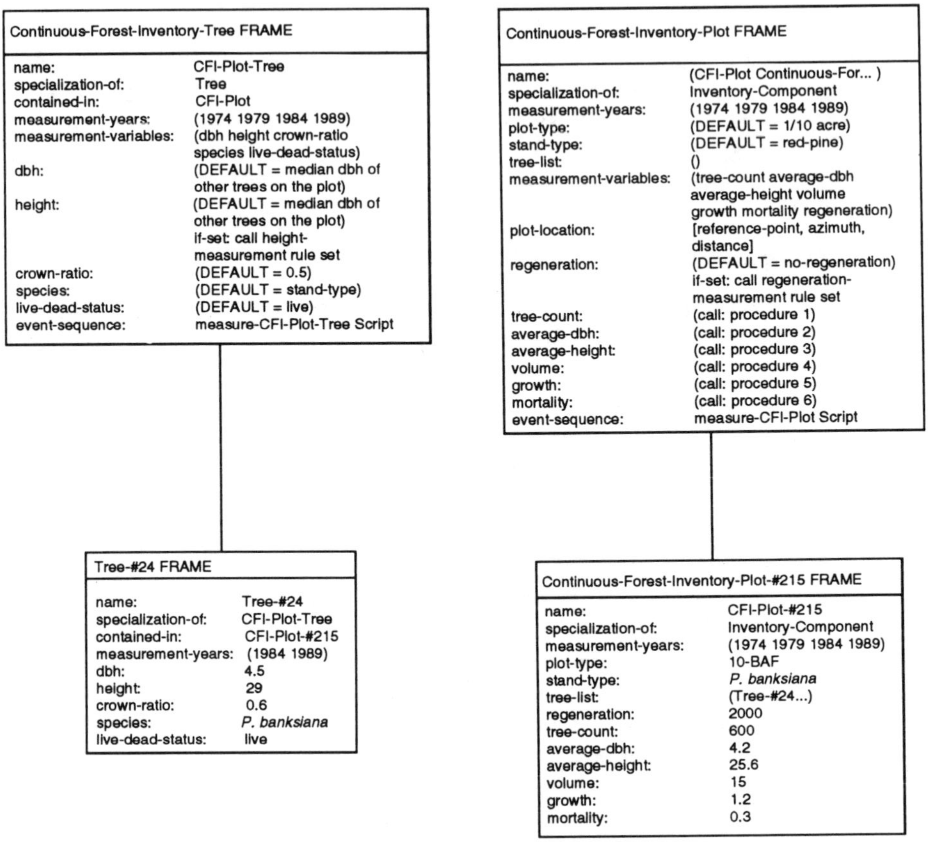

Figure 2–4. These four frames illustrate how a frame-based knowledge representation of Continuous Forest Inventory (CFI) plots might look. Included are two class frames with an instance frame for each. Many of the default values, procedures, and other properties of a class frame may be inherited by an instance of that class.

instance frames if an instance does not have specific values to override the defaults. A class frame may also contain procedures to be invoked when the value of a property is set, as in the regeneration property (if-set call *regeneration-measurement* rule set), or when the value of a property is calculated, as in the *tree-count* property (call procedure 1). While there exist inheritance relationships between classes and instances, other relationships may also exist between non-class frames, e.g., the plot instance #215 contains a reference to the tree instance #24 and visa versa. Each of the class frames in this example also contains an *event-sequence* property that describes the steps necessary for measuring a tree or plot. These slots reference a related representation called **scripts**, which are discussed briefly below.

A frame representation allows the user to express very sophisticated relationships between entities in a domain—relationships that may be difficult to capture by inference rules alone. The primary strength of frame-based knowledge representation is the ability to capture the structural arrangement of knowledge (Bowerman and Glover 1988). Often this structure becomes hierarchical in nature, proceeding from classes of entities to specific instances of a class.

A closely related technique to the idea of a frame is the idea of **scripts**. Representationally, a script is almost identical to a frame, but it contains sequences of events that describe one's understanding of a situation, or actions to be performed. Schank and Abelson (1977) first introduced this idea and, as an example, used a restaurant scenario complete with expectations of de rigueur actions and anticipated events. The CFI frame above could be extended to include scripts (Figure 2–5) for measuring a plot and a tree. Here, measuring a tree is a specialization of measuring a plot, so some of the properties of the *measure-CFI-plot* script are inherited. Each of these scripts refers to other scripts for more detailed action sequences.

Because of the highly-structured nature of frames, they have some very desirable properties for knowledge bases and their development. Some of these are:

1. frames are a way of grouping related pieces of knowledge into a single record that can be used to represent expectations and assumptions more succinctly than rules;
2. by their nature they force a more organized and modular content and development of a knowledge base;
3. temporal interdependencies can easily be accommodated and understood by using scripts;
4. frames can exploit the natural structure of knowledge and are very good in dealing with exceptions and defaults;
5. knowledge-base structure can be represented graphically, thereby increasing its understandability;
6. inheritance of class properties by instances permits a more generic storage of values, rules, and procedures so that the knowledge is more compact, understandable, and testable; and
7. because embedded rules are distributed and organized according to purpose, problems associated with unwieldy and convoluted rule sets are easier to mitigate.

On the other hand, their flexibility in control of the reasoning process can make their behavior hard to predict and understand (Jackson 1986). Much of their reasoning occurs via procedures distributed throughout the frame hierarchy

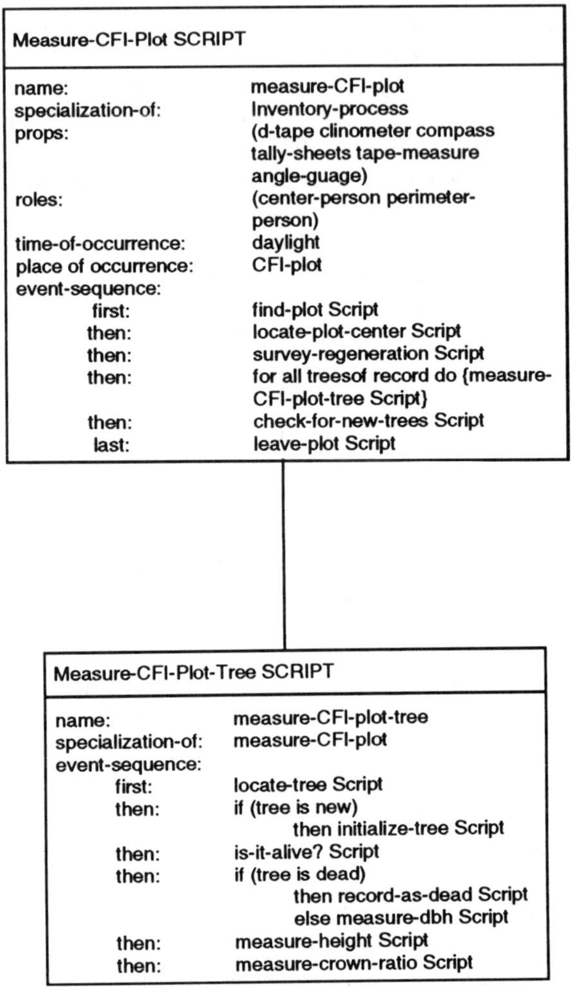

Figure 2–5. Where knowledge may consist of detailed sequences of events, a type of frame called a script may be very useful. Just like traditional frames, scripts may be either classes or instances. Property values are inherited in the usual manner. Because a script may refer to other scripts for more detailed steps, this representation can be useful for tailoring the level of information to the desires and needs of an individual user.

(Barr and Feigenbaum 1981). Thus, unlike the rule-based representation, there exists no common area where all inference methods reside. Also, the implementation of inexact reasoning, outside of attached rules, can be difficult. Frame-based systems are typically more difficult to design, develop, and implement than rule-based systems.

2.2.3 Object-Based Knowledge Representation

The idea of creating knowledge "chunks" that contain both declarative and procedural knowledge was pioneered in Minsky's use of frames. This approach received extensive work in the 1970's at Xerox's Palo Alto Research Center and eventually led to Smalltalk (Goldberg and Robson 1983), an **object-oriented programming** language. Object-oriented programming (OOP) is an outgrowth of the design and development of object-based techniques. In addition to the OOP programming style, objects are a valuable way to represent knowledge in knowledge-based systems.

The two most important concepts in **object**-based knowledge representation are *information abstraction* and *encapsulation*. First, by representing real-world entities as computational objects, program data and information become more abstract in a programming sense, but more concrete in an intuitive sense. That is, information is represented in a way that is more akin to its use and relationship to other entities in the real world. Second, encapsulation encloses both state and behavior in a single computational unit. These units are active in nature, they respond to *messages* from other objects and send messages to other objects. Each object is relatively autonomous because it has its own data and procedures to perform specific requested tasks. An individual object has no specific knowledge about other objects, but only that there is *some* object that can perform a particular operation. For example, in developing a simulation program for a manufacturing operation, a conveyor in the plant could be represented by a software object; and the object would respond to controls and would possess attributes similar to those of the real entity (e.g., speed, capacity, material conveyed, location, operating status, etc.).

As a more elaborate example, image that you are an employer with a large number of subordinate employees, each with their own office. Each of these employees has in their office access to certain pieces of information that are important for operating your business. Each employee also has the ability to perform certain tasks and to communicate, i.e. send messages to other employees. As the employer, you are not allowed direct access to any employees' information bases. The only way for you to get things done is to send messages to employees either: (1) requesting information or (2) asking them to perform some task. When you, the employer, send a message to an employee, you don't care what goes on in that employee's office as long as your request is completed satisfactorily.

For example, you may request a report of the last month's accounts receivable from one of your vice-presidents. Unknown to you, the VP sends a message to the Accounting Department requesting information on accounts. Then the VP may send the received information to an assistant who prepares a report. The report arrives on your desk without your exact knowledge of how it was produced. In this organizational structure, the sender of a message only knows that the receiver of the message can respond to it, the sender does not know, or care, how this is done. Neither does the sender have access to the internal affairs of the receiver. This is the essence of an object-oriented knowledge representation.

An object is a uniform representation of information. In this representation the emphasis is not on a global control structures that will invoke rules or navigate through frames, but rather, the focus is on designing objects and the messages they can send and receive from other objects. A major advantage of using an object-oriented representation is that it enforces modular design. The implementation of one object does not depend on the internal details of other objects, only on how they respond to messages (Goldberg 1984). By isolating each object from every other one, KBSs can be built up one object at a time, and objects from one system can often be reused in another system with little modification.

Object-oriented methods of knowledge representation share many similarities with frames. The terminology changes slightly, but the principal difference is that objects send messages to other objects to initiate actions while frames use a more open representation in which other frames or rules may execute a frame's procedures directly. *Local* data within objects are generally called "attributes", while frames use the term "properties". Both frames and objects can be organized hierarchically and both use inheritance to pass information and defaults to lower levels of objects. Objects execute "methods" when "messages" are received from other objects, while frames *can* execute "procedures" or call rules, most often only when they are specifically asked to. The methods executed by objects are actually pieces of program code, i.e. procedures, which are specific to a particular object and generally unavailable to other objects.

Because of their similarities, object- and frame-based systems seem to be converging rapidly (Parsaye and Chignell 1988). In fact, frame-based representations are becoming more infrequent, they are being replaced by object-oriented ones. Object-oriented representations possess all the advantages of frames and some others, as well:

1. object autonomy ensures that changes to one object do not create a cascade of required changes throughout the knowledge base;
2. many object-oriented programming languages are available that make the construction of these representations much easier (e.g. C+ +, Object Pascal, Object Common LISP, PROLOG+ +); and
3. computational objects and object-oriented programming offer a very natural way to represent a wide variety of real world situations.

The way in which objects compartmentalize information and knowledge means that they manifest many of the same disadvantages as frame-based systems. In particular, distributed reasoning makes object-oriented systems complex and makes understanding them difficult.

2.2.4 Network-Based Knowledge Representation

A **network** is a constellation of nodes and links in which each node stands for a concept, entity, or event, and each link represents a relationship between two nodes. Networks represent one of the earliest attempts at knowledge representation in AI (Parsaye and Chignell 1988). They were developed originally by Quinlan (1968) as a psychological model of human **associative memory** (Barr and Feigenbaum 1981). In an associative model of memory, closely related ideas or concepts are thought to be connected by direct links. Then, when a particular concept is activated, all linked concepts are also activated. This theoretical basis led to research and application of networks as one of the first means of knowledge representation.

In Figure 2–6 concepts like stand, red pine plantation #14, tree, leaves, and plant are related by *is-a*, *element-of*, and *has-attribute* relationships. Inferencing over this network could be accomplished with inference methods like (the term "rules" has been avoided here, so that there is less confusion with rules in a rule-based representation):

```
INFERENCE METHOD 1:
  IF ?X is-a ?Y
  AND ?Y has-attribute ?Z
  THEN ?X has-attribute ?Z,

INFERENCE METHOD 2:
  IF ?X is-a ?Y
  AND ?X element-of ?Z
  THEN ?Y element-of ?Z.
```

In the above, each of ?X, ?Y, and ?Z are variables which may have as their value any entity in the network. From Figure 2–5 and these inference methods we may conclude that "PLANT element-of CFI-PLOT", "RED PINE PLANTATION #14 has-attribute LOCATION", "TREE has-attribute LEAVES", and "TREE #23 ON PLOT #16 has-attribute LEAVES", among other things.

The major advantage of network representations lies in their spatial presentation of entities and those entities' interrelatedness. Many believe that this approach may mirror, to some degree, our own internal, associative mechanisms for storing and manipulating ideas.

Weaknesses of associative networks (i.e., semantic nets and procedural nets), which have led to their disuse, are: (1) they lead to a cascade of inferences and

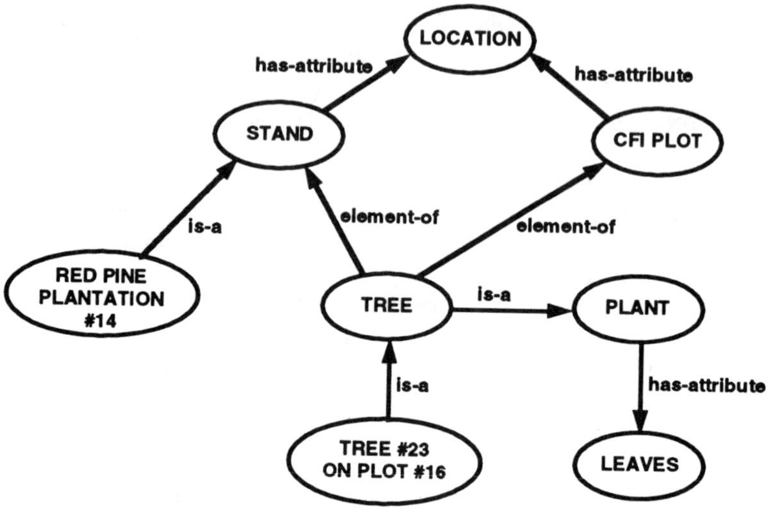

Figure 2–6. A network knowledge representation method uses nodes for entities of the domain and directed arcs between nodes for relationships between entities. Here, the arcs are labeled. In a concept graph (Sowa 1984), relationships would be displayed as nodes also and the arcs would be directed, but unlabeled. Inference methods are not inherent in a network representation but must be explicitly defined for the various relationships.

resulting deductions, which is difficult to control and navigate; (2) except for concept graphs (as noted by Luger and Stubblefield 1993), discussed below, they are logically deficient in expressing many of the distinctions that logic can make—in particular, *at least one, all, none*, etc.; and (3) there is no knowledge in such systems that tells you how to search for the knowledge you want (Jackson 1986), i.e. they contain static knowledge only without a mechanism for well-directed access to pertinent knowledge.

Despite these limitations, networks have some interesting implementations. Three of these are mentioned below.

Sacerdoti (1977) used *procedural nets* in the development of a general purpose planner to translate goals into implementation, i.e. a sequence of tasks. Instead of representing entities and their relationships, Sacerdoti's procedural nets contain *events* at various levels in a planning process. His networks are hierarchical and allow for the decomposition of any procedural step in a plan into a subordinate network of more basic planning steps. The result was a highly flexible planner, NOAH, that was able to construct the necessary component steps to solve a particular planning problem at any level of detail appropriate to the user.

Sowa (1984) used *concept graphs* as an explicit means for displaying statements of first-order predicate logic. He extended the earlier idea of semantic networks to include additional structures. These structures permit: (1) the distinction of concept types, individuals, and names; (2) operations on graphs to produce other graphs; and (3) the organization of graphs into propositions which may be treated as concepts themselves. His extension of semantic networks into very powerful conceptual graphs has created renewed interest in this method of knowledge representation.

More recently, *belief networks* were developed by Pearl (1986). His approach, based on probability theory, updates and propagates belief in a network of propositions (each proposition represents some statement about the domain of interest). Each link connecting two propositions represents a conditional probability between them. The belief in each connected proposition is modified (updated) as other propositions are modified or new ones introduced. In addition to providing a scheme for representing knowledge, belief networks also implicitly contain an inference mechanism and are designed to reason with inexact knowledge (see 3.5). This makes them a very powerful and complete system; however, the limitations and disadvantages of Bayesian probability are also present (these are discussed more in Chapter 3). One interesting advantage to belief networks is their ability to reason in both directions. That is, given some facts about a problem, the network can deduce hypotheses from those facts. A network can also "reason" the other way by deducing facts that must be true given belief about a particular hypothesis.

2.2.5 Logic-Based Knowledge Representation

Logic-based representations use formal logic, usually a variant of *first-order predicate logic*, to specify assertions and inferences. For example, we might want to represent the statement "all the trees are discolored yellow". In propositional logic, facts like this are often specified as a string constant or as an atomic symbol. In propositional logic there is no way to access the individual components of such a compound concept. It would have to labeled as follows:

trees_discolored_yellow,

or using a similar label, and it could have only 2 values: TRUE or FALSE. But that fails to capture the complexity of the relationship between trees and their discoloration and the fact that all trees of which we speak are so discolored. When we can refer to individual pieces of a statement, we have greater expressive capability.

First-order predicate logic is an extension of propositional logic such that access to the components of an assertion is permitted (Luger and Stubblefield 1993). We could represent the above fact as a predicate logic statement:

$$\forall X\ [\text{tree}(X) \Rightarrow \text{discolored}(X, \text{yellow})]$$

in other words:

for all X, such that X is a tree, X is discolored yellow.

This logic is labeled "first order" because quantified variables, such as *X*, are allowed to represent only entities in the domain of discourse not predicates such as "discolored". A major motivation for choosing logic as a knowledge representation method is that there exist good, efficient ways of reasoning with that knowledge (Boy 1991). Using first order-predicate logic to represent knowledge and well researched methods of theorem proving as inference procedures to reason about that knowledge, we can search computationally for solutions to problems (Buchanan and Duda 1983, Nau 1983, Harmon 1985).

When using a logic-based knowledge representation, a logic-based programming language, such as PROLOG, is utilized (e.g. Brown et al. 1985, Roach et al. 1985). PROLOG is convenient because a PROLOG program is, essentially, a set of assertions and inferences. Resolution (Robinson 1965) provides the underlying deduction mechanism. Logic programming's attraction stems from its paradigm of telling the computer *what is known* and letting the machine draw conclusions (Clocksin and Mellish 1987). This *declarative programming* style solves problems by describing what solutions should look like rather than how to compute them (*procedural programming*).

Logic-based representations force the constructed knowledge-based system to be quite rigorous. Logic-based systems have received less attention so far than frame- and rule-based ones, but PROLOG, as the primary example of logic programming, is being used quite extensively as a general purpose AI language. As we will discuss in chapter 7, because Prolog uses this method of knowledge representation and it contains inference and control methods, it is essentially a knowledge-based system **shell** as well as being an AI programming language.

The basic structure of a logic-based representation is one of rules, in the sense of logical *modus ponens*. Therefore, a logic-based system suffers from some of the same drawbacks as a rule-based representation (see Section 2.2.1). Unlike the other knowledge representation schemes, however, general-purpose programming languages have been designed around the use of logic, for example PROLOG. The downside of this is that one must learn a new programming paradigm in order to use logic-based representations easily. However, because these languages are also general purpose, they are much more flexible than typical rule-based representations, they are extensible, allowing a programmer to add whatever constructs he or she would like, and they often provide very useful, interactive execution and debugging environments. As we will demonstrate in Chapters 7-9, general-purpose programming features allow us to implement any of the other knowledge representation schemes within the logic programming language.

2.3 WORKING MEMORY

The **working memory** of a KBS contains assertions about the problem currently being investigated. These assertions may be obtained from the user via queries, from external programs, from a real-time process, from other knowledge-based systems, or from external data files (Figure 2–7). Assertions may be known facts, or they may be hypotheses which have been inferred from other facts already known. Because the ultimate goal of KBS consultation is to infer problem solutions, some of these assertions will be eventual solutions. All facts and hypotheses in the working memory together describe the current context, and the current state, of a consultation session.

When inferring a hypothesis from data, the necessary data items for each deductive leap must be present in the working memory. Some criteria must be established for determining when this state of "knowing" exists. For example, not all assertions may necessarily be absolutely true or absolutely false. Often, assertions have some uncertainty associated with them (see Inexact Reasoning and Uncertainty in Chapter 3). So, it becomes necessary to distinguish between assertions that have some degree of truth. Exactly how this is handled will vary depending on the knowledge representation scheme, the type of facts/hypotheses, and the method of handling uncertainty. When using KBS development shells, much of this work is done by the software; however, when coding in a language (even a high-level one such as Prolog), the developer must explicitly consider how facts and hypotheses are represented and stored, as well as how they are used in the knowledge.

In a rule-based system the knowledge base and the working memory are quite distinct types of structures. Assertions explicitly referenced in IF-THEN rules must be verified as true before a rule can fire. In frame- and object-based representations working memory and knowledge are contained in the same type of structure, either frames or objects. Also, assertions about some entity in a frame-based representation may not be explicitly stated in the frame where the knowledge is used. Due to some relationship with another frame (e.g., "inheritance" of attributes), however, it may be possible to assert some facts. In a logic-based representation, a particular proposition might require that all instances of a predicate be true, for example, universal quantification over the variable X in *tree(X)*. For the predicate *tree*(X) to be true, all possible instances of that assertion must be true in working memory. Consequently, different knowledge representations store and retrieve working memory items differently.

The method used to *match* working memory assertions and knowledge base assertions also varies. For example, simple symbolic statements of assertions must match exactly, e.g. *the stand is stressed*. The symbol, *the stand is stressed*, must explicitly appear in the working memory. Numeric assertions may also need to be exact, or a range of values may be permitted.

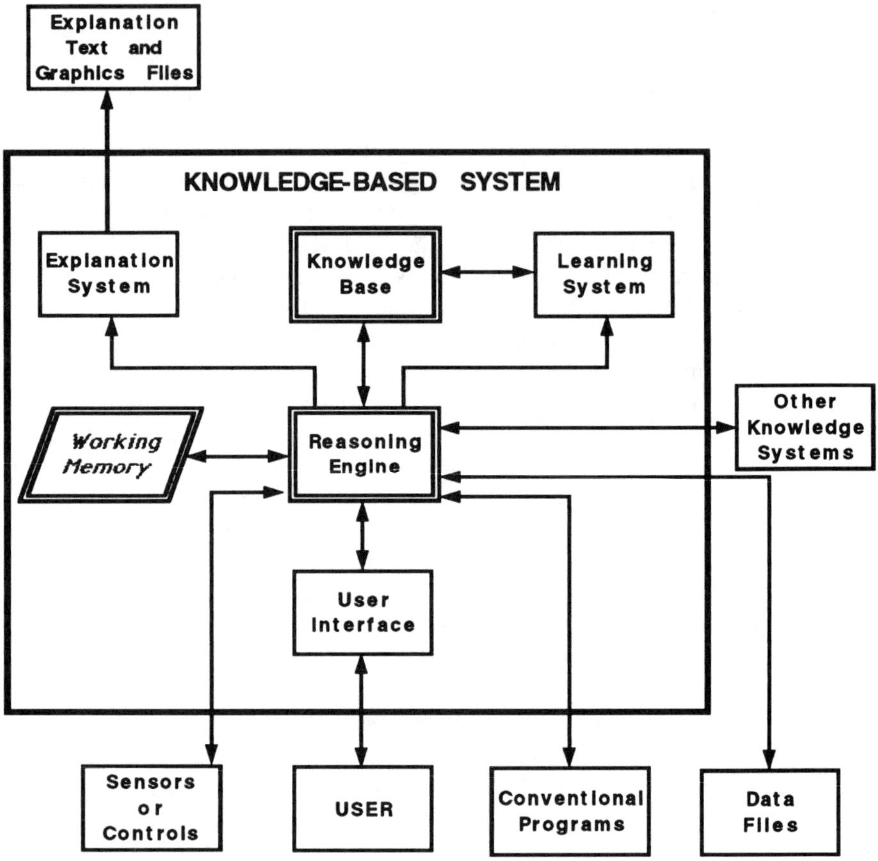

Figure 2–7. Working memory serves as a repository for all assertions that are pertinent for solution of the current problem.

Determining the truth value of an assertion varies among systems. Some systems assume all assertions are true with 100% confidence, or probability. Most systems, however, permit uncertain assertions and set some threshold value for asserting a fact as true (Harmon 1985). There are a number of methods for handling the inexact nature of what we know (or believe) as true. These techniques are collectively called **inexact reasoning** methods and are covered in section 3.5.

It may also be important to explicitly store in working memory assertions that are known to be *false*. Oftentimes, knowing that some assertion is false can be very important information for making other inferences. For example, in pest diagnosis, sometimes a particular symptom always occurs coincidentally with damage by a particular pest. Then, the absence of that symptom can indicate that the particular pest is *not* the cause of damage. In the following rule we can eliminate meadow mouse damage if we know that the bark symptom is not present.

IF NOT bark is removed at the base of the tree
THEN NOT meadow mouse damage.

Otherwise, the **closed world assumption** is typically used. This states that "we only know what is explicitly stated", and so anything that is not explicitly stated in the working memory is assumed to be false. In the above rule, a closed-world assumption would implicitly verify that the bark symptom is not present if we had not specifically indicated that it was present.

Furthermore, it is also important to state that something is *unknown*. An assertion that remains unknown (no evidence to indicate truth or falsehood) is usually treated as pseudo-false, i.e., it is not used to propagate any inferences. On the other hand, it may be valuable to explicitly state when some assertion is unknown, and then to base inferences on the *unknowing* state. In particular, when a closed world assumption is applied, it becomes more important to distinguish between assertions that are unknown and assertions that have been established as either true or false. For example, in the above rule when we have no information about the condition of the bark, a closed-world assumption would cause the rule to fire. Given our pest diagnosis rationale for the rule, however, this is clearly not what is intended. We would like to make the inference indicated in the rule above, only when we know that bark damage did *not* occur. By assigning all facts as "unknown" initially, the above rule will not fire unless we actually have some evidence to indicate that bark damage did not occur at the base of the tree.

2.4 SEARCH: INFERENCE AND CONTROL

A knowledge base is a passive entity; some means of executing its actions or navigating through its relationships is necessary. The process of reaching a problem solution from an initial problem description involves **search**. Search methods determine which states of the problem space will be visited, and in what order they will be visited.

In the most general sense, all human problem solving can be visualized as searching through some set of states, which describe some problem of interest. This search terminates when a state satisfying some solution criteria is reached. In this view, each state represents the total information known about a particular problem instance at any point in the solution process. As each new state is visited, more information is added to the total "understanding" of the problem.

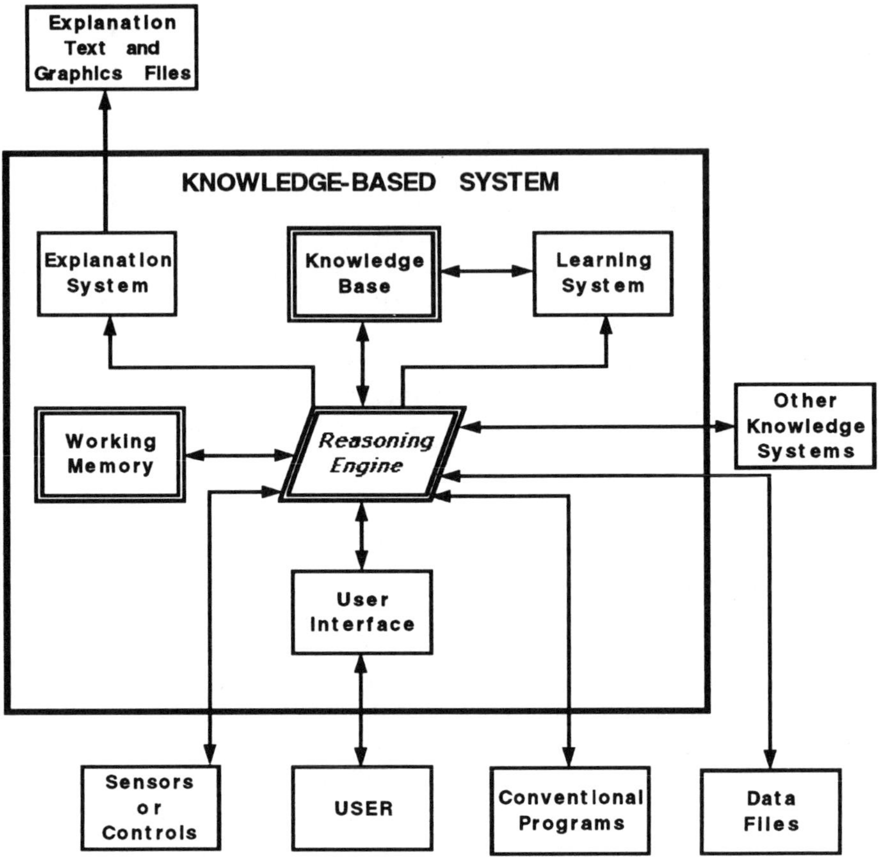

Figure 2–8. The reasoning engine of a KBS applies the knowledge present in the knowledge based to the problem description that is represented as assertions in working memory. Reasoning in this context has two components: inference and control. Inferencing allows the system to generate new assertions and control methods determine how the system goes about finding solutions.

When the current state contains a satisficing assertion (one that answers our primary question), then we can say that we have reached a solution state. This constellation of problem states is often called a *problem space* or a **state space**. Figure 2–9 depicts a very small problem space of unlabeled states; in most instances the cardinality (number of states) of the space will be thousands or millions of nodes (states). The game of chess has approximately 10^{40} possible board positions or states (Shannon 1950). In many problems the number of states becomes very large quite quickly (often referred to as a combinatorial explosion), so efficient search methods become very important.

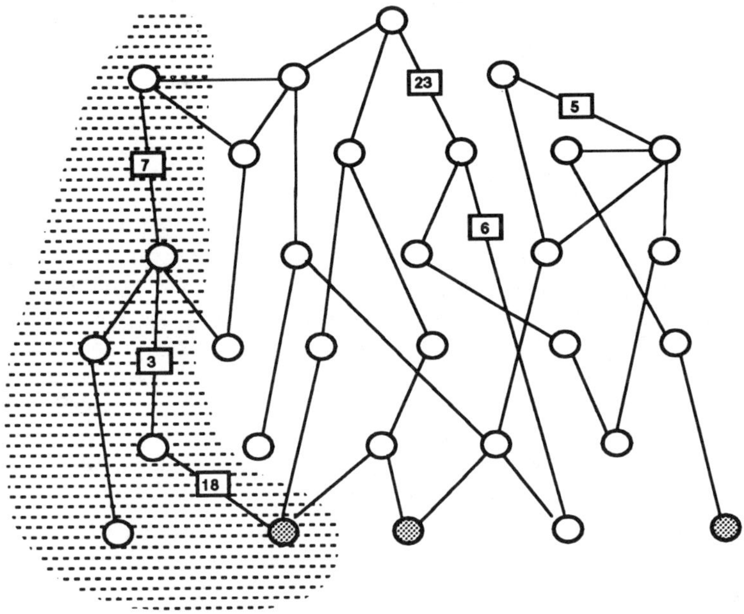

Figure 2–9. In general, problem-solving activities can be viewed as a collection of state nodes (circles) and inferencing transitions (connecting lines) between states. At least one state must be a solution state (shaded circles). Some of the transitions here are labeled with numbers corresponding to hypothetical rules that might be used to infer one state from another. Inference and control are the two components of searching for problem solutions. The shaded area of the figure represents one possible search control strategy, which has resulted in reaching a solution state in a relatively efficient manner.

Navigating through a problem space involves two mechanisms: (1) **inference** and (2) **control**. These two components of search determine allowable moves between any two states and the general strategy of moving through the entire space, respectively. The **reasoning engine** (Figure 2–8) performs these functions in a knowledge-based system. Often this component is referred to as an **inference engine**; however, because these algorithms provide both inferencing and control mechanisms for *reasoning*, we prefer the former, more encompassing term.

The model of a *state space*, as in Figure 2–9, can itself be a useful knowledge representation for a particular problem. All that is required are: (1) some state-based representation of the different stages of a problem (e.g., board positions for a tic-tac-toe game), (2) a set of legitimate operations for transforming one state into another, and (3) some criteria for determining whether a solution state (at least one must exist) has been reached. Exhaustively, visiting all possible states would be an inefficient, but sure-fire, method for reaching a solution. The set of legitimate moves are part of inferencing; selecting and ordering possible moves and establishing the solution criteria are aspects of control. The next section presents several general search strategies that have been applied to the state space problem representation.

2.4.1 Generalized Search

As noted above, generalized search (as in a state space) includes both inference and control methods. That is, they contain mechanisms to decide both the order that states will be visited (control strategies) and how new states are generated. After brief descriptions of several general search methods, we will examine some specific control and inference methods individually.

2.4.1.1 Exhaustive Search. One of the simplest and most basic techniques is brute force *exhaustive search*. Here, all possible solutions are enumerated and the best one selected. When a subject area is poorly understood or relatively small, this may be a desirable approach. However, the number of decision states, i.e. the size of the problem, increases exponentially as a function of the number of factors used in the decision process. Consequently, even small problems may take a long time to solve using this method. This technique does not represent a problem-solving method as much as the lack of any efficient method or the failure to use problem-specific information intelligently.

2.4.1.2 Heuristic Search. Whenever we use past experiences, i.e. empirical rules of thumb, to solve a problem, we are performing a **heuristic search** (Newell and Simon 1972). The more specifically a heuristic fits a particular problem, the stronger it is and the more rapidly an acceptable solution can be reached. Weak heuristics may result in combinatorial searches because they fail to reduce the search space significantly, hence, in the worst case heuristic search becomes exhaustive search.

Heuristics may take several forms. Scoring functions are sometimes used to calculate "distances" from an initial state to the current state and also from the current state to a solution state. The sum of these distances is the heuristic value. At any stage in the search process, it then becomes possible to select the shortest path to a solution by selecting the state with the smallest heuristic value. Search states can be placed on a prioritized list. In other instances a heuristic may consist of intuition or subjective judgment, in the form of an if-then rule, for instance. Tradeoffs exist between the power and the flexibility of heuristics. As they become stronger they become applicable to a smaller number of problems. Consequently, by combining both weak and strong heuristics a system can operate well for both easy and difficult problems.

2.4.1.3 Parallel Search. Another term often used for *parallel search* is **breadth-first search**. Basically, the idea is to investigate a number of solution paths simultaneously until some solution criteria are satisfied. When the only criterion for an acceptable solution is the best solution, this method produces an exhaustive search. This strategy does not use any domain-specific heuristics to organize its search, it blindly looks at all states at a given level before moving to the next level.

The investigative nature of this technique tends to be divergent and spreading, much like a forward chaining control strategy (see below). During any stage, parallel search and forward chaining try many new states or create many new hypotheses, respectively. As with forward chaining, multiple lines of reasoning or hypotheses are investigated concurrently, so it can be difficult to provide an adequate explanation of the reasoning process to an end user.

2.4.1.4 Serial Search. *Serial search* is, in some sense, orthogonal to breadth-first search and often goes by the name **depth-first search**. A single path is investigated until a solution or a "dead end" is reached. The path is determined by trying the next state immediately derived from the most current state. Although no heuristics are used, there exists a noticeable continuity in the search process because any state transition bears a close relationship to the previous step.

The focused aspect of serial search provides some measure of similarity to backward chaining (see below). Only a single state or hypothesis is generated at any stage in the reasoning process. When a dead-end is reached, that is, only previously visited states can be reached from the current state, then the search procedure backs up to the previous level and tries other unvisited states emanating from that node. Depth-first search can be more efficient on average than breadth-first search when goal states are many moves from an initial state. The directedness of depth-first search means that reasoning chains (i.e., the sequence of states to a goal state) will appear logical because each state is closely related to the previous state in the sequence.

2.4.1.5 Hypothesize and Test. Alternatively, this method has been called **generate and test** (Barr and Feigenbaum 1981, Kline and Dolins 1985). Here, hypotheses are generated one at a time by some process, and then an independent test procedure either accepts or rejects each one. This amounts to an exhaustive search for the first solution that passes a qualifying test. The hypothesized portion of this search process must incorporate some mechanisms for inferring new hypotheses (inference) and for deciding in what order to generate them (control). Generated hypotheses may be either final solutions or intermediate hypotheses (perhaps components of a final solution, as in problems that are solved by constructing solutions from smaller pieces). The *hypothesize and test* method is extremely general and, hence, can be applied to a wide variety of problems. Kline and Dolins (1985) note that application of this method to intermediate hypotheses results in a hierarchical generate-and-test process which possesses a greater degree of efficiency than hypothesizing and testing full-blown solutions.

2.4.1.6 Decision Tree Search. As its name implies, **decision trees** are represented as hierarchical structures. Their use assumes that the knowledge about a subject area can be organized into a taxonomy that has well-defined data and that partitions the different solutions. Each node in the tree represents a datum key whose value determines which lower nodes will be visited. Particular solutions are leaf nodes. In this type of search, the tree's organization forms the control strategy and the datum keys allow for inferences. Figure 2–10 depicts a typical decision tree formulated after a common taxonomic key.

While knowledge systems, particularly classification types of systems, can be created in a decision tree framework, this approach is not very elegant and not analogous to the way we think and solve problems. As a result, explanation capabilities are severely limited in their efforts to transfer an understanding of logical decision-making relationships. Additionally, not all subject areas can be neatly partitioned by data items in this manner. In some domains certain data keys may contain no information about a particular solution or contradictory information about some solution; it is unclear how these might be positioned in the tree. Rather than having the order and amount of information requested of the user determined and fixed as in a decision tree, it seems more flexible for it to depend on what the user knows or does not know. Decision trees may be useful for initial KBS prototypes, but their inflexibility and limitations make them poor candidates for problem solving in later prototypes.

2.4.2 Inference

Inferencing includes a number of activities:

1. allowable terms (statements about the application domain) used by the inference method(s) are specified;
2. combinations of terms that are permitted in the inference language are defined;
3. criteria are set for matching knowledge-base and working memory terms; and
4. the operation of actions on the working memory are determined.

Selecting which inference to apply at each stage in problem solving, and what to do when no more inferences can be made during a particular consultation are outside the jurisdiction of the inference methods. These activities are a function of the governing *control strategy*, which must take the appropriate actions under these conditions. This is discussed more below.

Given states, s_i and s_j (i, j unequal), inference refers to the method by which s_j can be reached from s_i. Most often, inferencing is accomplished by using the method of *modus ponens*. If-then rules are the most common example of this

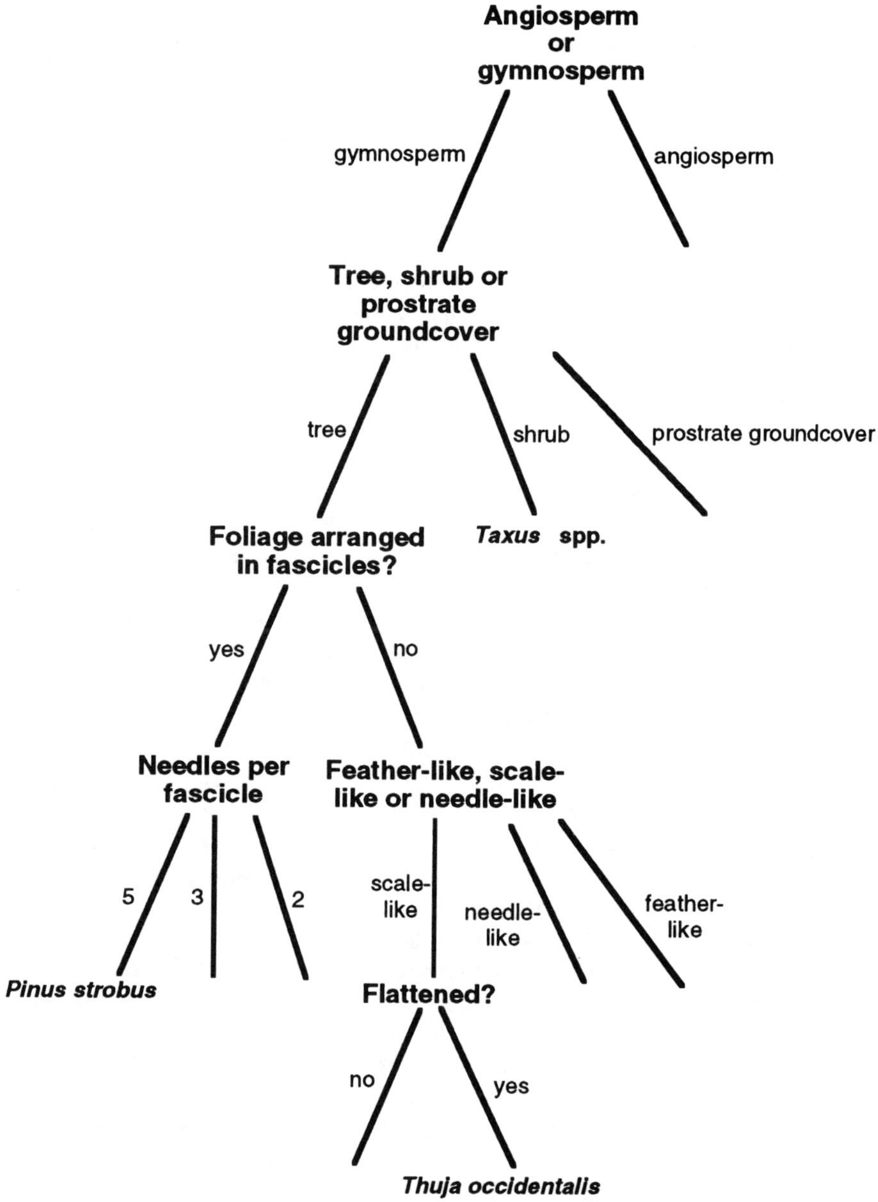

Figure 2–10. A portion of a taxonomic decision tree illustrates how different data items are used to partition the set of solutions (tree species in this case).

Table 2–1. Automated reasoning systems, AI programs and knowledge-based systems, use one or more of the following inference methods (some of these are found in Luger and Stubblefield 1993).

Name of Method	Inference Procedure
Abduction	From *if p then q* and *q* infer *p*.
Modus Ponens	From *if p then q* and *p* infer *q*.
Modus Tolens	From *if p then q* and $\neg q$ infer $\neg p$.
Inheritance	If *A* is a class, and *a* is a member of class *A*, and *p* is inherited from *A* by *a*, and *p* is true of *A*, infer *p* is true of *a*.
And-Elimination	From $p \cap q$ infer *p* and infer *q*.
And-Introduction	From *p* and *q* infer $p \cap q$.
Or-Exclusion	From $p \cup q$ and *not p* infer *q*.
Or-Introduction	From *p* or *q* infer $p \cup q$.
Universal Instantiation	From *for all x p(x)* and *a* is an instantiation of *x*, infer *p(a)*.

inference method. Here, when we are given the two statements *p* and *if p then q* we can validly assert *q* as a logical consequence under the *modus ponens* method. As universally applied as this method is, there are other inferencing techniques as well.

In the prior discussion of the frame representation scheme we saw that values for an entity can be inferred by **inheritance** from the values of another entity. This type of inference mechanism is built into the knowledge representation scheme itself. Still another related technique, *abduction* is the converse of *modus ponens*, i.e. from *if p then q* and *q*, we may infer *p* to some degree. This follows from the realization that whenever *p* is true, *q* is also true, therefore *p* and *q* are often true together—but not always. Stated more rigorously, if there exists a sufficiency relationship from *p* to *q* (i.e. *if p then q*) and we know *q*, then there also exists some evidence for *p* being true because *p* is sufficient for *q*. These and other common inference methods are summarized in Table 2–1.

Inference methods may be used explicitly in the reasoning engine or implicitly as an underlying model of reasoning in the domain. For example, in a system that uses frames or objects (e.g. Reinhardt 1987), inheritance between a class and a member of that class is explicitly defined by links within the knowledge representation. PROLOG (Clocksin and Mellish 1987) explicitly uses *modus ponens*, universal instantiation, "and-introduction", and "or-introduction" as built-in inferencing methods. On the other hand, Schmoldt and Martin (1989) use abduction and *modus tolens* as part of a diagnostic model for pest problems of forest trees. In their pest diagnosis system, the underlying inference method would be:

pest model: cause (pest) *implies* effect (symptom)

$$p \Rightarrow q$$

However, in real-world diagnostic circumstances, it becomes necessary to reason in the opposite direction,

diagnosis model: effect (symptoms) *implies* cause (pest)

$$q \Rightarrow p$$

which is abduction. Also, the absence of certain symptoms, *not q*, may imply that a particular cause (pest) cannot be the source, *not p*, of the observed pathological state—this represents *modus tolens*.

Although, in this circumstance, these inference methods are used to model diagnosis, the actual knowledge representation implementation may still use *modus ponens*, "and-introduction", and "or-introduction" to represent these problem-solving methods. That is, the abductive method, *if p then q* and *q*, would be implemented as, *if q then p* and *q*. Referring back to our scleroderris canker example rule above, our pest model would say:

Rule in the pest model:

IF G. abietina (scleroderris canker) is the causative agent,
THEN Green discoloration is present beneath the bark on
affected stems.

However, from our knowledge of pathogens we have good reason to diagnose scleroderris canker (*p*) from green discoloration (*q*)—*q* implies *p*, abduction. So, our implementation rule would appear as:

Rule in the implementation of diagnosis:

IF Green discoloration is present beneath the bark on
affected stems,
THEN G. abietina (scleroderris canker) is the causative agent.

Therefore, our model is abductive, but our implementation of the knowledge still uses *modus ponens*.

There are problem instances, however, where we would like to make inferences about causal relationships. Functional relationships between entities in the domain are sometimes known. In the biological sciences there often exists some knowledge of how certain agents influence other agents. As an example, a rise in soil temperature results in greater micro-organism activity. When these sorts of relationships are known and represented, it becomes possible to exploit the cause-effect associations to draw inferences and to simulate a biological system and perturbations to it. This is the intent of qualitative models (e.g. Schmoldt 1991, Whittaker *et al.* 1991) and the basis for model-based reasoning (e.g. Plant and Loomis 1991). A convenient characterization for such an interrelated set of entities and processes is some sort of network knowledge representation scheme. It may then be possible to reason from cause to effect or reversibly from effect to cause depending on the directionality of the connections. Because *causal inference* is oriented to the way we sometimes think about difficult and unfamiliar problems, it can produce very intuitive and understandable explanatory support about how solutions are reached.

Some arcs in the state space representation of Figure 2–9 are labeled with numbers, these correspond to particular rules, or other inference steps, such as inheritance. In general, the application of all possible inference methods will eventually lead to a solution, provided that the necessary supporting evidence can be confirmed. However, a control strategy that directs the inferencing methods to operate efficiently in the exploration for solutions can be more pleasing for the user and intimates a more intelligent agent.

2.4.3 Control

In addition to the inferencing process that defines legal moves between states of a problem space, a reasoning engine must also contain some overall strategy for moving about the entire space. The methods by which this is accomplished are referred to as *control strategies*. They are called strategies because they outline a course for the solution search. The overall search activity is not complete, however, without the ability to actually move to a new state, i.e. inference, which we have covered previously. Hence, the search for solutions includes both control strategies and inference mechanisms. The shaded area of Figure 2–9 contains some states that might be visited using a particular strategy. Less efficient strategies might visit far more problem states, in the worst case an exhaustive search might take place. Controlling the sequence in which problem states are examined in a KBS is a significant problem.

A number of control strategies are outlined in the following subsections. Each method possesses some advantageous characteristics that make it useful under certain circumstances. These control methods are useful for systematically or efficiently finding solutions for problems, they should not to be confused with knowledge representation methods or methods for inferencing. Neither are they necessarily mutually exclusive nor do they represent an exhaustive list. Often, it will be desirable for a system developer to utilize several techniques simultaneously, perhaps one or more for moving through the problem space and another for ranking or representing solutions. Each of these techniques mirrors some general method of problem solving that has been found useful in human reasoning. Because each KBS application possesses its own peculiarities, it may be necessary, in some situations, to create a new method tailored to the needs of a particular application. This is where scientific discovery and research diverge from straightforward technology transfer.

As one reads through the AI literature, one frequently comes across the control strategies of the following subsections. By briefly mentioning some of them here, we have attempted to identify the major ideas employed for controlling the search for solutions. Readers interested in more detailed implementation issues should consult general AI texts (e.g. Barr and Feigenbaum 1981 and 1982, Cohen and Feigenbaum 1982, Barr et al. 1989) and any specific sources cited below.

2.4.3.1 Forward Chaining. When employing a **forward chaining** (data-driven, antecedent reasoning) strategy, the problem solver begins with some known facts about a problem. New facts are then obtained by the inference "rules" or methods. These new facts are used with the old facts to produce more new facts. Eventually (and hopefully) this terminates when a solution state is reached, that is, when a final conclusion is inferred.

For example, in a rule-based knowledge system, rules are selected for application when their antecedent conditions are satisfied by the current facts in working memory. Because several rules may be selected in this manner, there must be some method for selecting which one will be used at any point in time. **Conflict resolution** criteria are applied to the rules already selected until one has the highest priority. Luger and Stubblefield (1993) mention the following set of possible conflict resolution mechanisms:

1. immediacy: the first such rule found is used;
2. refraction: after a rule has fired, it may not be used again until the values of its antecedent conditions have changed, this prevents infinite loops;
3. recency: the rule whose antecedent matches the most recently modified working memory terms is given priority; and
4. specificity: rules with more terms in their antecedent are given higher priority.

Let us assume that we are given as a starting point two facts:

tree height = 2 ft.
the distribution of needle damage is throughout the crown

Also, we have the following rules available in our knowledge base:

RULE 1:

IF the size class of the stand is seedling
AND the distribution of needle damage is throughout the crown
AND the needles are discolored
THEN the entire seedling is discolored

RULE 2:

IF tree height < 4.5 feet
THEN the size class of the stand is seedling

A forward-driven inference chain would try each rule and see whether there are enough facts in the working memory to prove each consequent true. In this case, the second rule "fires" and is therefore able to place the true fact:

the size class of the stand is seedling

into working memory. After this rule is applied and working memory is modified, the data-driven procedure is applied again to select the next rule. This continues until the necessary termination conditions are satisfied or, alternatively, the user may be asked to provide information that may allow more rules to be selected.

Because rules that are candidates for application are selected on the basis of the current context, the working memory must be searched often. As working memory search requirements become extensive, a great deal of processing time may be expended for each cycle. Another disadvantage of this data-driven, fact-finding process is that it can give the user of the KBS the impression of random questioning with little sense of direction toward a solution.

2.4.3.2 Backward Chaining. The second major strategy for controlling the search for solutions is **backward chaining**. In contrast to the forward chaining strategy above, a backward chaining (goal-driven, consequent reasoning) approach begins with a goal to be solved. Then by examining the legal moves that are able to generate this goal, the strategy can determine which conditions must be true to satisfy the goal. These conditions then become *subgoals* for **recursive** investigation. Search continues, working backward until known facts are reached; the trail of subgoals then specifies the solution path. A rule-based example of backward chaining, from forest pest diagnosis (Figure 2–11), demonstrates this goal-to-data chaining method. Backward chaining is a simple, but effective, strategy for solving many kinds of problems.

The decision to use one type of strategy or the other depends greatly on the type of problem being solved by the KBS. This decision is discussed further in Chapter 6. It is also influenced by the system development environment available, i.e. the software used to implement the KBS. Goal-driven strategies are suggested (Luger and Stubblefield 1993) if:

1. A goal or hypothesis can be easily formulated or is given in the problem statement.
2. There are a large number of inference "rules" that match the facts of the problem and, hence, produce an increasing number of hypotheses or goals. Early selection of a goal to pursue can eliminate much needless search.
3. Problem data are not given, but must be acquired by the problem solver. Data acquisition will be guided by the search for solutions. This has the advantage of requesting only that information that is essential to solving the problem at hand.

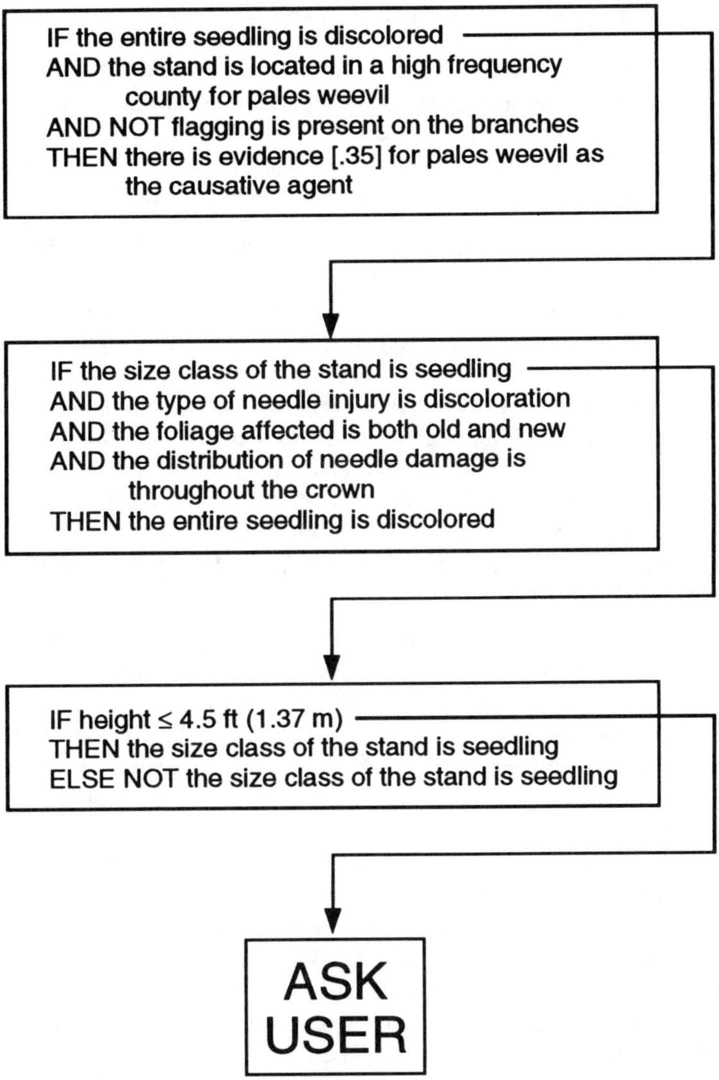

Figure 2–11. In a backward chaining strategy, a goal to be confirmed is identified and then a recursive search procedure seeks evidence to support that goal. A pest diagnosis example from Schmoldt and Martin (1989) shows how the search for evidence in favor of pales weevil injury (the consequent of the first rule) causes other subgoals to be investigated (i.e., *the entire seedling is discolored*). These other subgoals are antecedents in the rule that contains the current goal of interest. At some point, necessary atomic (basic) information (i.e., *tree height* in the last rule) must be known or the user is asked to provide what cannot be inferred by rules.

Data-driven control methods are appropriate for problems where:

1. All or most of the data necessary are provided in the initial problem statement.
2. The number of potential goals is large, but there are only a few ways to use the available facts of a problem instance.
3. Solutions or goals are difficult to formulate. Problems, whose solutions are constructed from many separate pieces, would be an example.

If an accomplished human expert is available, that individual could provide a good example, via his or her problem-solving methods, of which strategy to employ. How one might go about eliciting this type of knowledge from an expert is discussed in Chapter 5. Regardless, the system designer should consider the availability of data and the ease of determining potential goals.

Variations on these two approaches usually result in a mixture of forward and backward chaining. Intermediate hypotheses in the data-driven phase may be used to select a goal for investigation in the goal-driven phase. When the goal-driven phase reaches a large number of dead ends, the data-driven phase is applied again to suggest additional hypotheses for investigation. This approach results in a search from both data and also goals, hoping to meet somewhere in between.

2.4.3.3 Confirmation by Exclusion. Under certain circumstances, solving a class of problems may require the elimination of all infeasible solutions in order to select the most viable one. This **confirmation by exclusion** approach is essentially exhaustive, but with a negative criterion. That is, we eliminate the worst, and the best is what's left. Buchanan and Feigenbaum (1978) used this approach with generate-and-test in the DENDRAL system. Pople (1982) found confirmation by exclusion to be a useful technique in the face of absent or limited medical data. When the elimination process of confirmation by exclusion does not reduce to a single solution, we are left with a partial solution consisting of several alternatives. Some other technique may then be applied to further discern a single solution. On the other hand, with incomplete data or uncertain data, a set of likely solutions may be the best answer possible.

As an example of confirmation by exclusion, the organization of rules in Figure 2–3 shows *Elimination Rules*. These rule were used to exclude a pest from consideration prior to expending any effort to find evidence confirming that pest. Pests that were not excluded in this manner were then included in the final solution set of pests.

2.4.3.4 Intermediate Hypotheses. It is often impossible or confusing to try to infer a final solution from the data comprising the initial specification of a problem. We don't generally solve problems this way because often there exists no clear and direct connection between observations and solutions. Several stages of intermediate inferences usually lie between initial data and a final solution. Because these intermediate statements about a problem are not observable facts, we label them hypotheses.

Intermediate hypotheses allow the aggregation of smaller pieces of a problem into progressively larger pieces. These larger pieces contain generalizations that may, in turn, indicate greater generalizations or maybe possible solutions. In Figure 2–11, we are trying to infer a damaging agent in the first rule. Some of the antecedents for that rule are intermediate hypotheses and can, therefore be inferred by using other rules, i.e. the second and third rules in the figure. Intermediate hypotheses strategically organize a subject area hierarchically, allowing us to create a tree of inferences. Each hypothesis may be used for more than one hypothesis above it in the tree, so technically this is not a tree but rather a graph. In actuality, producing a graphical or tabular representation (i.e. dependency tables in Chapter 10) of this inference graph may be quite useful during system development. One graph could be constructed from facts and hypotheses used to confirm each hypothesis and solution; another graph could be constructed similarly, but to refute each hypothesis and solution. The result would be a complete description of all possible inferences for the problem domain. It would then remain to represent this knowledge and include additional control strategies to navigate the inference structure.

Intermediate hypotheses are quite useful apart from organizing knowledge and guiding system development. As explanatory rationale is provided to an end user by a system, the structuring of hypotheses enables the user to see greater detail of the logical steps involved in a decision. There are fewer leaps of faith from data to conclusions. The benefits of intermediate hypotheses make them an almost indispensable strategic tool for any good KBS.

2.4.3.5 Divide and Conquer. The idea of breaking a problem up into smaller, easier to manage subproblems is not a new one. We were taught to do this through most of our mathematical training; although it wasn't formally named as such. Sometimes it had a name like "integration by parts". *Divide and conquer*, for example, is the basis of recursion (see Chapter 7) and has been an established technique in computer programming for a long time (Horowitz and Sahni 1978). Backward chaining and intermediate hypotheses techniques are both strategies for subdividing a larger, more difficult problem into manageable units. Here we are giving this general, divide-and-conquer approach a particular name and suggesting that there may be other ways of applying it. For example, we might want to create a hierarchy of decision trees, where each one addresses one aspect of a larger problem. When working with constructed-solution problems (i.e. problems whose solution consists of several component pieces), each of the components of a complete solution can be tackled separately. The divide and conquer technique assumes a high degree of independence between subproblems. Any dependencies should be few and well understood.

2.4.3.6 Group and Differentiate. Classification types of problems are characterized by the need to select among a group of alternative solutions when recommending a solution to a particular problem (Schmoldt 1989). One strategy for accomplishing this classification is to *group* alternative solutions into categories according to some measure of similarity among the representative solutions within each group. Then, an additional strategy can be utilized to *differentiate* between those solutions within a group. This process corresponds to a two-stage hierarchical classification. First, a particular problem is classified as belonging to a particular group, and a second classifier further distinguishes it as most closely similar one of the members of the group. Sherman (1981) found group and differentiate to be superior for diagnosing birth defects as opposed to scoring and ranking hypotheses on evidence alone.

Another advantage of this approach is that it produces, in the worst case, a partial solution consisting of a group of possible solutions when no single, best solution can be produced. When a system implements this process it becomes more robust in response to a wide variety of incomplete inputs (Schmoldt 1989). Combined with confirmation-by-exclusion during the second stage, this becomes a very effective control strategy.

2.4.3.7 Matching. Resorting to some sort of *matching* procedure may indicate that a good model of problem solving is lacking. Matching was first described by Newell (1973) and used extensively in R1 to configure VAX computer systems (McDermott 1982). In this control strategy, the data describing a problem are matched against possible solutions using some comparison criteria; the best match indicates the best solution. The matching mechanism employed may require either exact patterns or approximate patterns. In the first case, all necessary data are expected to be available for the matching procedure; the second case implies that partial matches are allowed and useful. Unfortunately, except when the matching criteria are very intuitive, it will be difficult to provide a user with any reasonable account of the decision-making behavior.

2.4.3.8 Evidence Accumulation. In a court of law, jurors often use *evidence accumulation* to determine the guilt or innocence of a defendant. Similarly, in certain problem-solving situations selecting a final solution on the basis of the total weight of evidence favorable to that solution may be a desirable strategy for controlling the search for acceptable solutions. Individual pieces of evidence (facts and hypotheses) may contribute a certain amount of support for (or possibly against) one or more hypotheses. These hypotheses are often final solutions, but not necessarily. After all evidence has been collected and applied through inferencing, the solution with the most support is accepted as the best answer. This is one of the ideas behind MYCIN's implementation of certainty factors (Shortliffe and Buchanan 1975), and it was applied to pest diagnosis by Schmoldt and Martin (1989).

2.4.3.9 Truth Maintenance. A system that represents entities, values of those entities, and relationships between those values, and makes inferences by maintaining consistency within this representation is referred to as a **truth maintenance system**. Early work by Doyle (1979) dealt with revision procedures for modifying knowledge non-monotonically. Bayesian belief networks (Pearl 1986) constitute one type of truth maintenance system in which the values of entities are the probabilities of particular events, and relationships are conditional probabilities between those events. Truth maintenance systems are well equipped to deal with incomplete information and are able to identify incompatible inputs. Explanatory rationale then becomes more like a set of consistency constraints, rather than a sequence of deductive decisions like in rule-based systems.

2.4.3.10 Constraint-Based Reasoning. Constraint-based methods identify problem solutions by developing a compatible set of parameter values that satisfy certain problem-specific constraints. Because these methods need to maintain compatibility among the parameters describing a problem, constraint-based methods can be viewed as a variant of truth maintenance systems. However, instead of defining constraints *a priori* that must be adhered to, constraint manipulation attempts to reach a solution by dynamically constructing constraints that a viable solution must meet. Also, the final solution may contain specific values subject to certain constraints. Because constraints may change as the solution procedure proceeds, it may be necessary to revise previously accepted parameter values and assertions.

This approach is useful when a solution can be represented as a set of parameters. For instance, managing wildlands often requires that fires be introduced to main wildlife habitat, reduce dangerous wildfire fuels levels, etc. Planning for a prescribed burn may require setting certain limits on particular burn parameters. The set of all burning parameters and their ranges of values might constitute a solution to such a planning problem.

Under this general technique a number of individual techniques can be used nonexclusively. *Least commitment* maintains parameter bounds as wide as possible for as long as possible, only constraining when absolutely necessary. In this way constraints may not have to be relaxed later. *Constraint propagation* represents solutions exclusively by parameter constraints and, hence, only the value ranges for parameters need to be propagated through a system. *Constraint relaxation* applies some procedure to select constraints that can be relaxed. This relaxation may be necessary at the end of a consultation to produce a final solution from an overly constrained problem, or it may be required during constraint propagation to remove conflicting constraints as they occur.

2.5 KNOWLEDGE-BASED SYSTEM ARCHITECTURES

Knowledge-based system *architectures* are a means for organizing larger knowledge-based systems. The architecture provides the framing or context within which KBS modules operate. As a knowledge base grows larger, three distinct problems appear (Parsaye and Chignell 1988):

1. the system becomes harder to understand
2. different types of knowledge and different knowledge representation and inference methods need to be used, and
3. response time begins to deteriorate as the amount of computation increases

Issues which KBS architectures need to address are (Kahn 1982, Stefik and Bobrow 1985):

Information hiding. Information hiding restricts certain information to only those modules that need to use it and hides information from others. In some cases information between modules may be contradictory so it is important to provide some insulation between those different versions of reality.

Modular entities. When multiple paradigms of knowledge representation and inferencing are used, there are no globally uniform mechanisms for dealing with entities in the domain (Leung and Wong 1990). Therefore, these entities can only fit together if they are modular.

Extensibility. Large knowledge systems grow larger and more complex because there is incentive to support the large initial investment by expanding and improving the original. Therefore, the system must be extensible and allow for the introduction of new or better knowledge sources without drastically changing the way that they are used in the system.

Data abstraction. Data abstraction helps simplify many information changes by hiding data storage and manipulation details from higher level algorithms.

Powerful computational semantics. Problem-solving models, inferencing, and control methods must be capable of operating on very different data. Data may contain temporal or spatial relationships or may be uncertain, and finding solutions may require causal as well as heuristic models.

Some common architectures are: multiple contexts, model-based reasoning, and blackboards (Walters and Nielsen 1988). The basic requirement is that the modules which make up the knowledge-based system be constrained as follows:

1. the code in each module is not directly accessible to any other module;
2. each module has a local memory space which can be manipulated by its own code but not by other modules;

3. there exists a centralized, carefully controlled way for modules to communicate with each other.

2.5.1 Multiple Context Architecture

In section 2.1 we presented 5 methods of representing knowledge within a single *context*. A context refers to a knowledge base with knowledge representations that are used for solving problems within a single universe (Harmon et al. 1988, Walters and Nielsen 1988). For some problems, it is necessary to solve the same problem in different contexts (universes) and then decide which one is most relevant to the present problem. For example, suppose we have an aspen forest for which we need to find a forest management prescription. If timber management were our only "context", we could provide a single system to determine some plausible set of treatment recommendations, as was done by Rauscher et al. (1990). However, if we were interested in multiple-use management, we might consider analyzing the aspen forest in "multiple contexts" such as wildlife, recreation, esthetics, as well as timber. A final recommendation might then result from a conflict resolution knowledge system that identifies conflicts and uses some procedure to ameliorate the disharmonic solutions.

2.5.2 Model-Based Reasoning Architectures

Model-based reasoning refers to an architecture that combines aspects of traditional modelling methods and AI techniques. For example, a typical KBS approach might "reason" about the silvics of a forest stand by using heuristic rules, whereas a traditional model-based approach might treat the different components of the forest stand, e.g. trees, understory growth, soil, climate, wildlife, as separately modeled compartments with quantitatively described flows of matter and energy between them. When the heuristics in a typical knowledge-based system fail, the system normally fails as well. Under the best circumstances the system may ask the user to "teach" it a new case to incorporate into the knowledge base (see chapter 3), but this still scores as a failure when viewed in the context of the current problem to be solved.

In the latter case of the traditional modelling approach, model results are "inferred" from the model structure, i.e. the organization of compartments, and the mathematics of each component. But there is little opportunity to understand the operation of the forest system, because cause-effect is hidden within a set of mathematical relationships. A model-based reasoning architecture, however, combines these two approaches by using the overall structure of a traditional systems model with the heuristic reasoning and inferencing capability of KBSs to implement individual components. This marriage of KBSs and systems modelling results in a system with greater explanatory power, but less precise solutions (Schmoldt 1991).

Furthermore, by combining a typical KBS and a model-based architecture KBS into one system, it is possible to reason at both a shallow and a deep level. When the "shallow" knowledge, represented by the heuristics, fails (Walters and Nielsen 1988), the deep knowledge model can be invoked to reason from first principles. Torasso and Console (1989) refer to these types of systems as "second generation" expert systems. In effect, this architecture simulates how a human expert responds when stumped by an unfamiliar case. That is, when heuristics fail, humans often resort to a basic model of the domain to arrive at a solution. A system capable of applying basic principles about a domain enables us to build more robust KBSs. Unfortunately, developing these domain principles, building the model-based reasoning system, and linking these systems into the KBS are all difficult problems that have confined this method largely to the research laboratory.

2.5.3 Blackboard Architecture

The **blackboard** knowledge system architecture originated with the HEARSAY speech understanding system of Erman and Lesser (1975). A blackboard framework is a scheme for organizing reasoning steps and diverse domain knowledge to construct a solution to a problem (Englemore *et al.* 1988). The central issue is what, when, and how knowledge will be applied. In contrast, a traditional problem solving model consists of a program acting on a database (Figure 2–12a), where problem-solving knowledge and control structures are embedded in the procedures. For a "classical" KBS structure, the input to the model and the results of computations are kept in working memory (Figure 2–12b). The reasoning engine interacts with the working memory and the knowledge base until a satisficing condition is detected. One major weakness in this structure is that the knowledge representation is dependent upon the nature of the reasoning engine which requires that all knowledge must be represented in a uniform way (Englemore et al. 1988).

The blackboard architecture (Figure 2–12c), is a collection of independent entities communicating only via a global database, the blackboard. In addition, a "supervisor" agent, with one or more problem solving strategies, is used to explicitly control the order in which the other entities gain access to the blackboard. The entities *never* communicate directly with each other. Thus applications can be implemented using different combinations of knowledge representations, reasoning schemes, control mechanisms, and programming languages. Blackboarding makes the modular organization of knowledge easier, provides a natural model for extension into a distributed or parallel-processing computing environment.

The blackboard framework can be used to implement the multiple-context architecture through its strictly enforced knowledge modularity. In theory,

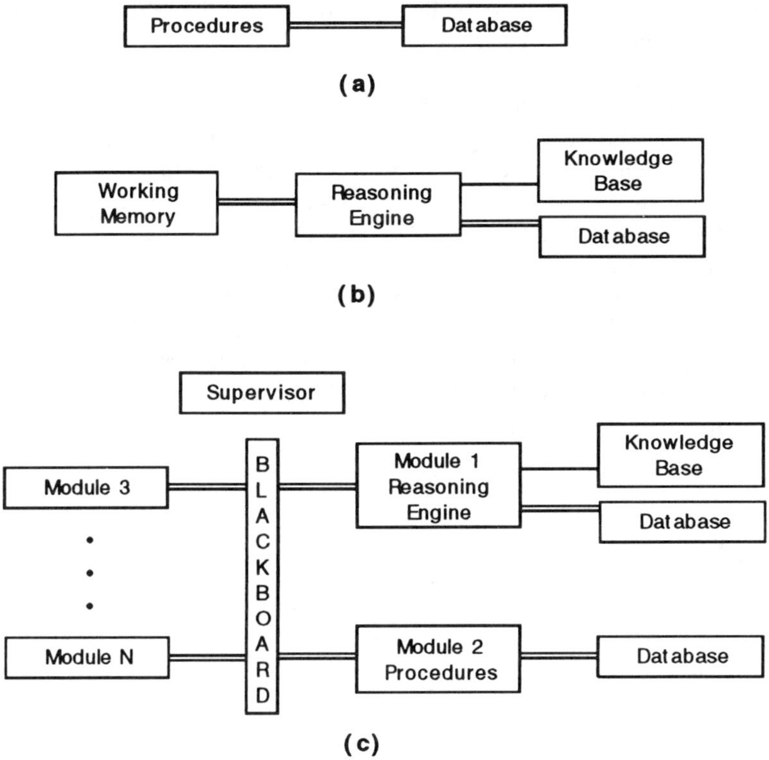

Figure 2–12. A schematic representation of three common problem solving architectures indicates three possible methods for structuring knowledge: (a) simulation model architecture; (b) "classical" knowledge-based system architecture; (c) blackboard framework architecture.

model-based reasoning may also be adapted to the blackboard concept. An implementation of the blackboard architecture is presented in chapter 8.

2.6 SUMMARY AND ADDITIONAL READINGS

Knowledge representation is one of the most challenging tasks in the design of knowledge systems. It is here that the development team of knowledge engineer(s) and domain expert(s) will spend most of available resources and it is here that formal guides can help the least. Representing knowledge is an art, not a science, and can best be learned by doing, preferably under the guidance of an experienced teacher. Additional discussion about knowledge representation may be found in Luger and Stubblefield (1993; Chapters 8–9), Walters and

Nielsen (1988, Chapters 13–14), Parsaye and Chignell (1988, Chapters 4–5), Harmon and King (1985, Chapter 4).

The reasoning mechanism in a KBS finds new assertions from a set of established assertions. Control of the inference process deals with issues of when to begin, which knowledge representations and inferencing procedures to call up and in what order, and when to stop. Further readings on these topics may be found in Parsaye and Chignell (1988, Chapter 7), Luger and Stubblefield (1993, Chapters 2–5), and Carrico et al. (1989, Chapter 2).

Knowledge-based system architectures are knowledge-system organizational models rather than knowledge representation methods. Nevertheless, architectures have as much to do with actual implementation as do knowledge representations. Further reading may be found in Walters and Nielsen (1988, Chapters 15–17), Parsaye et al. (1989, Chapter 4), Englemore and Morgan (1988), and Carrico et al. (1989, Chapter 7).

CHAPTER 3

Other Knowledge System Components

The knowledge we have acquired ought not to resemble a great shop without order, and without inventory; we ought to know what we possess, and be able to make it serve us in our need.

Leibnitz

In chapter 2, we introduced the important concepts of knowledge representation, reviewed inferencing methods grounded in symbolic logic, discussed methods to control the inference process, and introduced several knowledge system architectures. These issues make up the core of a knowledge-based system (Figure 3–1). To increase the utility of KBSs, we must also provide for explanations, interfaces both with the user and other computer software, and user-guided learning at run-time so that the domain knowledge available to the user can expand. These topics are discussed in this chapter. To round out our presentation of knowledge system components and techniques, we also review some ways to reasoning with uncertain knowledge. Following these topics, the remainder of the text examines how to apply those ideas to design, develop, and implement knowledge-based systems.

3.1 EXPLANATION AND JUSTIFICATION

The purpose of an explanation facility is to enable a program to display an understandable account of the motivation for all of its actions (Davis and Lenat

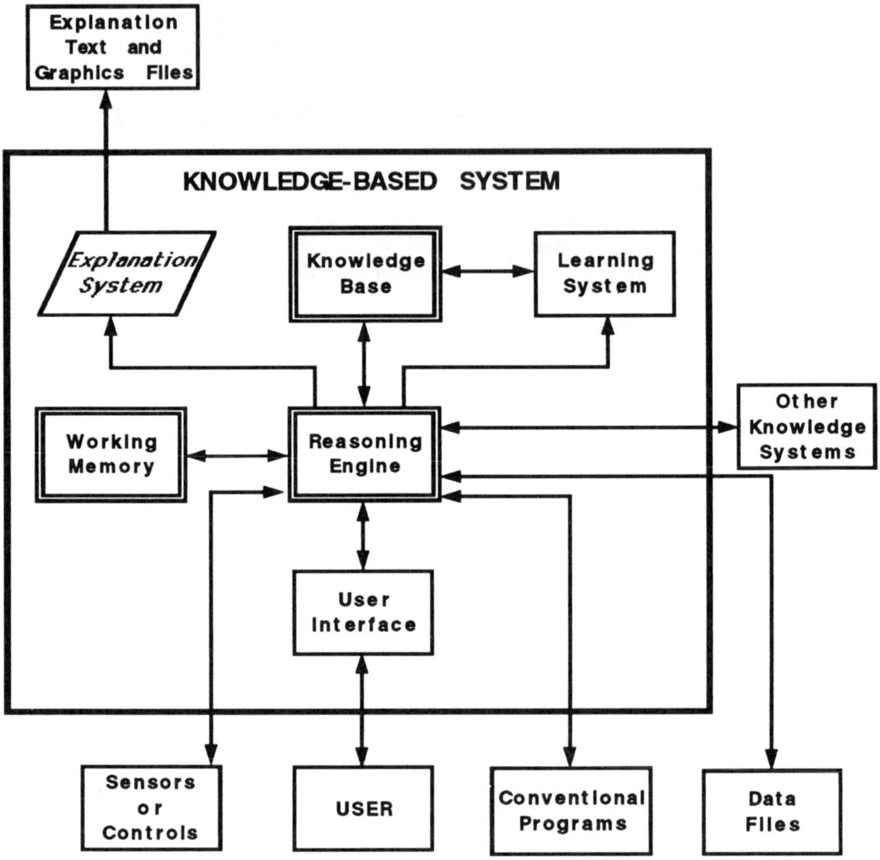

Figure 3–1. Explanation capabilities are an important add-on feature to the basis components of a KBS.

1982). Explanation is part of the larger issue of **human factors engineering**, which also includes the user interface (see 3.2 and Chapter 6). Explanation systems are not involved with the correct execution of knowledge-based systems. Instead, their focus is on helping the user understand what is happening within the system. It is important to write programs that can explain their own actions for three major reasons.

First, it is important to convince users that a system's conclusions are reasonable. Explanation and justification are essential to convince users that the reasoning of the KBS is correct (Jackson 1986). Explanation provides a basis for the user to either correct his or her own judgment or to reject that of the KBS (Sell 1985).

Secondly, users should understand how a system reached its conclusions. A knowledge-based system should not appear like a black box with its inner workings hidden from view. If a user is able to ask the system why certain information is needed, or how a particular conclusion was reached, or why a particular conclusion was *not* reached, results become more acceptable, even to those less knowledgeable in the domain. Typical consultations with human experts involve dialogues between advisor and advice seeker; a similar capability is highly desirable for knowledge-based systems.

Third, a good explanation facility can help the developers debug the system. Knowledge base complexity grows as size increases. Without some explanation facility, faulty reasoning is difficult to locate and correct. The system developer can use explanations to better tune the knowledge base to have more realistic dialogues with the user as well as to correct misbehavior (Merritt 1989).

According to Sell (1985), there are at least three kinds of explanations:

1. *interpretive* – explains what; gives definitions and details for terms that appear in questions to the user and in the knowledge representation language (e.g. rules);
2. *descriptive* – explains how; states facts, relations, criteria, purposes, and how these are derived; and
3. *reason-giving* – explains why; states reasons, causes, laws, principles, to validate program behavior.

3.1.1 Interpretive Explanations

Interpretive explanation is designed to help the user understand any questions that are asked of him or her. This can be accomplished by giving more background information about the question or by defining any specialized vocabulary contained in the question. Interpretive explanations are usually available only upon direct request by the user. Some examples of interpretive explanations are:

1. explain further- restate the question and amplify the available background information
2. define 'Word'- what does 'Word' mean?

In the PREDICT expert system for diagnosing pest damage in red pine stands, a user might be asked to provide information about branch flagging, i.e., "Has flagging been observed on the branches of affected trees?" Figure 3–2 displays the type of explanatory material that a user can request when presented with the above query. This explanation contains both types of interpretive material. That is, Figure 3–2 both defines the term "branch flagging", and it explains the significance or implication of flagging.

FLAGGING

When an entire section of foliage on a shoot or branch is discolored (flagged), the causal injury is usually within or at the base of that section. Individual shoots (shoot flagging) or entire branches (branch flagging) may be affected.

Figure 3–2. This explanation screen for the term "flagging" from the knowledge base of PREDICT provides the user with both interpretive and descriptive information. Terminology is defined and their importance to the problem at hand is described.

3.1.2 Descriptive Explanations

Descriptive explanations, also called *justifications* (Kline and Dolins 1985), provide a trace of, or a rationale for, the steps used by the program to arrive at a solution. Justifications may also include rationale for not choosing certain alternative reasoning branches.

When descriptive explanations for decisions are provided by a system, it may be necessary for that system to maintain a *trace* of each decision step. A trace is a record of which decisions were made and in what order they were made, e.g. "fired" rules in a rule-based system. Two subcategories of justifications may be usefully identified here:

1. *Introspection* (trace dependent): examination of the trace of the problem solving process. This method relies on building a proof tree, i.e. a record of details of how conclusions were obtained;
2. *Concoction* (trace independent): producing a justification that is only partially dependent on the actual decision-making trace. This method relies on a well designed series of textual statements that when strung together give a plausible explanation of the conclusions produced by the system (Lee 1986). These filtered explanations allow the creation of an explanatory theory that complements rather than strictly repeats the trace (Sterling and Shapiro 1986, Riordan and Carden 1990).

In both trace-independent and trace-dependent explanations, it is desirable to include explanations of the problem solving strategy that was actually used by the reasoning engine (Chandrasekaran et al. 1989).

3.1.2.1 Trace-Dependent Explanations. Descriptive explanation often only involves explicating heuristic knowledge and can be performed via a backtrace of the heuristics (often in the form of domain rules and meta rules). For example, in the pest hazard rule:

IF the average diameter of trees in the stand is greater than 2
AND the condition of the stand is stressed,
THEN there is MODERATE HAZARD for Red Turpentine Beetle,

only the facts used to conclude hazard for red turpentine beetle are conveyed to the user. There is no biological causation to indicate why these factors are important to a particular hazard assessment for a particular pest. Some systems, such as taxonomic classification systems, are based strictly on empirical knowledge, also referred to as "surface level" knowledge. In some classification systems, no underlying theory exists while in others the underlying theory is unknown (Merritt 1989). When a system reasons using heuristics, it may be sufficient to relate motivation to the user at that "surface" level. On the other hand, the user may need a "deeper", more detailed level of reasoning to understand the rationale involved in a decision (see section 3.1.3 Reason-Giving Explanations). In the latter case, a system must use or, at least, possess a detailed model of the domain. Which level of explanation is provided will be determined by the type of use a system is designed for, general knowledge level of the end users in the domain, and existence of a well-understood and detailed model of the subject area.

There are a number of trace-dependent explanations that are of interest and should be included in a well rounded knowledge system development toolkit (Lee 1986):

1. why is this question being asked?
2. how was this conclusion reached?
3. why was a goal NOT concluded as true?
4. browse through a truth maintenance system
5. turn on a full trace showing each rule attempted along with a short explanation

Most current explanation facilities in KBSs are typically rather limited (Wick and Slagle 1989). Two types of explanations are almost universally offered: *why* and *how*. The origin of this type of explanation can be linked to the discovery made by the originators of MYCIN that a trace of the problem-solving activity can yield useful explanations (Davis and Lenat 1982). Implementation of these two facilities is graphically-based in more sophisticated systems and is text-based in most others. The other trace-dependent explanation facilities (3-5 above) are infrequently available.

The problems with trace-dependent explanations are:

1. Explanations are heavily influenced by what is easily and naturally produced by the KBS structure rather than by what the domain experts are familiar with (Millet and Gilloux 1989).
2. For any but the smallest knowledge base, trace-based explanations typically produce too much output (Sterling and Shapiro 1986). The overwhelming amount of text is not well organized and is simply not intelligible.
3. Often a knowledge base contains control structures, e.g. meta rules. Because this control knowledge is encoded in the cryptic, difficult-to-read format of the knowledge base language, it often has little or no meaning to the user when displayed as part of an explanation trace.

An abbreviated meta rule from PREDICT appears in Figure 3–3. When a user is presented with a trace of reasoning by the system, these sorts of rules appear along with the more intelligible domain rules. Without going into any explanation of this meta-rule, it is apparent that this type of control information remains incomprehensible to all but the most knowledgeable user.

3.1.2.2 Trace-Independent Explanations. Trace-independent explanations, also known as knowledge base analysis (Chandrasekaran et al. 1989), are implemented by developing a system which analyzes the knowledge base and allows the user to browse through the resulting summary knowledge. Explaining the knowledge base does not depend upon having an execution history of the reasoning engine (a trace) available.

Trace-independent explanations may explore the knowledge base along with the user or create an entirely new knowledge base depicting the causal theory supporting a problem-solving knowledge-based system (see section 3.1.3). Some functions that are important for trace-independent explanations are presented below. Ross (1989) and Wick and Slagle (1989) present these and many more ideas for knowledge base analysis:

1. list all the rules in the KB
2. list a particular rule
3. list all the facts in the working memory
4. view a hypertext system of the entire KB
5. activate a DBMS w/ SQL query access to KB
6. list the set of all predicates used in either the premise or conclusion of any rule
7. list the set of all predicates used in any rule premise
8. list the set of all predicates used in any rule conclusion
9. list the set of all predicates that can be asked about (i.e. for which questions exist)

```
RULE [ Set order using flagging ]
IF Area Selection = OFF
AND NOT the type of needle injury is defoliation
AND flagging is present on the branches
AND investigate possible damage by Saratoga spittlebug
AND investigate possible damage by Pales weevil
AND investigate possible damage by Root tip weevil
AND investigate possible damage by Winter injury
AND investigate possible damage by Red pine needle midge
AND investigate possible damage by Red pine shoot blight
AND investigate possible damage by Scleroderris canker
THEN Select Order CF 95
AND CONF(investigate possible damage by Pine bark beetle) := 100
AND CONF(investigate possible damage by Needle cast) := 100
AND CONF(investigate possible damage by Pocket gopher) := 100
AND CONF(investigate possible damage by Meadow mouse) := 100
AND CONF(investigate possible damage by European pine sawfly) := 100
AND CONF(investigate possible damage by Red pine sawfly) := 100
AND CONF(investigate possible damage by Red-headed pine sawfly) := 100
AND CONF(investigate possible damage by White grub) := 100
AND CONF(investigate possible damage by Pine tussock moth) := 100
```

Figure 3–3. This meta rule from the PREDICT knowledge base demonstrates how cryptic and uninformative meta rules can be for anyone unfamiliar with the rule structure of the knowledge base. Basically, this control rule says that if there is no defoliation but there is branch flagging, then the pests listed in the premise should be considered further. The pests listed in the conclusion of the rule have their "investigate" clause set to a confidence factor (CF) of 100 explicitly, so the reasoning engine will not consider them further.

10. list the set of rule names of rules that can conclude each premise
11. list the set of rules which use each predicate in their premise
12. for each predicate, list the set of all values it can hold according to the usage in the KB

3.1.3 Reason-Giving Explanations

Reason-giving explanations, also called *warrants* (Kline and Dolins 1985), describe the authority (usually causal) for any judgments made by a system. Warrants generally require a much deeper representation of the domain to defend the knowledge contained in a KBS. Often both dynamic generation of text and rules from deep knowledge may be used. Static text is very easy, but time-consuming, to include within a system. Dynamic text and rule generation

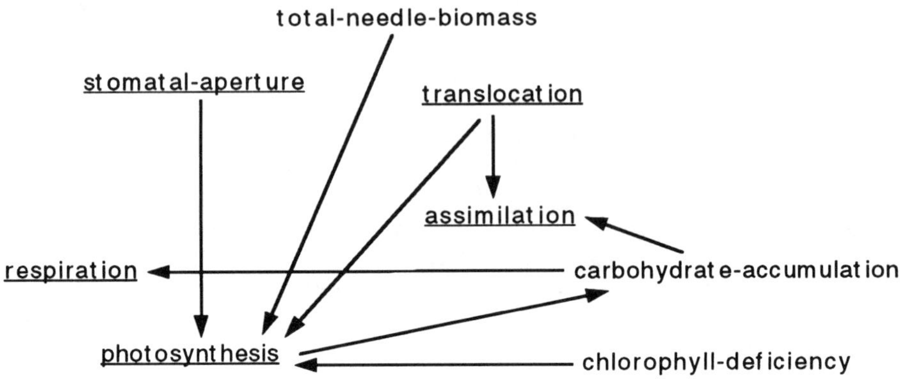

Figure 3–4. An influence diagram like this one shows relationships between model parameters. Links indicate that one parameter exerts an influence on the value or on the change of another parameter.

(Swartout 1983) is much more difficult to accomplish, but, on the other hand, large volumes of text then do not have to be entered into a knowledge base.

In some cases, it may be desirable to develop a causal model as a second knowledge base. This would parallel the problem-solving knowledge base but would focus on explanation. This task is very demanding because it requires the availability of a theory of the domain which can then be modeled (Millet and Gilloux 1989). Unfortunately, a causal-model theory is not available for many domains of interest.

As an example of a system that develops a causal model of deep knowledge Schmoldt (1991) created a qualitative simulation model of atmosphere-plant-soil interactions for ponderosa pine (*P. ponderosa*) in an attempt to understand tree responses to ozone and drought stresses. Figure 3–4 show part of an "influence" diagram in which links between various simulation parameters are labeled. Each influence represents a dependency between parameters. At the most basic level, each parameter changes its value over a particular time frame, e.g. hour, day, week, etc. Temporal heuristics can be used to relate parameters that vary over different time frames, for example:

```
IF often(day, photosynthesis(hour, LOW))
THEN carbohydrate_accumulation(day, MODERATELY-LOW).
```

This heuristic states that if the hourly value of photosynthesis is LOW often during a day, then carbohydrate accumulation for that day will be MODERATELY-LOW. As longer time frame changes occur, they can be logged and used later to produce interpretive causation explanations for simulation results.

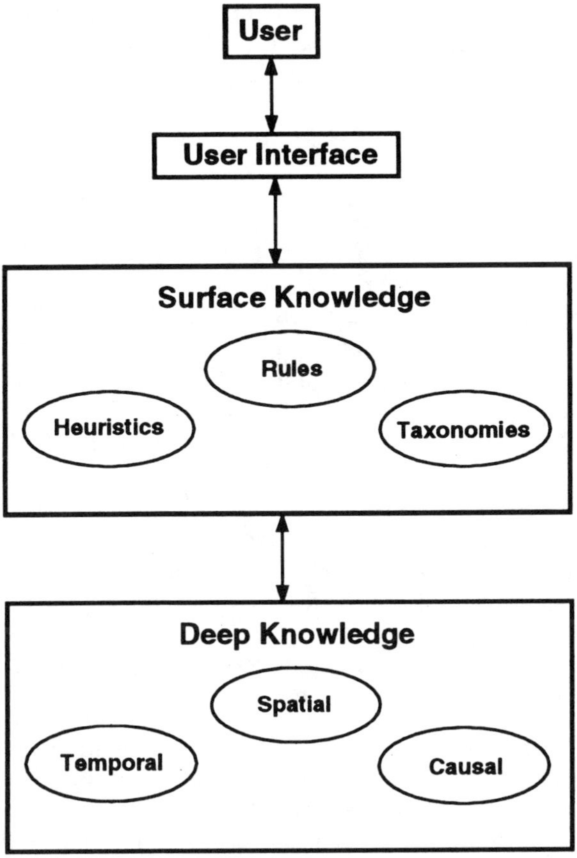

Figure 3–5. Second generation knowledge systems include causal models beneath the surface of the heuristic and compiled knowledge. These "deep knowledge" models are useful for providing reason-giving explanations and for solving hard problems that the surface knowledge is unable to solve.

So called "second generation" knowledge-based systems incorporate "deeper" domain knowledge in their explanation systems (Torasso and Console 1989). These more advanced systems are not designed to represent only associations between hypotheses and observational data ("surface" knowledge). They are extended to include some understanding of the modeled system as well ("deep" knowledge). Additional layers of complexity are added to the traditional, heuristic domain models to provide this enhanced level of understanding (Figure 3–5). These layers may include temporal, spatial, and causal knowledge. Heuristic and interpretive layers may be used to insulate the user from much of the detailed knowledge until it is needed for solving harder problems or for producing explanations. This architecture can be very useful for accommodating variable user understanding and explanation needs.

3.1.4 Important Issues in Explanation Management

The ability of knowledge-based systems to explain their results and for their reasoning chains to produce these results is of crucial importance. At the same time, there is a wide-spread consensus that the explanatory power of most current systems is rudimentary and generally unsatisfactory. It is fair to say that the art of giving good explanations is not yet well developed. Most current approaches to knowledge-based system development require expressing knowledge in such cryptic ways and at such low levels that it is hard to build meaningful high-level explanations (Chandrasekaran et al. 1989). Consequently, much research focuses on developing new concepts and methods to improve explanation in knowledge-based systems and to better understand what type of explanations people deem acceptable in everyday life (Schank 1986).

Several issues are summarized, below, which make it difficult to implement and generalize more unified and expressive explanation systems.

1. *Explanation Requires Knowledge*—Research in explanation systems has highlighted the fact that explanation requires knowledge *beyond* problem-solving knowledge (Swartout 1983). However, given explicit encoding of problem-solving strategies, explanation systems can use this knowledge to offer users explanations of the overall problem-solving strategy of the system (Hasling et al. 1984). Also, when given the availability of "deep" knowledge, explanation systems can provide higher quality explanation of the shallow, problem-solving knowledge (Chandrasekaran and Mittal 1983). Most system builders do not commit the additional and significant resources to add non-problem solving knowledge to a knowledge-based system.
2. *Explanation Query Types*—Most KBS offer only two types of explanations—*why* and *how* (Neches et al. 1985, Wick and Slagle 1989). We need to expanded the standard queries to include the *what* type of query, designed to give insight into definitions of objects and events used in a knowledge-based system (Rubinoff 1985). One of the most difficult problems in the explanation area is to answer *why not* questions (Millet and Gilloux 1989). Schnupp et al. (1989) provide a knowledge-based system that shows one method for answering the *why not* question.
3. *Explanation Levels*—It has been found that explanations should vary in the amount of detail presented to the user as well as in the type of content (Paris 1987). Naive audiences need explanations that focus on "process" whereas expert users require explanations of "structure". Making explanations user dependent requires a model of the user—a current research area. State of the art

adventure-type video games pass the user through a personality test whose results determine the permissible actions of the user's "hero". A similar strategy might well be used to assess a KBS user's level of domain expertise. Research by McKeown et al. (1985) has been working in this direction. Another approach is to give the user a "dial" or other selector that he or she can adjust in order to specify the level of explanation desired for "why" or "how" questions.

4. *Explanation of Uncertainty*—Problems arise when trying to explain knowledge-based system conclusions involving inexact reasoning (Wick and Slagle 1989). This should not be surprising because a human expert exhibits similar difficulty. Often, "Doing it this way just feels right" or "I had a hunch" are the best explanations that an expert might provide. Fuzzy sets (discussed below) offer a way to interpret KBS levels of truth and falsehood in an intuitive vocabulary that is more easily understood by users than likelihood or probability values.
5. *Explanation Granularity*—If system behavior is represented at a higher level than rules and frames, say at the task level (Clancey 1983), explanations can be produced at the task level. The task level is closer to the concepts and symbols easily understood by domain users. Such explanations are very difficult to achieve when the system grain level consists only of rules and frames.

3.2 USER INTERFACE

The user interface is a component of human factors (see Chapter 6). A user interface refers to the physical and sensory interaction between computer and user (Figure 3–6). *Human factors engineering* tries to understand and model the total human-machine environment. In addition to a user interface, human factors issues also include social/psychological aspects of the user group and the organizational and physical climate within which the knowledge-based system is expected to operate.

The manner in which a KBS consultation appears to the end-user carries a great deal of importance. Systems may fall into disuse if they lack good user-interface facilities (Walters and Nielsen 1988). For most consultations with a KBS the user must supply some problem-specific information based on requests from the system. A number of techniques are used to accomplish this. First, more sophisticated systems that have natural language front-ends accept English-like responses. Second, it has become relatively standard for users to input data using on-screen graphic objects, i.e., a graphical user interface. Using a pointing device, such as a mouse, the user may turn an "on-screen dial" and receive visual feedback on the value that he or she is selecting. Third, on most systems, the user is presented with a query in the form of a menu of items from which answers may be selected, or line input where the user fills in labeled data cells, in a dialog box. Natural-language interfaces are the most flexible, graphics user interfaces are the most natural

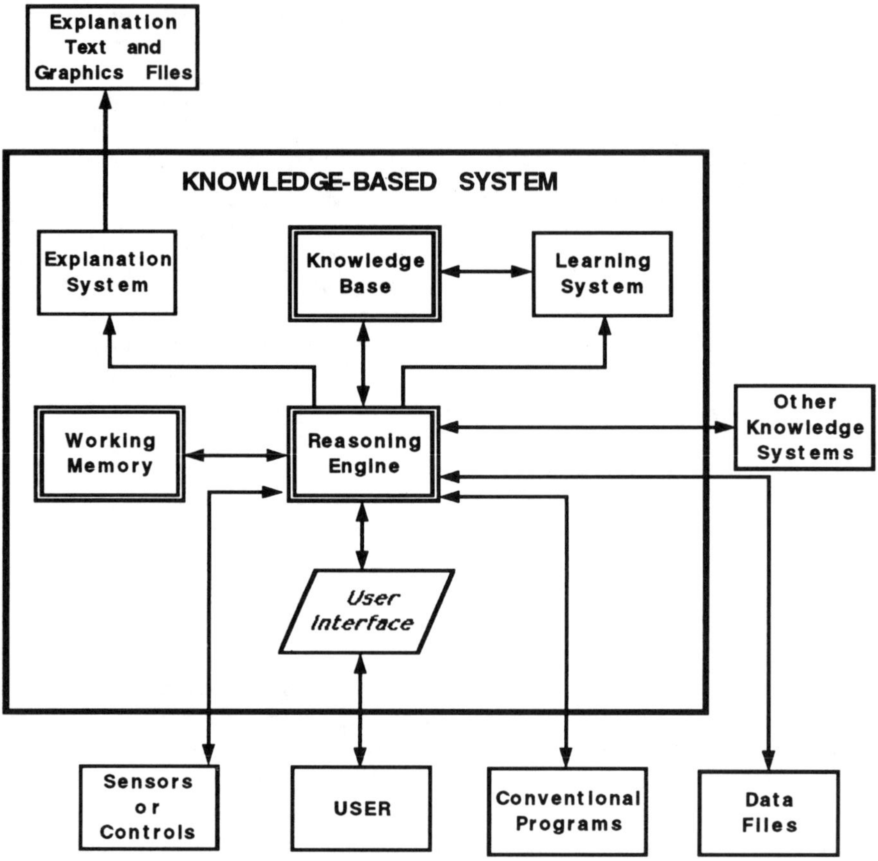

Figure 3–6. The user interface is part of a comprehensive model of human-machine interaction. How a KBS interacts with a user can often be just as important as how well the system actually performs its task.

to use, and menus restrict the values that may be specified, but are more helpful for users less familiar with the domain.

Reports of results can be displayed in a variety of ways, including customary textual descriptions of the solutions or graphically oriented presentations . These results may be displayed on a monitor screen, printed, or sent to a file for later use. As computer hardware advances arrive on the technology scene, video and voice output will become more commonplace.

Many KBSs also permit users to store the context (i.e. all facts and hypotheses) of the current session. The list of hypotheses also includes the solution(s) reached during a consultation. This snapshot of the system constitutes a record of the current decisionmaking strategy and also provides a "paper trace" of the user's problem-solving effort. By storing this information regularly, currently accepted decisions and decision strategies can be reviewed and updated over time.

It may also be desirable to provide the user with an experimentation (*what-if*) facility; this permits a **sensitivity analysis** of the solution. For example, at the end of a session, one or more answers to investigative questions could be changed; the session would be run again and the results compared with those previously obtained. In this way, the user can be aware of how sensitive a solution is to incomplete information or information which contains some uncertainty.

Suppose we have a KBS to recommend silvicultural practices for enhancing wildlife habitat. An exploratory, what-if capability could provide the user with the ability to modify characteristics of the problem (e.g. population levels of various species) and compare answers. Maybe several alternative scenarios, covering differing management objectives, could be examined simultaneously. The system could identify where differences in reasoning occurred and how varying the input changed the solution.

3.3 LEARNING CAPABILITIES

Currently, there are few KBSs that have a learning component as depicted in Figure 3–7. Most that do, have been specifically designed to learn as a primary part of their functioning (e.g. the Automated Mathematician program of Lenat, 1980 or the scientific discovery programs of Langley et al., 1987). At the present time, learning-based KBSs and expert-based KBSs are separate entities (Weiss and Kulikowski 1991). We've included learning in our comprehensive KBS design because we feel that it represents an important future step for improving the "intelligence", and at the same time functionality, of KBSs.

If we can write programs that treat themselves as data and that can rewrite themselves during run-time, then such programs can evolve and grow, becoming richer with use (Waldrop 1987). It is just such capabilities that make the AI languages such as LISP and PROLOG so different from other computer languages. These programming languages treat programs as data and are able to generate new program code during run time and then execute that code; they are dynamically extensible. Therefore, programs are able to modify themselves at execution time, and in particular, they can modify their knowledge bases and, in effect, learn.

3.3.1 Fallible Systems

In solving real world problems the question is often not so much whether an answer is "the best correct answer possible" but whether the answer is **satis-**

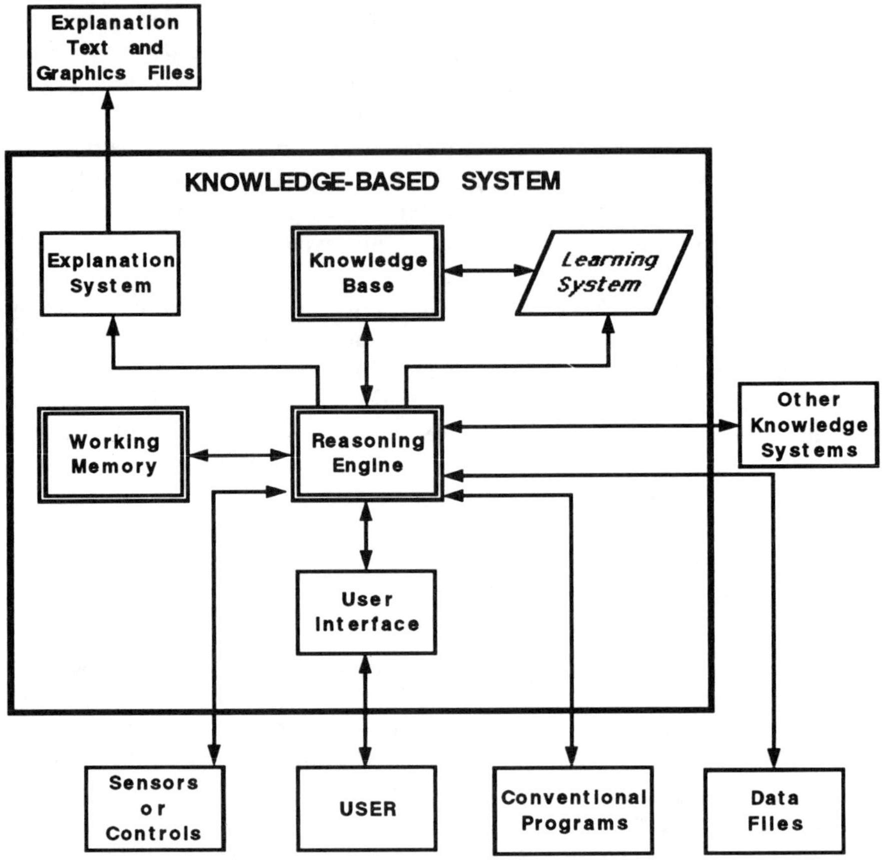

Figure 3–7. The ability of a KBS to acquire new knowledge or to record heuristics about unfamiliar cases enables it to keep current with advances in the field and to increase its breadth in dealing with exceptions.

ficing (Simon 1969). That is, does the system provide an acceptable answer to the problem in question? For example, suppose a forester has been given a goal to offer 500 cords of aspen for sale in a working district this year. In theory, an optimization algorithm, such as linear or dynamic programming, might be used to calculate some "best" selection of aspen stands to harvest among the numerous candidates. If done so, the answer might be called an optimizing solution. In practice, optimization is very expensive and often it is not realistically possible due to the large number of elements that must be considered in the solution space and the requirement to quantify all relationships including those that may be critical but inherently not quantifiable. So, our forester most often simply chooses among the candidate aspen stands, using formal or informal heuristics—selecting a set of stands that while not strictly optimal, are satisficing, i.e. they will solve the problem satisfactorily.

KBSs are absolutely different than algorithm-based systems in this sense. Because algorithm-based systems are constrained to work correctly only in neat, controlled special-case situations, they can be guaranteed to provide the correct answer when used correctly. As a consequence, their results are considered correct and so learning is not an issue for these systems. In KBSs, on the other hand, we only require that solutions are "good enough", that they satisfy some minimal criteria for acceptability. This change in philosophy allows us to address many more situations because we relax the constraint that the answers must be guaranteed to be correct in all situations or to be optimal. Because acceptability criteria may change and more situations can be addressed, KBSs can be prone to errors in judgment. That is, they will eventually make mistakes, and they must learn from those mistakes. In the most desirable case then, a KBS must somehow try to recognize when it has broken down, report to the user where it has broken down, and finally, help the user direct the KBS in repairing itself—all without the computer program failing or "bombing."

3.3.2 Learning Situations

There are several general ways that a KBS can learn. Learning is basically of two different types: supervised and unsupervised—the difference being the presence of a formal teaching element. Teaching (supervised) may consist of knowledge provided by an expert or it may consist of examples or other predefined patterns. In unsupervised learning, an unaided system creates new knowledge during the course of solving problems. In the learning approaches listed below, the first three are supervised and the remaining ones are unsupervised.

Knowledge acquisition. Knowledge is explicitly deposited in the knowledge base either through direct modification of the knowledge base or by computer-assisted knowledge acquisition. Chapter 5 covers knowledge acquisition in detail.

Learning by example. Generalized knowledge (in the form of rules) is induced from specific example problems. Michalski et al. (1983) have demonstrated substantial success in rule induction from examples for classification type problems. Considerable work has also been done by Quinlan (1986, 1990).

Learning by instruction. Given a descriptive specification for a solution, the system generates knowledge structures to implement that solution. The system takes advice from an expert or a user and incorporates it into its existing knowledge base. Advice may be either at the problem level (e.g., specify an actual solution to a new problem) or at the strategic level to suggest a general rule-of-thumb applicable to many different problems.

Rote learning. The system memorizes specific examples and re-applies them when they recur. Because of the large number of instances that must be stored and searched, this learning mode is only useful in situations where standard, or pat, solution steps are required—often at the beginning or end of a solution process.

Learning by analogy. A current problem instance can be compared to other, similar problems. The familiar problem and its solution are then used as a template to find a solution to the new problem. This new problem and its solution can then be added to the existing data base of solved problems and can be applied to other new problems in the future. *Case-based reasoning* systems rely on analogy to associate patterns, inherent in problem descriptions, with likely solutions.

Parameter modification. Weights assigned to inference rules or coefficients used in heuristic functions can be modified by a learning program as it receives feedback about its success. Weiss and Kulikowski (1984) used success rates as feedback to generalize and specialize rule-base rules—this testing and revising, however, occurred outside of their system's normal mode of operation. Samuel (1963) dynamically modified coefficients in a polynomial heuristic function in his checkers playing program based on how advantageous a board position it held several plays later.

Concept formation. Examples which possess particular attributes (e.g., chemical substances and their attributes, such as bitterness, reactivity, salinity, etc.) are used to generate higher-level concepts about classes of entities and their relation to other classes. For example, attributes about chemical substances can be used to infer the notion of acids and bases. Concept formation is used extensively in BACON and the related programs of Langley et al. (1987).

As noted previously, systems are most often designed exclusively either as learning systems or as expertise-based systems. Little, if anything, has been done to merge the two types of knowledge acquisition methods. Expertise-based systems work well for specific, narrowly defined cases and fail miserably when pushed behind this familiar territory. The ability to learn exceptions and still maintain a general knowledge of the domain would enable expert-based systems to maintain their currency for problem-solving by improving themselves as they work.

3.4 INTERFACING TO CONVENTIONAL PROGRAMS AND DATA FILES

Because KBSs are just one component of decision support software systems, we expect that most useful KBSs will have to interact with files and programs outside the KBS. On the one hand, a KBS may need to access database, spreadsheet, or text files and retrieve specific data needed by the knowledge base. On the other hand, a KBS may need to run conventional computer programs, that execute well defined algorithms, and which may use the results in its knowledge base or provide data for the knowledge base. The perspective we have proposed in Figure 3–8 places the KBS in a central position with

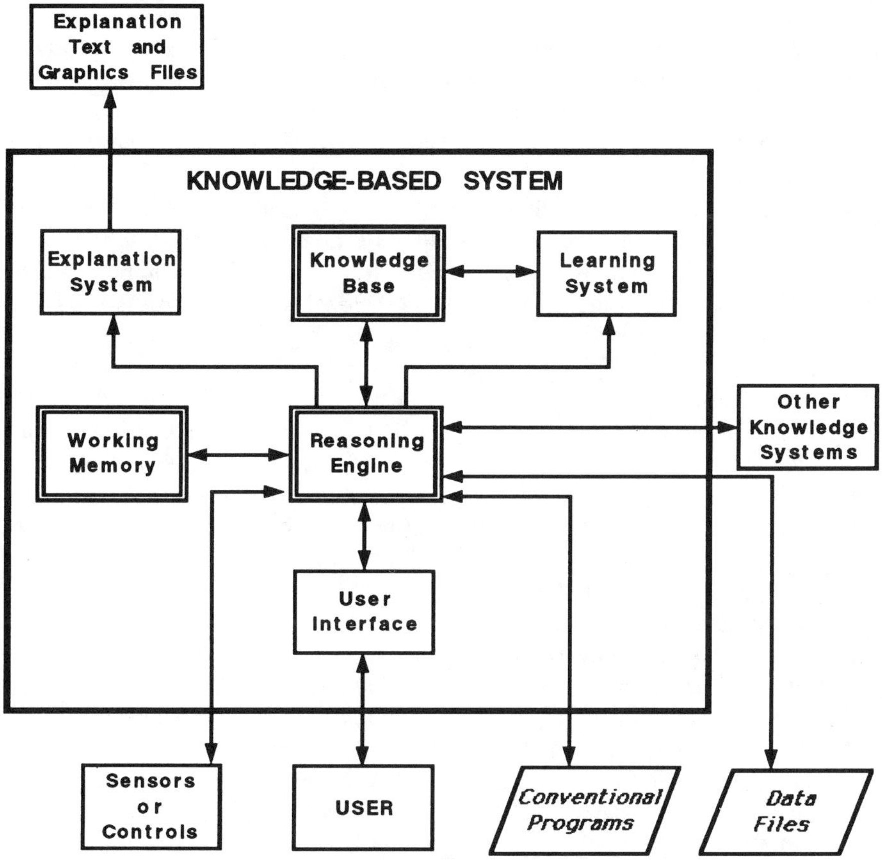

Figure 3–8. KBSs often need to be integrated with existing computing resources, such as conventional programs and data files. The reasoning engine often contains explicit calls to external software and data.

conventional programs subordinate to it. However, the programs need not be organized in that way. It is entirely possible for the main focus of a development project to be "traditional" algorithmic software with small KBSs providing key information at various points during processing. Nevertheless, to operate in either scenario, however, KBSs must be able to merge with existing programs and data files.

A later stage in the adoption of any new technology is its seamless integration into previously existing technologies and ways of doing things. So, it is likely that KBSs will become just another computer programming tool and that most popular programs will be a mixture of conventional and knowledge-based programs. In fact, we are seeing this already, even popular personal computer game programs are using AI and KBS methods. One of the most important application areas for KBS may be as Intelligent Front Ends (IFE's) (ÒKeefe 1986). An IFE is a KBS that sits between a conventional program and the user, generates the necessary input instructions and data for the conventional program following a dialogue with the user, and interprets, displays, and explains output to the user.

Geographic information systems (GIS) and data base management systems (DBMS) are both useful information storage, retrieval, and display tools. However, because of their extensive functionality, they are also very complex, which makes them difficult for inexperienced or occasional users to access. By accepting English-like phrases from a user and translating them into the necessary data manipulation and display commands of a GIS or DBMS an IFE could make these complex software systems easier to use. Several such interface systems have been developed for GIS's (e.g., Friedl and others 1988, Ripple and Ulshoefer 1987, Usery and Altheide 1988). Also, IFE's can make simulation models much easier for novices to use (e.g., Lemmon 1986) and, at the same time, guard against misuse of the model (applying it to situations outside its area of validity) and misinterpretation of the results. These are two serious problems with computational models.

When conventional programs (CP) and KBSs are combined and must interact, often they must share some data items. There are several methods for passing this information back and forth between programs. A data base program or spreadsheet program might serve as an intermediary store between a KBS and a CP. This approach might be useful when a large number of programs (KBSs and CPs) need to access and modify information. In some circumstances, it might also be necessary for a user to modify data independent of the KBS and CP, an intermediary store would easily permit that type of intervention. An example of this scenario would be the sharing and modification of corporate (organizational) data.

Another method of sharing uses the RAM memory of the machine to temporarily store data. The calling program writes the shared data to a specific bank of physical memory locations; the called program then reads that data into its own variables for processing. This technique is useful when speed is important and there are only a small number of data items to be shared. Its major disadvantage has been machine dependence. However, a variant of this approach has surface recently. This new approach, often called dynamic data exchange, is due to the movement by computer operating system architectures toward interchangeable and shared software modules rather than completely independent applications. Now many computer operating systems permit programs to send data directly to other concurrently executing programs. The operating system handles the underlying details of how the exchange actually occurs. Because many different operating systems support this data exchange mechanism, a programmer sees only the high-level, machine-independent interface.

The simplest and most machine-independent method shares data using a flat ASCII files. All programs read and write textual information to a selected file that serves as a sort of global memory for all programs in the larger system.

Because a large amount of our existing information and many traditional methods for manipulating data reside in other formats, e.g., data bases (relational and spatial), spreadsheets, etc., it becomes imperative that our newly developed systems (KBSs) interact fairly seamlessly with our old environments. The blackboard framework architecture discussed in Chapter 2 offers an excellent organizational mechanism to accommodate the complex interfacing needs of these hybrid, CP and KB systems.

3.5 INEXACT REASONING AND UNCERTAINTY

To make the best decisions possible in an imperfect world, we must often use the best available resources at our disposal. Often desirable resources are unavailable, sometimes the ones we do have are unreliable, and sometimes our understanding is incomplete and can only provide us with vague answers. For KBSs to be useful, they, like us, also must possess an ability to reason in this type of "dirty" environment. Some of the methods used for **inexact reasoning** are described in the sections that follow.

3.5.1 Uncertainty

When solving quantitative or qualitative problems, people are often faced with imperfect information, and often rely on incomplete knowledge or approximate knowledge (e.g., heuristics). Some data items may also be missing; this results from poor observation or from unavailability of the data. Additionally, not all of the data that is available about a specific problem may be known for certain; more often it is believed true with some subjective level of confidence, i.e. some belief. Also, decision procedures for the knowledge applied to a problem do not usually lead to conclusions which are absolutely certain; they are often subjective judgments or "good guesses". Therefore, the limitations of imperfect information are further compounded by the approximate nature of the heuristics relied upon for inferencing.

For any particular problem, if we are able to express all possible combinations of findings and hypotheses in a mutually exclusive manner, heuristics may not be required. This is because only one rule would be satisfied for any given case (Weiss and Kulikowski 1984). The taxonomic decision tree of Figure 2–10 is an example of such a situation in which data and conclusions are each mutually exclusive. Here, at each branching stage of problem solving, the user is presented with a set of mutually exclusive possibilities. This dichotomous method of traversing the decision space results in a single path to each possible solution. The problem space of Figure 2–6, on the other hand, has several paths to each solution (shaded nodes) and consequently allows greater flexibility in problem solving. Even so, uncertain answers to questions may still occur. Most knowledge systems incorporate some technique for coping with either or both types of uncertainty, information and knowledge. Users may be allowed to specify values indicating levels of confidence in their data, and inferences in the knowledge base may contain values to indicate the relative strength of their conclusions.

Consider the situation in which a fisheries biologist must deal with the effect that runoff from a relatively new fertilizer chemical might have on fish populations. Laboratory data may exist that suggest deleterious effects on at least one fish species. However, long-term effects on that species' life cycle are unknown, effects on other species can only be inferred, translocation of the chemical within the environment is not understood, injurious concentrations of the chemical are not precisely known, etc. The decisions that this biologist is faced with contain many reasoning steps that can only be handled by educated guessing. Not only is the available rationale vague, but pertinent information remains incomplete and, if obtainable, probably uncertain as well. In many ways, this situation is typical of decision scenarios that arise daily in natural resources management. All the different possible courses of action in a decision-making situation do not organize themselves into neatly structured decision trees with obvious road signs indicating which branch to take next. To address these difficulties, researchers have developed methods of incorporating incomplete and uncertain knowledge and data into intelligent systems.

3.5.2 Reasoning with Uncertainty

The following pages are intended to provide the reader with a flavor for some of the approaches used to address inexact reasoning. Our descriptions are not intended to explicate all the details of these methods or provide a recipe for their implementation. It is assumed that interested readers will search out the included citations and those at the end of the chapter for more details.

3.5.2.1 Bayesian Probability. One of the earliest attempts at dealing with uncertainty used likelihood as described by probability theory. PROSPECTOR uses Bayesian methods to update the belief in a hypothesis in light of evidence obtained (Duda et al. 1976). Belief in a hypothesis is equated with the probability of that hypothesis being true. Belief is expressed as a single point value or probability; there is no additional number to set the precision of this point value. Rarely, does anyone indicate their belief in something by providing a precise numerical value. Usually terms, such as "quite certain" or "very unlikely," are used. Those vague terms imply a range of belief and not a single value. So, while any inexact reasoning approach that uses point values for belief or truth is not *de facto* useless, it does not mirror human reasoning in the strict sense.

Two pieces of information are necessary before Bayes' Theorem can be used:

1. some prior belief (or likelihood) that a particular hypothesis is true and
2. the likelihood that the current evidence is true given that the hypothesis is true.

The second item is a conditional probability, wherein the probability of one event (in this case a piece of evidence) is conditioned on (or viewed in the light of) the truth of another event. Then it becomes possible to calculate the likelihood of the hypothesis given the current evidence.

The following discussion outlines a mathematical description. The use of Bayes' theorem requires an a priori belief (also called a *prior probability*), $P(H_i)$, in each hypothesis and it also requires conditional probabilities, $P(E \mid H_i)$, of the current evidence, E, given each possible true hypothesis. This relationship between evidence and hypotheses can be seen in Figure 3–9. Here, we have three mutually exclusive hypotheses H_1, H_2, H_3. Given some evidence, E, each of these hypotheses has some probability of being true. Ultimately, the value we seek is called a *posterior probability* (it results from an updating of the prior with the conditional information) and is represented by the expression, $P(H_i \mid E)$. In an actual KBSs, the posterior probability is usually an inference of some sort relating the evidence to a particular hypothesis, e.g. an if-then rule. Mathematically, this can be calculated as:

$$P(H_i \mid E) = \frac{P(E|H_i) \cdot P(H_i)}{\sum_{k=1}^{n} P(E|H_k) \cdot P(H_k)}$$

where the denominator is precisely $P(E)$. Using Figure 3–9, $P(H_i)$ is the percentage of the total area of the big rectangle occupied by the rectangle H_i, and $P(E|H_i)$ is the percentage of the area in H_i that is shaded.

When someone is asked to assign probabilities, they are often coached to give their answers in terms of odds likelihood values, e.g. 3-to-1 or just 3. This is often easier for the individual making assignments, and the results can be directly translated into probabilities by $P(A) = O(A)/(O(A)+1)$.

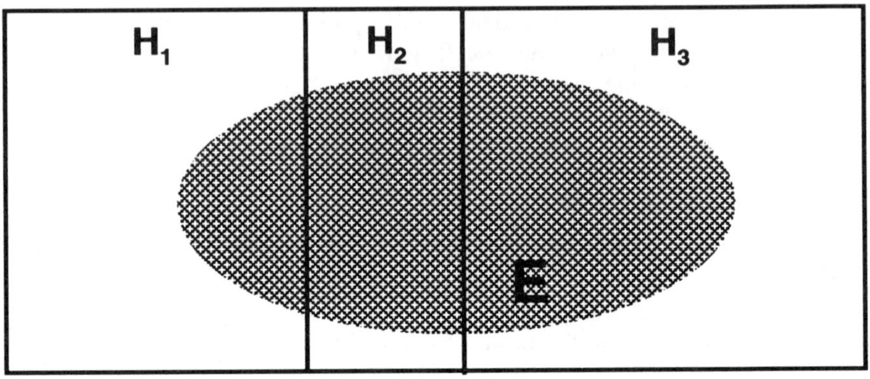

Figure 3–9. Given three mutually exclusive hypotheses H_1, H_2, H_3, each has some probability of being true a priori, i.e. before E has been observed. The a priori probability of each hypothesis is reflected by its relative area within the total area of the rectangle. Each of these hypotheses has a different probability of being true under the condition of the evidence, E, being true. Therefore, with the observance of evidence, E, the focus is no longer the relative area of each hypothesis within the large rectangle, but rather the relative area of each hypothesis within the shaded oval.

As an example, we can estimate the probability of H_1 given the evidence E in Figure 3–9 by using the values in Table 3–1.

$$\sum_{k=1}^{n} \mathrm{P}(E \mid H_k) \cdot \mathrm{P}(H_k) = .20(.36) + .55(.20) + .4(.44) = .362$$

$$\mathrm{P}(E \mid H_1) \cdot \mathrm{P}(H_1) = .20(.36) = .072$$

$$\mathrm{P}(H_1 \mid E) = .072/.362 = .199$$

When the evidence E may also be uncertain and if we allow for evidence against hypotheses, i.e. $\mathrm{P}(\neg H \mid E)$, then the use of Bayes' Theorem becomes very complicated, both to carry out calculations and to ensure that all numerical estimates are mutually consistent. For those wishing to pursue those details, we suggest Duda et al. (1976) or Leonardo et al. (1985).

Table 3–1. Probability assignments associated with Figure 3–9 that are necessary to estimate $\mathrm{P}(H_i \mid E)$.

	H_1	H_2	H_3
$P(H_i)$	.36	.20	.44
$P(E \mid H_i)$	.20	.55	.40

Three drawbacks to this approach result from (1) the need for extensive historical or subjective estimates for all the conditional probabilities, (2) statistical assumptions of independence for the hypotheses, and (3) Bayesian probability's inherent reliance on the principle of the excluded middle (i.e. some event is either true or false). All of the prior probabilities $P(H_i)$ must be calculated from historical records or be subjectively estimated. Also, the conditional probabilities of every possible combination of evidence, that might be anticipated, must be determined for all possible hypotheses. Very few disciplines have such a wealth of well-recorded data. This first problem aside, an assumption of the Bayes method requires that each of the $P(E|H_k)$ must be independent for all k. This means that the probability of seeing evidence E given H_i must not be affected by the observance of E given H_j. In general, this assumption is rarely justified.

The principle of an "excluded middle" says that either A must be true, or $\neg A$ must be true. Probability assigns A and $\neg A$ "likelihood" values, but once this is done for A, $\neg A$ has a complementary value, i.e. $P(\neg A) = 1 - P(A)$, regardless of whether any evidence pertains to $\neg A$. Both A and $\neg A$ cannot be true with a probability greater than .5.

A fourth disadvantage is more philosophical and less technical, but none-the-less important. There exists a large amount of social science literature to indicate that people do not think in this probabilistic way (von Winterfeldt and Edwards 1986). In fact, many people are downright ignorant with regard to statistical probability—innumeracy abounds. They exhibit a strong penchant to over-estimate the likelihood of rare events and a corresponding lack of confidence in the notion of central tendency (Paulos 1988). Although rare events do occasionally occur, average values (or central values in the distribution of possible values) are much more likely to occur. And, of course, the "gambler's fallacy" is a widely held belief, wherein one believes that following a string of outcome A for a series of events, outcome B is more likely, even though each event occurs independently of the other. Many of these types of problems have been researched and some of the results are present in Section 5.9, Expert Biases and Eliciting Uncertain Knowledge.

If a Bayesian approach aims to model human reasoning as probabilistic, it fails because probability theory does not appear to be a realistic model. The millions of people who think, yet nevertheless have little understanding of probability and, in fact, reason counter probabilistically, are ample proof of this. On the other hand, as a model of reasoning with uncertainty probability is limited only by the former three drawbacks. Nevertheless, Pearl (1986) uses Bayesian probability in a model of inexact reasoning in the form of a network of probabilistic beliefs. His implementation of Bayesian probability eliminates some of the above difficulties. Olson et al. (1990) present the details of Bayesian belief networks as applied to ecosystem management problems. Such a belief network includes the sum total of knowledge about a problem, and enables one to reason about that knowledge globally and simultaneously. This seems to have substantial intuitive appeal.

With regard to probability theory or any other inexact reasoning method, we have to decide whether we are modelling human reasoning or modelling a rational decisionmaking technique. In the case of probability theory it would seem to be the latter. There are, however, other approaches to inexact reasoning that focus on more subjective and heuristic methods for incorporating less than certain information and knowledge, and that address the former, that is, modelling human reasoning.

3.5.2.2 A Theory of Certainty. Certainty Theory or Confirmation Theory (Shortliffe and Buchanan 1975) is used in MYCIN, a medical diagnostic system developed in the mid-1970's. This approach uses probability-like measures for belief. There is one value for belief in a hypothesis and a second value for disbelief in the hypothesis. Belief and disbelief in each hypothesis are treated as two separate values. A measure of increased belief (MB) for each hypothesis is manipulated distinctly from a measure of increased disbelief (MD) in the same hypothesis, each lies in the range $[0 \leq \text{MB}, \text{MD} \leq 1]$. Unless there is an inference rule that sets MB or MD to 1, each approaches 1 asymptotically as new evidence is obtained. The measure of belief in a hypothesis following the application of a rule, MB_{new}, is calculated from the current belief value, MB_{old}, and the belief associated with the rule, MB_{rule}, as:

$$\text{MB}_{new} = \text{MB}_{old} + \text{MB}_{rule} \cdot (1 - \text{MB}_{old})$$

The measure of disbelief, MD, in the hypothesis would be calculated in a similar manner.

A combined value, called the *certainty factor*, CF, is calculated as the difference (MB − MD); this represents the sum total of all evidence presented for and against a hypothesis. As the reader may have noticed, the range for this certainty factor becomes $[-1 \leq \text{CF} \leq 1]$. Here, −1 indicates absolute disbelief, 1, absolute belief, and, 0, either a lack of evidence or contradictory evidence.

Let's use our example from Table 3-1 and assume the following:

$$\text{MD} = 0$$
$$\text{MB}_{old} = \text{P}(H_1) = 0.36$$
$$\text{MB}_{rule} = \text{P}(H_1 | E) = 0.199$$

Then our updated measure of belief in H_1, MB_{new}, can be calculated as:

$$\text{MB}_{new} = 0.36 + 0.199 \cdot (1 - 0.36) = 0.487$$

Because each rule used for assigning MB or MD assumes no prior evidence for that hypothesis, the pieces of evidence in the premises of each rule should be independent. If it is suspected that the premises of two rules are not entirely independent, the system developer can reduce each rule's confidence values (i.e. MB_{rule} or MD_{rule}) or, alternatively, add an additional clause to each rule that would ensure that they are mutually exclusive, and so, only one of them is ever applied in any situation. It should be noted that combining all belief into one value obscures the separate contributions of favoring and disfavoring information and also, as with Bayesian methods, there is no interval for this belief value, only a single point estimate.

The above discussion has not specified how one goes about determining the measure of belief (or disbelief) for a hypothesis (i.e., conclusion) during the evaluation of a rule. In the general structure of a rule-based representation, we have rules like A: IF E THEN H [CF(A)], where we want to calculate the confidence factor, or belief, in H when the rule fires successfully. Because E is often composed of a number of premise clauses connected by *and*, *or*, and *not*, confirmation theory provides an algebra for aggregating their separate certainty factors. For premise clauses, P_1 and P_2, in a rule:

$$\mathrm{CF}(P_1 \text{ and } P_2) = \min(\mathrm{CF}(P_1), \mathrm{CF}(P_2))$$

$$\mathrm{CF}(P_1 \text{ or } P_2) = \max(\mathrm{CF}(P_1), \mathrm{CF}(P_2))$$

$$\mathrm{CF}(\text{not } P_1) = -1 * \mathrm{CF}(P_1).$$

Once we have obtained CF(E) then we can obtain CF(H) by multiplying CF(E) by CF(A). The rationale for this calculation is that the confidence factor of the rule assigns H a belief value of CF(A) assuming that E is known with certainty. When E is known with less than certain belief, then CF(H) is reduced proportionally.

Certainty factors have been used almost exclusively in rule-based systems. This has been both the cause for their widespread use and the basis for their limitations as a model of inexact reasoning. Not all human reasoning can be modeled as independent rules that are loosely related to other pieces of pertinent information. This emphasizes the advantage of using multiple knowledge representation schemes.

Like probability theory, certainty factors produce a single final estimate of truth (or belief) and the evidence in each rule must be independent (as noted above). But, unlike probability theory, certainty factors treat evidence about a hypothesis and its negation separately, hence, the excluded middle assumption is not used. Despite the criticisms that this method lacks rigor and uses localized reasoning, this approach remains the most widely applied method for handling uncertainty in knowledge systems. It is difficult to criticize a technique, whether it be certainty theory or any other approach, as lacking theoretical bases, when the process it simulates (i.e. human reasoning with uncertain knowledge) is inherently subjective and not always rational itself (Luger and Stubblefield 1993).

3.5.2.3 Possibility Theory. Zadeh (1979) takes a different perspective of the concept of uncertainty. Whereas probability theory treats uncertainty as randomness or likelihood, it may be more appropriate to consider the *meaning* of information instead. Many of our descriptions of what we know and how we use our knowledge can only be specified in *vague* terms. For example, "short person" and "tall person" are vague descriptors because they are non-explicit and subjective. Zadeh developed **possibility theory** as an alternative to probability theory's treatment of random events; however, because it is analogous to probability theory, it suffers from some of the same problems (Quinlan, 1983a). Possibility theory is an extension of probability theory wherein the "excluded middle" principle is relaxed.

Fuzzy sets were introduced by Zadeh (1965) to represent possibility measures. Let S be any set (e.g. the possible heights of women, in inches), s an element of S, and F a *fuzzy subset* of S (e.g., heights of tall women). Then the possibility that s is in F (i.e. that a particular height, say 66", is in the set of heights that are interpreted as "tall" for women) is defined by a membership function q. The function q maps all heights onto the interval [0,1]. "Tall" is then referred to as a *fuzzy number* representing the fuzzy set F. A tabulation of $q(s)$ for some values of s in S is presented in Table 3–2. In keeping the idea of a fuzzy number, we can add, subtract, multiply, and divide fuzzy numbers that are defined over an arbitrary integer base.

Table 3–2. From possibility theory, all possible heights of women have an associated possibility, defined by q, that indicates each height's membership in the fuzzy set "tall". All heights less than those listed are assumed to have a membership value of 0 and, similarly, all heights greater than those list are assumed to have a membership value of 1.

height s in S	membership $q(s)$ in "tall"
63	0.0
64	0.1
65	0.3
66	0.4
67	0.6
68	0.7
69	0.8
70	0.9
71	0.9
72	0.9
73	1.0

The calculation of belief in a hypothesis H based on a rule of the form A: IF E THEN H CF(A) can proceed in a similar fashion to confirmation theory, above. There are a number of differences, however. First, any premise P in E, will be of the form X is F, which means that the variable X is in the fuzzy set F, for example, X might be the height of Jane and F the fuzzy set *tall*. The interpretation of that clause would be that "Jane is tall" and the membership value q associated with Jane's height, i.e. X, and the set *tall* would represent the belief that Jane is tall, i.e. $q(\mathrm{X})$. Second, the clause conjunctions *and* and *or* are treated exactly as above, but *not* is 1 minus the membership value of the clause. In general, if F is the fuzzy set *tall* and F' is the fuzzy set *not tall*, then *not tall* has the membership function $q(F') = 1 - q(F)$.

Two drawbacks to possibility theory are: (1) it assigns a single value to the belief of any hypothesis (e.g. the possibility that 66″ is tall equals 0.4), and (2) as with probability, the distinction between evidence *for* and *against* hypotheses becomes lost.

There are also some advantages to this approach. The ability to interpret and utilize natural language terms such as "many", "large", "tall", or "more or less" seems consistent with the way that people communicate information and knowledge to each other. Also, fuzzy sets are an example of what is referred to as a *multi-valued logic*, that is, a single term can have several values simultaneously, each one with its own belief level. For example, in addition to the fuzzy set "tall", we may also have a fuzzy set "above average". The membership of 66″ in the fuzzy set "above average" might be 0.8, while it is 0.4 in the fuzzy set "tall." So, the height, 66″, could be interpreted as both "above average" and somewhat "tall". By capturing vagueness and permitting multiple-valued entities fuzzy sets mirror some of the inherently inexplicit ways we reason about our world.

3.5.2.4 Dempster-Shafer Theory of Evidence. The *Dempster-Shafer* theory of evidence makes a distinction between *probability* and *ignorance* (Shafer 1976). Rather than limiting belief in a hypothesis to a single number, Dempster-Shafer takes a more informative and statistically conservative approach to place upper and lower bounds on belief. In their method, they separate belief in a hypothesis, belief in its negation, and ignorance about the hypothesis. Mathematically, ignorance is the remaining belief that has not been assigned to an assertion and its negation.

By using "belief functions" to represent the uncertainty of hypotheses, they relax some of the axioms of probability theory. The resulting system becomes a super class of probability theory (Luger and Stubblefield 1993). Dempster-Shafer and probability theories coincide when there is sufficient evidence for the former to produce point estimates. However, because Dempster-Shafer theory is fundamentally very similar to probability theory, it suffers from the same need for large numbers of probability assignments and from the need for independence assumptions. However, by considering "ignorance", they've avoided the inherent *excluded middle* principle (either A is true or $\neg A$ is true) of probability theory's treatment of randomness. Because the specifics of this approach are extensive and complex, we do not consume space here; instead we refer interested readers to a very complete description by Tanimoto (1987).

3.5.2.5 Non-Monotonic Reasoning. While all of these numeric methods may have some particular advantages for one application or another, it is extremely doubtful that people actually use any of these for everyday reasoning. Certainly we don't use them consciously and we don't use them as an exclusive means of reasoning. The experts in natural resource management that we have worked with find it very difficult and unnatural to attach certainty values to their heuristics in such a way that the system is consistent. This means that if certainty factors are used at all, the system developers must make adjustments to the initial factors in order for their conjunctions to provide correct conclusions. Schmoldt (1989) used ideas from communication theory to modify the initial certainty factors of a rule base. While these inexact reasoning methods may propagate uncertainty to some acceptable degree, they don't provide an explanation of why particular statements might be uncertain because they lack a structure of the relationships between objects (Luger and Stubblefield 1993).

Also, as these methods are usually applied, they only provide for continual additions to what is known. There is only a partial capability for retraction of hypotheses in the light of new information, i.e. they reason monotonically. One exception is the mechanism in Bayesian belief networks which acts like a truth maintenance system, i.e., when the probability of one belief is changed, all of the beliefs that depend on it are automatically updated. A partial exception is certainty theory, but only when measures of belief and disbelief are updated separately.

People, on the other hand, are able to revise their belief in hypotheses when previous evidence is discovered to be false or when new, contradictory evidence comes to light. As evidence is retracted, the belief in hypotheses based on this evidence must be revised also, i.e. reasoning is *non-monotonic*. A major difficulty with non-monotonic logics is that retracting only a single hypothesis can result in a cascade of other retractions or modifications to beliefs. Early work by McCarthy (1980) on *circumscription* and work by Reiter (1980), de Kleer (1986), and Davis (1990) have tried to make non-monotonicity easier to implement and intuitively consistent with the way most people think.

3.6 SUMMARY AND ADDITIONAL READINGS

We have now finished our presentation of the major components of knowledge systems and associated concepts. It is important that the reader understands this background information even though these chapters have focused more on generalities and overviews of KBS methods rather than on their implementation. Starting with chapter 4, we dive into how to define, design, and implement knowledge-based systems. The reader is likely to experience a rather flat learning curve because the planning, design, and implementation of a KBS is filled with choices and trade-offs, which make the process confusing. There is no single best way to build a KBS system, so we have attempted to illuminate as many different options as possible. Nevertheless, without a thorough reading of the background chapters 1-3 and some of the additional references, where needed, a reader unfamiliar with this material is likely to be ill-prepared.

For further reading about explanation, user and software interfacing, and inexact reasoning you may consult the following sources: Bowerman and Glover (1988, chapter 8), Luger and Stubblefield (1993, chapter 8). An excellent treatment of uncertainty and its computation in KBSs can be found in Giarratano and Riley (1994) and in Tanimoto (1987).

CHAPTER 4 Planning the Application

Perhaps it is just a matter of working out the details, but as any scientist would agree, the details are everything.

G. Johnson

We have just completed the first section of the book. This included: motivation for the application of AI and KBSs to help solve resource management problems (chapter 1) and a look at the major components of knowledge-based systems (chapters 2 & 3). Now, in chapters 4–6 we turn our attention to the detailed steps of planning and designing a knowledge-based system, and acquiring and organizing the necessary knowledge to implement it. System implementation and evaluation issues are covered in chapters 7–10.

The process of developing a knowledge-based system, from conception to final delivery product, is a long and complex task. It can involve a number of different people with vastly different skills, possibly each with different perceptions of what the goals and potentials of the development project might be. In many ways this process is different from traditional computer software development, where the users, system outputs, algorithms, and data are often more easily identified from prior experiences or expectations. The development of knowledge-based systems is less structured than conventional programs, application problems tend to be more complex and ill-defined, and there is less relevant prior experience to draw upon as a guide.

While there exist no widely accepted standards to apply to the development of a KBS, several approaches have been proposed and used. This chapter focuses on the early stages of system development, i.e. defining and planning an application. We introduce an overall design methodology which combines elements of traditional software engineering and trial-and-error exploratory programming.

4.1 PARALLELISM AND CYCLING IN THE DEVELOPMENT PROCESS

Many factors influencing the design of a KBS must be considered in parallel because each stage in development interacts with both previous and future stages, this necessitates continual revision. True parallelism is difficult to achieve; however, we can simulate it by cyclical iteration (Walters and Nielsen 1988). Each pass through the steps of KBS development may require additional knowledge acquisition efforts and some modifications to system design and implementation methods. Our presentation of system development in the following pages will have a sequential flavor to it. This recipe-like formula, however, only works for a general orientation of broad topics. As actual systems are created and many details need to be specified, there often exists a need to step back and reconsider previous commitments as one proceeds. Or, it may be necessary to leap forward and wrestle with issues whose impact could cause drastic changes if left to a later stage.

Ideally, we would like to be able to juggle all the important development issues simultaneously. Unfortunately, this is nearly impossible to do except for the most trivial applications. By resorting to a cyclic approximation of the parallelism ideal, however, we may pay a price for time and effort lost on ideas that will need to be scrapped later and replaced with better designs and implementation. A certain amount of this cost is reasonably accepted, but the combination of exploratory programming and careful system design can help reduce the number of dead-ends and can increase efficiency.

Prototyping loops that constitute the core activities of *design*, *implementation*, and *evaluation* effectively simulate parallelism. Iteration allows the development team many opportunities to include all necessary requirements as construction proceeds. This flexibility becomes important because not all problem characteristics and their effects on system design and implementation can be foreseen in the early planning stages of a KBS project.

All the different KBS development paradigms, that have been proposed, employ the same core activities of design, implementation, and evaluation. Different paradigms are distinguished, however, by the manner in which these activities are carried out, and the amount that each activity is emphasized. In the next section, we will present and discuss several alternative methods for developing knowledge-based systems.

4.2 ALTERNATIVE DEVELOPMENT METHODS

Any attempt to synthesize and organize complex knowledge about some subject is necessarily difficult and time consuming. When that knowledge must be organized in such a way that it can also be *applied* in a problem-solving manner to a variety of real-world situations, then the development task becomes even more challenging. With this additional stipulation of functionality, the knowledge must not only *convey information* about the subject of interest, but it must also contain some instructions about *how that knowledge can be applied.*

Two very different general approaches to system development have been described in the literature. They differ in the manner and timing of when decisions are made about system characteristics. In the earlier years of AI work, system development methods emphasized early coding of system prototypes with the *a priori* knowledge that several such prototypes would be replaced in favor of others that were better able to handle an evolving vision of the full system. In contrast, traditional software engineering principles resolve most decisions up front to permit well-directed project development. So in one case development decisions are made as they become important during the project, and in the latter case all decisions are made in an initial design stage.

Whenever there exist diametrically opposed ideas like these, there also tends to be a movement toward a more middle ground. More recent development methods have produced a third method, a compromise between the above two methods. The following paragraphs present: these two distinct and traditional approaches, a hybrid approach that is a compromise of these two, and also a second type of middle-of-the-road method that represents the authors' perspective of the development process.

4.2.1 Rapid Prototyping

Exploratory programming methods were the first to be applied in knowledge-based system development. Here, system developers explore different implementation approaches by programming pieces of a system. The pieces that seem to work well are retained and less successful ones are scrapped. Exploratory programming was further encouraged by AI programming environments that supported this type of effort. Exploratory methods dominated because they seemed to fit the needs of AI applications, which tended to be quite different from traditional software applications. This means of programming led directly to the **rapid prototyping** approach to knowledge-based system development,

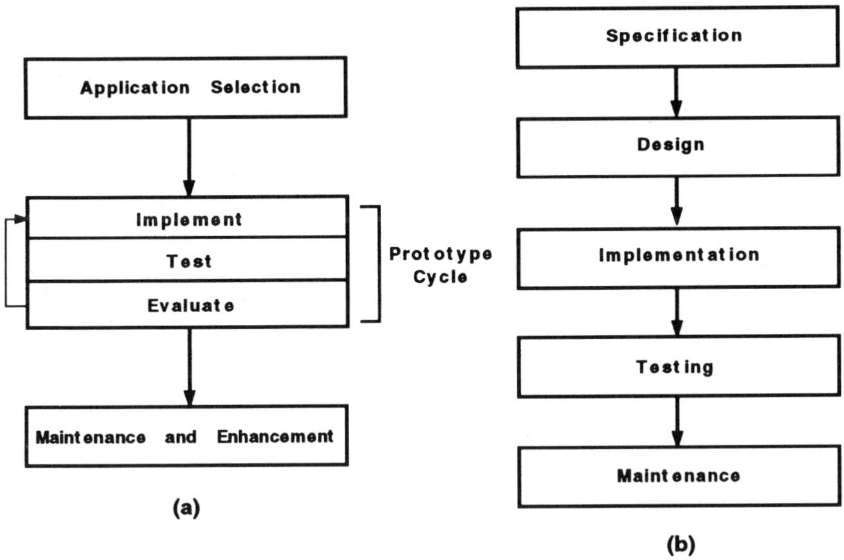

Figure 4–1. A rapid prototyping development process using exploratory programming (a) centers mostly about the prototyping cycle of implement, test, and evaluate. Traditional software engineering methods (b) emphasize a more systematic analysis of system requirements and their implementation. Although the diagram (b) does not depict it, there exists ample opportunity for cycling back to previous stages in the software life cycle.

which uses trial and error as a primary mechanism for obtaining results (Walters and Nielsen 1988). Some of the characteristics of AI applications, in general, and KBSs, in particular, that imply rapid prototyping system development include (Luger and Stubblefield 1993):

1. application problems are poorly specified;
2. solution methods are heavily domain specific;
3. heuristics are empirical and so is programming them;
4. programming must be incremental because the knowledge of solving a problem is generally acquired by degree; and
5. selecting and implementing knowledge representations and search methods have not been well established theoretically or empirically.

In an analogous sense, one could say that these early methods of programming used heuristic-like system development methods—similar to the heuristics that are applied to finding solutions within an actual application.

Figure 4–1a illustrates an example of a traditional rapid prototyping development cycle. Selection of an application is quickly followed by an initial prototyping effort. Cholawsky (1988) notes that this first prototype often serves as a combined feasibility study, design document, and functional specification effort. After prototype implementation, testing, and evaluation, this exploratory programming loop is repeated iteratively. At some undetermined time and under poorly-specified circumstances, a mature prototype surfaces when the nth prototype appears to operate satisfactorily and does everything that the developers feel is reasonable. Then the developers must address system maintenance and system enhancement issues as part of delivering an operational system. This final stage involves providing for adaptability as knowledge changes, producing companion documentation, and easing this new software into the already existing organizational computing environment—which may include hardware, software, and user conventions.

Under this description of a system development life cycle, Floyd (1983) distinguishes between three functions for prototyping:

1. *exploratory* – for identifying and clarifying system requirements;
2. *experimentation* – for investigating possible alternatives of system design, including knowledge representation and search methods; and
3. *evolution* – where prototypes represent and are converted into actual versions of the system.

Several drawbacks exist to the *exclusive* use of rapid prototyping as a development style and these have resulted in its decreasing popularity. Rapid prototyping fails to accommodate deployment issues such as data problems, human-computer interfaces, hardware/software incompatibilities, and potential for scaling up future designs and program code. Inordinate amounts of time may be spent prototyping an application with a low cost-benefit ratio or working on a system design that is poorly matched to the problem and its solution. Also, not all the principal players with interests in the system are consulted at the outset. While the idea of gradually phasing through multiple prototype enhancements toward an operational system seems attractive (e.g., the three purposes of prototyping noted above), for many applications too many issues and concerns may not be specified nor included in system development at the most advantageous time.

Despite these limitations, there are some circumstances where rapid prototyping may be acceptable or desirable. Quick investigation of the feasibility of a KBS approach to a particular problem may be accomplished by constructing a small prototype (Oliver 1986). An initial prototype could be part of a larger feasibility study or as a proof of concept to secure funding. For some applications a cursory development effort may produce something of immediate use—this immediate-solution system may be all that is required as a final system. In some research applications where the user *is* the developer and system use does not extend outside of the lab, a prototype constructed quickly might be expedient and sufficient. An example of this situation might be a small scale simulator for laboratory use to test, or possibly generate, experimental hypotheses about protein synthesis.

Rapid prototyping is generally not appropriate where institutional or organizational policies prohibit discarding prototypes and re-tooling. When this is the case, more systematic approaches are required to ensure that large portions of an initial prototype may be utilized in subsequent system revisions.

4.2.2 Software Engineering

The lure of exploratory programming practices and the availability of sophisticated AI programming languages (e.g., LISP and PROLOG) to support this trial-and-error style initially seemed to promise programmers freedom from the more rigid specifications of traditional *software engineering* (Walters and Nielsen 1988). While this chimera might be real for small applications, as system complexity grows there exists a need for more systematic approaches (Buchanan et. al. 1983, Cholawsky 1988, Gudes et. al. 1986, Oliver 1986, Schmoldt 1987, Schmoldt 1989, Walters and Nielsen 1988). In contrast to the rapid prototyping development method, the software engineering development method has been designed for writing traditional software programs and many books exist on the subject (e.g., Pressman 1982).

Software engineering provides a highly structured and sequential progression through the steps of a software development project (Figure 4–1b). The major difference between software engineering and rapid prototyping, described above, is the incorporation of formal planning and design phases. These additional stages of development address definition of the application problem, software requirements, and possible solution methods. The complete cycle includes: an analysis of system feasibility and system requirements, an outline of problem-solving methods, a description of implementation methods, some provision for testing and refining the prototype, and specific details of delivery and maintenance. While Figure 4–1b shows a strictly unidirectional process, the need often arises to cycle back to a prior stage for redesign, re-specification, etc. (Parsaye and Chignell 1988).

In addition to a software life cycle, software engineering also offers some rules-of-thumb for implementing these stages. Traditional wisdom suggests that the allotment of time and effort for specification/design, software construction, and testing/evaluation should be distributed approximately as 40–20–40, respectively (Pressman 1982). Proponents of software engineering reason that through the use of a comprehensive design phase, actual programming can be a relatively short task.

However, software engineering deals with data and information processing as opposed to the knowledge processing of AI applications. Requirements and techniques for data and information processing have been firmly established due to the more precise nature of the concepts being manipulated. In contrast, knowledge acquisition, knowledge representation, and knowledge processing are relatively new ideas. Proven methods for their application and implementation are not well established. Hence, it becomes difficult to transfer all the methods and guidelines used in software engineering directly to the development of knowledge-based systems.

Some techniques borrowed from software engineering, such as top-down design, program modularity, and testing/evaluation can translate into very valuable lessons for KBS development. Top-down design involves the analysis of the problem to be solved into, first, sub-problems of a general nature and, second, into more and more specific details of problem solving. In doing so, there exists less opportunity for one to lose sight of larger goals while working on the detailed aspects of a problem. Program modularity follows closely from the top-down paradigm by creating many separate and independent program modules. Each interacts with only certain other modules in clearly specified ways. Any non-trivial application that one programs necessarily contains unforeseen errors and omissions. So, testing and evaluation are important steps to help guarantee software correctness and usability. The general approach suggested by the methods in Figure 4–1b, subtleties like the 40–20–40 rule, and good programming methods (e.g. data abstraction and program modularity) can form a useful basis for and be adapted to knowledge-based system development.

4.2.3 Knowledge Engineering

Buchanan et al (1983) propose a KBS development approach that forms a compromise between rapid prototyping and software engineering. Hence, we have referred to their method as constituting a *knowledge engineering* life cycle analogous to a traditional software engineering life cycle. These ideas have been widely adopted because the knowledge engineering approach incorporates the need for exploratory programming within the framework of a well thought-out design. So, you can see that this represents a mixture of rapid prototyping and software engineering. Here, we present their approach as a de facto knowledge engineering methodology; in the next section we introduce our variant of it.

Five distinct steps are included in this methodology (Figure 4–2). Several cycles through these steps may be necessary to accommodate multiple prototypes on the way to a final system. These steps are: (1) identify the problem and the resources available (planning), (2) conceptualize the elements of the problem, i.e., formalize the data and the solutions, (3) define the problem solving model, (4) implement a knowledge base and search methods, and (5) test the prototype. Depending on the results of the testing phase, it may be necessary to reformulate concepts or redesign some structures or refine some

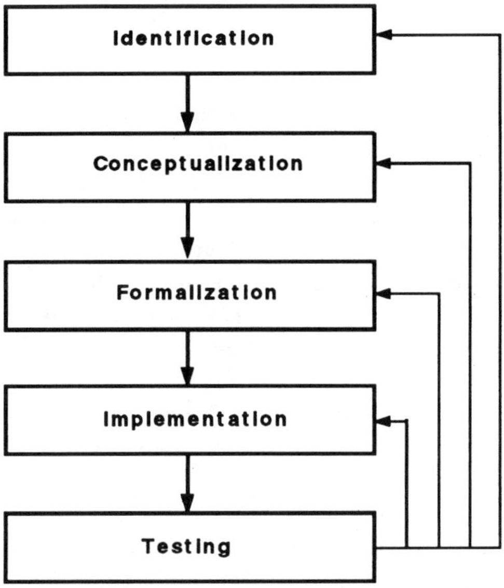

Figure 4–2. Buchanan et al. (1983) describe a knowledge engineering life cycle that borrows from both of the approaches in Figure 4–1. One serious drawback is the absence of delivery and maintenance concerns.

implementation components (e.g. rules, frames, or problem-solving techniques). This life cycle description, however, omits any provision for delivery of the operational system and no mention is made of maintenance issues. There are a number of variations on this approach by other authors (e.g. Cholawsky 1988, Parsaye and Chignell 1988), but the general stages of the method appear to be relatively invariant.

There are two important ideas emphasized by the formulation of Buchanan et al (1983):

1. system development contains two primary phases—first, planning and conceptualization, and second, construction and testing of an appropriate architecture;
2. knowledge acquisition occurs throughout the different development stages and does not constitute a stage in and of itself.

The first point seems to be fairly well accepted within the literature on KBS development and workers in the field have come to realize this sequence is important. However, the second idea diverges somewhat from what some authors consider as the general sequence of activities (e.g., Parsaye and Chignell 1988). Many knowledge engineers don't especially enjoy knowledge acquisition because it is difficult, time consuming, and delays the more interesting (from their perspective) aspects of design and construction. Consequently, knowledge acquisition often becomes relegated to a one-time phase of system development, rather than an ongoing activity. This attitude is unfortunate because it is often essential to return to the source of knowledge, particularly as a sequence of prototypes are advanced through the development process. It is very important to acquire additional knowledge and restructure the knowledge based on a more refined understanding of the domain knowledge—whether that knowledge originates from documents, people, or other software.

Our system development approach (Figure 4–3) follows the basic principles of the knowledge engineering life cycle. The first phase is *application planning* (Chapter 4) which leads to a formal *problem definition.* The processes of *knowledge acquisition* (Chapter 5) and *application design* (Chapter 6) lead to a written specification of a *knowledge model* and a *human factors model.* Once the problem has been defined and the knowledge and human factors models designed, our concern shifts toward implementation where we make some *pre-implementation decisions* (Chapter 6), introduce the selected *AI programming system* (Chapter 7), and *implement* an initial *prototype* (Chapter 8) and an advanced *prototype* (Chapter 9). *Testing and Evaluation* (Chapter 10) is then performed on each prototype to decide whether the knowledge-based system adequately meets the goals established in the problem definition phase. If it does not, then the cyclical iteration of the rapid prototyping method is activated by returning to different stages of the development process depending upon the results of the testing and evaluation phase (Figure 4–3). If the knowledge-based system is satisfactory, then the system is *delivered* (Chapter 10) and the development process is finished.

4.3 PROBLEM DEFINITION

This section details the initial phase of system development—*problem definition* (Figure 4–3). The initial phases of a project constitute what is often termed, a "feasibility" study. Essentially this means "Are we doing the right thing?" and "Can we do what we say we can do?" Exactly the types of things this entails is greatly influenced by the particular application. For example, when the user is the developer, feasibility may mean a few hours programming several different ideas on the computer. In other instances, an organization may only require a problem definition as a feasibility study for obtaining project go ahead. In more demanding establishments where costs and benefits are closely scrutinized, a problem definition stage and one pass through an initial prototype may be necessary for a proof of concept. Such a preliminary prototype should be able to demonstrate some rudimentary capability in the subject area. Sometimes, however, the look and feel of the prototype's user interface may carry more

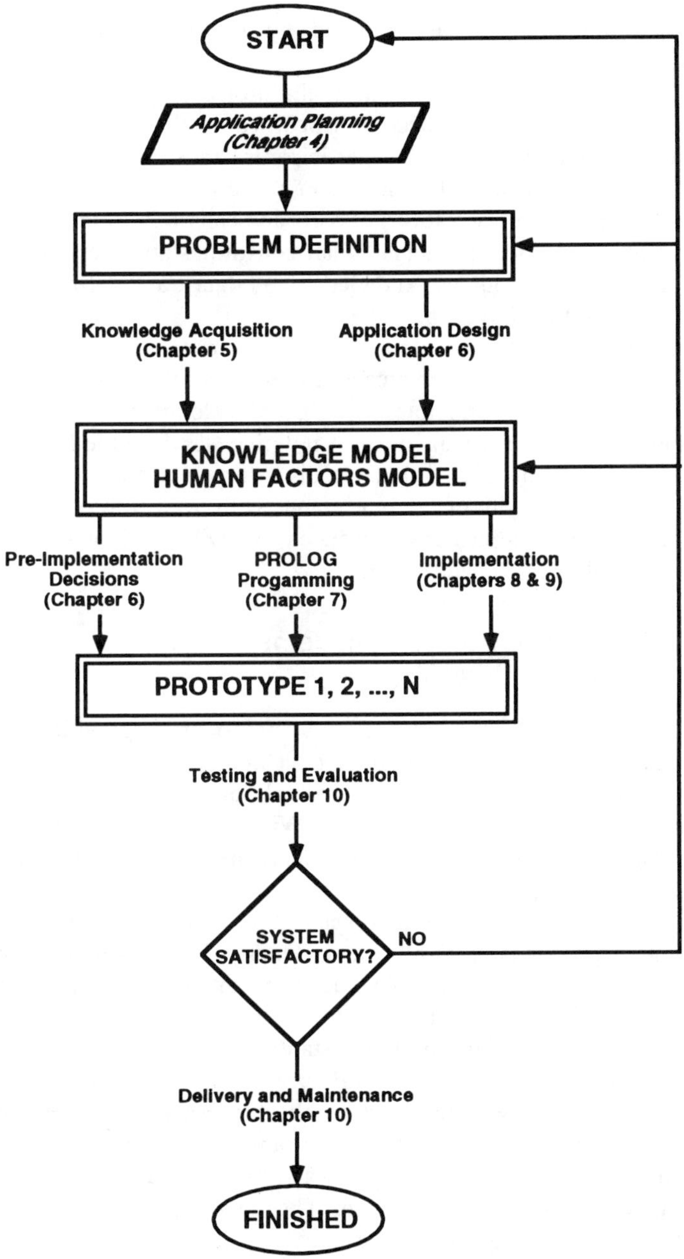

Figure 4–3. Our modified knowledge engineering life cycle borrows heavily from the design in Figure 4–2. This description uses some different terminology and associates chapters in this book with knowledge-based system development processes.

weight than any competence exhibited by the knowledge in the system (Cholawsky 1988).

Feasibility does not just determine if the project is possible, but also helps craft a plan of how it will be accomplished. To do this, the people concerned with system development (i.e. the participants) must understand as much as possible about the project and each person's role in it. These personnel form a "team" (Walters and Nielsen 1988) of sorts and can be instrumental in detailing problem solution characteristics (i.e. what constitutes an acceptable solution to a problem), locating resources (knowledge and data), and defining the requirements of an operational system.

Before we present the problem definition/identification phase more thoroughly, it may be useful to introduce an example problem which will enable us to make the important principles of KBS development less abstract. This application concerns the management of red pine stands in the Lake States area of the U.S.A. Later in this section we also discuss a companion application which addresses the health of red pine stands with respect to pest management. As KBS development topics are introduced, we will continually refer to these applications to illustrate specific points.

4.3.1 A Red Pine Forest Management Advisory System (RP-FMAS)

In the problem definition phase, it is important to provide a context or background that serves to orient both developers, funding organization, and users of the target system. A century ago, red pine (*Pinus resinosa*) was found on about 7 million acres in Minnesota, Wisconsin, and Michigan, the Lake States region of the United States. Today there are only 1.2 million acres of red pine forests in the upper Great Lakes region, including both natural and planted stands (Bassett 1984). It is a favored species in the region because it has many uses and high productivity (Kallio and Benzie 1980). This accounts for the more than one million acres that have been replanted with red pine since 1930.

Knowledge gained from many years of research on the establishment, composition, and growth of red pine forest stands, was organized and published in a manager's handbook for red pine (Benzie 1977). The manager's handbook provided field foresters with the latest and best information available in a useful format. It gave foresters a synthesis of red pine research results covering many aspects of silviculture in an explicit way that helped focus discussion and identify areas needing more research. The handbook found immediate and widespread acceptance as a reference aid for experienced decision makers and as a training tool for students and newcomers to red pine forest management.

One lesson learned from the experience of creating a manager's handbook was the importance of organizing and synthesizing research knowledge in a form that not only conveys knowledge but also provides instructions on how to apply it. Nevertheless, several weaknesses of the handbook format for delivering knowledge soon became apparent. First, a printed publication was expensive to publish and difficult to update and keep current. It was out of date before it was even published. Second, giving general advice for only a few "average" stands is the best that a printed handbook can accomplish. The field forester must utilize great technical skill and sound judgment to adapt prescriptions for actual stands. Developing a handbook to give advice for the full range of specific situations would result in such a large and unwieldy document that its use would be discouraged. Third, it was not possible to integrate modern dynamic simulation tools, such as growth and yield models, in a printed publication. Therefore, the most sophisticated computer models of forestry could not be utilized by revising the manager's handbooks in their present printed format. To overcome these weaknesses, Artificial Intelligence (AI) methods were proposed to update the red pine manager's handbook as a computer-based advisory system.

4.3.2 The Participants and Their Roles

Next, let us turn our attention to the individuals that might be involved in a knowledge-based system effort. As many as five different groups of participants may need to be part of system development (Walters and Nielsen 1988). In the minimal case there may only be a knowledge engineer and no one else (engineer = user). A **knowledge engineer** is an individual that has some understanding of the acquisition, structuring, representation, and manipulation of knowledge, i.e. someone that builds knowledge, much like a civil engineer might build bridges. In developer-equals-user situations, many of the knowledge engineering development steps that address organizational issues or end-user considerations may not be necessary. A single-user system, however, still requires careful planning and design for successful results.

Often there are other people involved in the project in some way. A knowledge engineer and some user group will generally be the minimal team. *Users* are the people that have a need for advice or knowledge to perform well in a specialized field. When some of the knowledge captured by the system resides in human **experts** (i.e. people with specialized knowledge), then those specialists become part of the development effort. Here, the system may often be referred to as an expert system, however we strongly resist the use of that terminology for reasons mentioned in Chapter 1. In certain situations there may be multiple experts or multiple knowledge engineers or both.

Depending on the resources of the development group, *programmers* may also be available. They are responsible for implementing (coding) the actual system. Programmers can be of great benefit to a project, because their presence often allows various development activities to proceed in parallel. For example, while programmers are implementing a prototype version, a knowledge engineer may be organizing resources or personnel for testing the prototype or the engineer may be engaged in additional knowledge acquisition activities. In large projects, in particular, programmers can be essential cogs in the development wheel.

Because overall organizational considerations may be important, the managerial level may also provide some input. *Managers*, acting for the people who provide facilities and pay bills, often insist—and rightly so—that their voice be heard, at least in the planning stages.

Roles, i.e. activities and duties, assigned to all these different participants will be determined by domain specifics and the presence or absence of any of the participants. In the absence of certain development team participants, one individual may be asked to assume several roles. For example, a programmer with some systems development background (not necessarily knowledge systems) may be asked to work on some knowledge system design tasks. Or, experts may serve both as experts and as members of the user group, if they are the people for whom the system is being built. Or, as is often the case, when programmers are unavailable, the knowledge engineer becomes the programmer. Regardless of the cast of characters on the development team, operations move most productively if everyone understands their assigned tasks within their role.

The development team organized for building the Red Pine Forest Management Advisory System, designated RP-FMAS, included: two red pine management experts, three pest specialists, three potential end users, and the authors, who served as knowledge engineers and programmers. This knowledge-based system example is meant to distill these experiences so that readers can better understand how the important steps in system development fit together.

4.3.3 Understanding the Application Problem

Forest management is a complex activity which focuses research knowledge and practical experience on the accomplishment of some goal(s) to produce some valuable product(s). The quantity and quality of the supporting knowledge for forest management is highly variable, where often the majority is qualitative rather than quantitative in nature. Managers' handbooks have proliferated (e.g., Basset 1984, Benzie 1977), but attempts to provide comprehensive automated systems to help understand and guide the forest management process are few and of limited value (Seal 1966, Bell 1971); prospective developers of these automated systems have been waiting for a more powerful technology. AI methods promise to make comprehensive forest management advisory systems practical and useful for the first time.

4.3.3.1 A World View. A major conceptual tool to help understand the application problem is a strategic, or world, view. A **world view** provides a vision of the context in which the application problem exists and leads to the identification of the major modules and their interrelationships. For example, RP-FMAS is viewed by the development team as only the first step in creating for forest science a testable, complex hypothesis of the forest management system. The current version of RP-FMAS is designed to provide advice and structure for the simplest of forest management systems, the even-aged management system. The next step is the development of the aspen forest management advisory system, AS-FMAS (Nute et al. 1995), to gain more experience in knowledge representation and advisory system implementation for even-aged forest systems. It will then be possible to combine the two even-aged systems into a general even-aged forest management advisory system that treats red pine and aspen merely as special cases.

Further work must then determine how we can incorporate an uneven-aged system, such as northern hardwoods species, into our world view. Eventually, it would be desirable to integrate the multi-storied forests of the western United States into this framework, also. Following this process will result in a precise definition for what we mean by "forest management", and will produce a forest management advisory system of national scope and utility. Such a comprehensive and detailed world view, or paradigm, can then be the focus of debate within the forest research community leading to rapid evolutionary improvements. The most immediate gain of such a strategic vision for forest management research is to provide a way to prioritize the vast array of sub-problems and thus concentrate resources on those with the highest pay-off or that have the greatest social importance.

More recently, there has been an ideological shift, of sorts, away from *forest management*, in the traditional sense, toward a more encompassing view of management, *ecosystem management*. Although, this change will affect the way in which landscapes (particularly public lands) are managed, it does not change the basic intent or structure of our previous world view, only how it is implemented. Now, instead of viewing forest management as manipulating stands of trees, it consists of inventory, monitoring, and adaptive management of ecosystems. Implementationally, this means that the health and functioning of many individual aspects of the forest will be carefully supervised.

As an independent and self-interesting module, forest health may have its own world-within-a-world view. The origin of the PREDICT project was research oriented like RP-FMAS. Unlike RP-FMAS, however, PREDICT was not designed primarily to understand and solve a difficult problem in forestry. Its primary goal was to demonstrate and evaluate the applicability of AI methods and KBS structure to solve forestry-type problems. Consequently, project emphasis focused on the technology more than on the pest management application and a forestry-related world view.

MYCIN (Shortliffe 1976) was extremely successful in utilizing AI methods

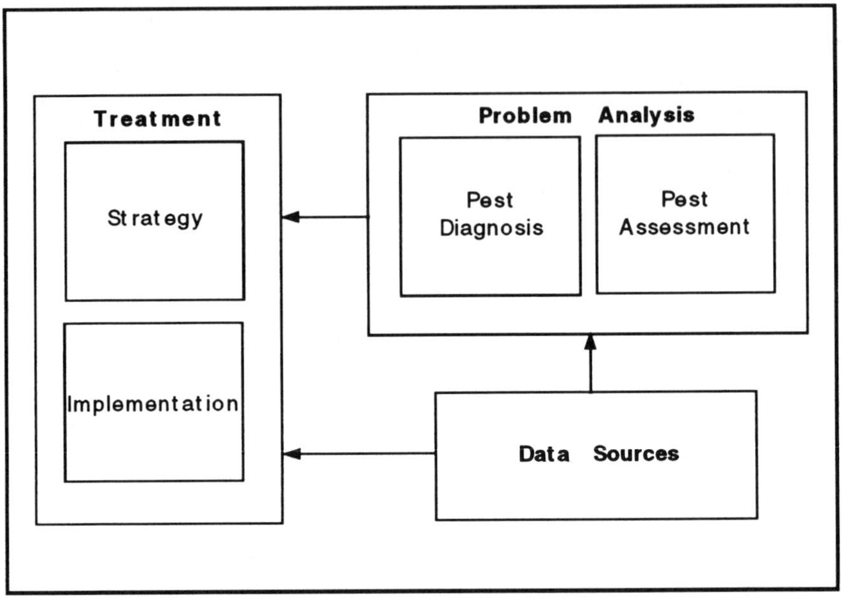

Figure 4–4. A world view of pest management of red pine consists of both problem analysis and treatment recommendation as the two major components. At the present time only the problem analysis section has been implemented.

for diagnosis and treatment of bacterial infections. The extension of AI methodology to pest management in forestry seemed like a natural transition (Note: in this and subsequent discussions, the term "pest" refers to both insects and diseases.). This simplistic observation, however, proved to be an over-simplification of the transfer of technology. Substantial differences exist between diagnosing/treating forest pest problems and doing similarly for human bacterial problems (Schmoldt 1987). Figure 4–4 depicts the world view of pest management as composed of two separate decision areas: (1) problem analysis and (2) treatment. Problem analysis may identify a forest stand as predisposed to injury by particular pests—pest hazard; or the analysis may diagnose a current stand problem as caused by one or more pests—pest diagnosis. Subsequent treatment recommendations will incorporate several factors: (1) whether the analysis is diagnostic or hazard assessment, (2) the pests identified, (3) how plausibly pests are identified, and (4) other management or silvicultural considerations. Because the aim of this entire project was extension of KBS technology, problem analysis was selected as a non-trivial application, with treatment recommendation postponed for future research.

Because PREDICT deals only with one aspect of forest stand management, it lies within the sphere of overall stand management and, hence, fits nicely into

RP-FMAS. PREDICT identifies hazards or diagnoses that must be considered as part of the overall treatment of any stand. Any silvicultural prescriptions that are made by RP-FMAS can incorporate PREDICT's knowledge about current and anticipated stand health problems.

4.3.3.2 Is a KBS Viable?. After a world view or strategy for the problem has been formulated, then the methodology for solving it can be selected. An overview of the broad range of possible software solution methodologies is depicted in Table 4.1. Much published scientific and practical knowledge resides in natural language systems (e.g., books, pamphlets, articles, handbooks) that have high generality and ambiguity, but low problem-solving power. Quantitative models, on the other hand, have high problem-solving power, but are not generally applicable to situations outside of their explicit models. More recent methodologies (e.g., hypertext, KBSs) attempt to strike a middle ground between these extremes.

For the purposes of this chapter (and for this text), the question is: Should we use a KBS approach to solving this problem or is another solution methodology better? Many people have made lists trying to categorize problems that are amenable to AI methods. We've reiterated some of those ideas in Chapter 1. Does an algorithmic solution to the problem exist, and if so, is it feasible? Generally an algorithmic solution, if it exists, is to be preferred over an AI approach. Certainly this does not prevent embedding an algorithmic solution method in a knowledge-based system, where the KBS serves as a "front end" to mediate between program and user. Maybe an algorithmic and symbolic solution method can be combined for a better result (Rauscher et al. 1989). On the other hand, the current natural language systems may be quite adequate for problem solving, with some updating and reformatting of the delivery mode—this presents another option. These are critical questions that need to be asked prior to building a KBS.

Table 4–1. Language-based knowledge representations vary according to their problem-solving power and their generality.

Natural Language Systems		Artificial Language Systems Logic		Mathematics
NLT[a] Documents	Hypertext Documents	Knowledge-Based Systems Qualitative Simulation	Numerical Equations Neural Networks	Analytical Equations
High	←	Generality	→	Low
High	←	Ambiguity	→	Low
Low	←	Problem-Solving Power	→	High

[a] Natural language text

Because red pine forest management knowledge had already been highly organized (Benzie 1977) and the human experts who did much of this organizational work were interested and available, we concluded that a KBS in this domain was possible. In fact, the major questions were not about the domain knowledge, with which even the knowledge engineers on the project were well acquainted. The major question was whether the current AI technology had matured enough to handle the complexities of forest management. We operated, at first, largely on faith that the AI technology was capable enough for our purposes.

Other KBS developers may require a more extensive analysis of an application problem. Later in this chapter, we look at a highly structured methodology by Laufmann et al. (1990) for evaluating potential KBS applications. Before presenting their ideas, however, we examine some of the issue that will need to be addressed prior to making such an analysis.

4.3.3.3 Application Environment. The environment under which the system is expected to function will determine many aspects of design. Systems that must operate in a hostile environment or in real-time decisionmaking will require special considerations different from a traditional desk-top application. Because end users can be considered part of the operating environment, if they have not previously been identified, they should be now. Even in non real-time situations the decision time frame may vary considerably from several minutes, to hours, to days. At this stage, these types of things are only identified; later they will be tied to system specifications via requirements analysis.

The RP-FMAS was intended for use by field practitioners, forest managers, researchers, and educators. In planning for the system, much time was spent on identifying the features that would be most needed for each of the four major user groups.

4.3.3.4 Cost-Benefit Analysis. A cost-benefit analysis for a KBS examines the various factors, and their costs, involved in developing a system and also anticipated benefits from system deployment. Both costs and benefits are given some numerical score, and the two quantities are compared. The importance of this step varies drastically between situations. For example, in a research organization there may be many qualitative, or intangible, benefits associated with developing a KBS. These might include increasing the amount or quality of knowledge in a subject area (Rauscher 1987b) or demonstrating that a KBS approach can be useful for problems of a particular type (Schmoldt 1987) or meeting academic needs, such as training graduate students. These kinds of benefits may not be easily quantified because they are not immediately tied to the productivity of an organization or the generation of some product. When benefits or costs are difficult to quantity, any comparative analysis based on those values carries less weight for project go-ahead. Also, in situations such as the research environment there is a tendency to ignore a balance sheet of costs and benefits because many aspects are intangible and net costs are accepted as part of the business of research.

On the other hand, in more financially conscious organizations there is a need to be accountable for expenditures and for returns on investments. As a result, both costs and benefits become much more tangible and explicit. Shareholders expect dividends and the payback period must be kept short. Therefore, the comparison of benefit and cost values can be an important part of project justification.

When a cost-benefit analysis seems indicated, one should identify and document the anticipated benefits, both immediate and long term. Then place some sort of monetary expected value on these benefits, in terms of improved productivity, better job performance, etc. Expected costs should include hardware and software for development (and delivery), personnel involved, and other costs (e.g. routine disruption, travel, hidden costs). Both costs and benefits should be computed in current dollars by using some expected discount rate. Because benefits are usually delayed until project completion, while costs are incurred now and until project completion, benefits are more heavily discounted. Therefore, \$1 of benefit when it is received is worth much less than \$1 of cost when it is incurred.

How closely the present value of costs and benefits balance should indicate whether the development effort will have significant payoff. Usually a benefit-cost ratio can be calculated by dividing the total value of benefits by the total value of costs. If this ratio is near 1, a payback at least equal to the investment could be expected; as the ratio becomes larger, a greater payoff can be anticipated. Some companies like E. I. duPont[1], who have invested heavily in KBSs, report average benefit-cost ratios of near 15. Walters and Nielsen (1988) recommend a benefit-cost ratio of at least 2 for project success. Walters and Nielsen (1988) also provide some guidelines for estimating the value of benefits and costs, both tangible and intangible ones.

Few reports in the scientific literature describe the use of cost-benefit analyses. This situation exists largely because the focus of KBS examples in research is often not on the deployment implications of system development. Technical issues are usually the primary consideration. That is, how can we effectively solve problems of this type?, or can we codify and apply human expertise in this subject area? Many of the guidelines for selecting KBS applications, presented in Chapter 1, are designed to implicitly contain cost-benefit and risk assessments. We expect, however, that many business enterprises use an explicit cost-benefit analysis as part of the feasibility study.

The first version of RP-FMAS was intended to be primarily a research and educational tool to test out the concept. Thus resources were committed to the project without a detailed cost-benefit analysis.

---

[1] This information was provided in an unpublished 1987 report by Ed Mahler of duPont.

4.3.3.5 Define the Solutions. At this point, it becomes necessary to understand an application problem in more detail. In a very basic sense, we must ask, "What are we trying to do?", that is, what is the problem we are trying to solve? To do this, we must explicitly characterize valid solutions to typical problems. There are at least three attributes that categorize an application's solutions:

1. *specificity* – the degree to which possible alternatives can be listed a priori;
2. *number* – how many alternatives may constitute an answer to a problem; and
3. *plausibility* – some score that represents the likelihood or usefulness or value of each alternative.

Figure 4–5 displays some types of application problems according to solution specificity. This figure is a one-dimensional variant of Kline and Dolins (1985) two-dimensional description. We identify three levels of specificity: pre-enumerated solutions, constructed solutions, and simulated solutions. For pre-enumerated, a possible list of alternative solutions may be known a priori, i.e. we may be able to specifically list each possible alternative solution. In the second category, solutions may need to be constructed and described via other parameters within a problem description—there exists a little less specificity. In the third case, symbolic simulation, a solution would be the final state of all or some part of the system parameter set after some simulated time interval. In this latter case, not only are we unable to make a list of possible answers, we are unable to determine ahead of time which parameters will be indicative of a final state.

Readers may note that truth maintenance has been included here, whereas it was also listed as a control strategy in Chapter 2. While truth maintenance can be viewed as a control strategy, when it is used as the primary method of finding

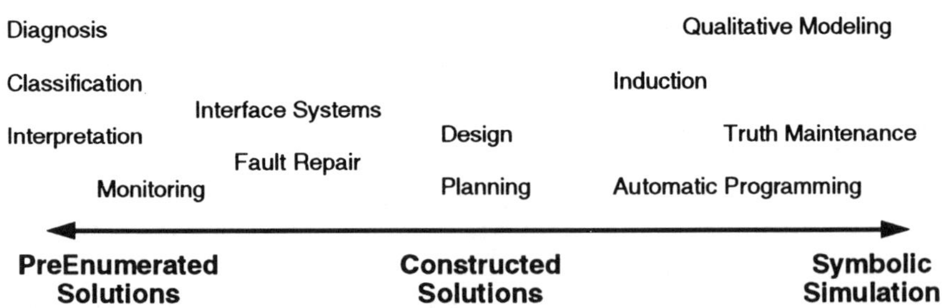

Figure 4–5. Different application problems have different levels of solution specificity. For those problem types toward the left of the diagram a list of possible alternative solutions is known in advance. As we move to the right into other types of application problems, enumerating solutions a priori is less easily accomplished. To the far right solutions are often represented by the state of an entire system.

problem solutions, that system is often referred to as a truth maintenance system.

As we move from pre-enumerated to simulated solution problems we retain less and less ability to easily recognize valid solutions. At least in this sense, pre-enumerated solution types of problems seem to be less complex and easier to understand and develop systems around. These are also often referred to as **classification** problems because problem solving entails classifying a particular problem instance as belonging to one of a set of pre-enumerated categories. Most KBSs, including RP-FMAS, are of this type and high specificity may be one of the reasons for their popularity.

Regardless of solution specificity, the number of alternatives that constitute an answer may also vary. Some applications have a single best solution which can be determined from the data for each problem instance. On the other hand, there are also many applications where the structure of the problem does not always permit a single conclusion to be reached. For RP-FMAS (Figure 4–6) and the diagnosis component of PREDICT (Schmoldt and Martin 1989), multiple alternatives may be given to the user depending on the data for a particular case. This multiplicity may result from conflicting expert opinion, from inexact heuristic knowledge, or from uncertain or incomplete user input.

Multiple-alternative solutions may also result from incorporating multiple opinions in the same system. Different human experts sometimes reach different conclusions from the same data. It would not be unreasonable to build a system to incorporate such a bifurcated reasoning structure; in fact, it probably should be constructed that way if system knowledge is to reflect the current state of scientific knowledge about that problem. To determine whether multiple-alternative solutions are part of the application (or should be), current experts in the field and users need to be consulted. As a rule of thumb, though, there are very few applications in which the state of the science and the certainty of the data used to make decisions is of sufficient quality to give a single solution as an answer.

When multiple-alternative solutions exist, there will often be some measure of plausibility attached to each answer. This indicates to the user that the data suggesting one conclusion may be preferred over others. Methods that can be applied to accomplish this were presented in Chapter 3. Under some conditions it may not be possible to identify one answer as preferable; this is most often the case in simulation problems. D'Ambrosio (1989), however, has implemented a simulation mechanism that does include belief estimates for different conclusions. Heuristic methods are by nature inexact; this fact suggests that plausibility measures should be used for all applications whenever possible.

The next two sections contain examples of an application's operational structure as a way to understand an application problem. These application-specific perspectives are formulated once it has been decided to pursue a KBS

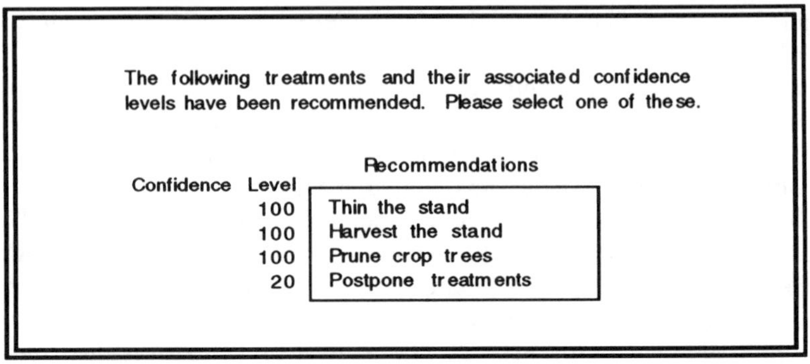

(a)

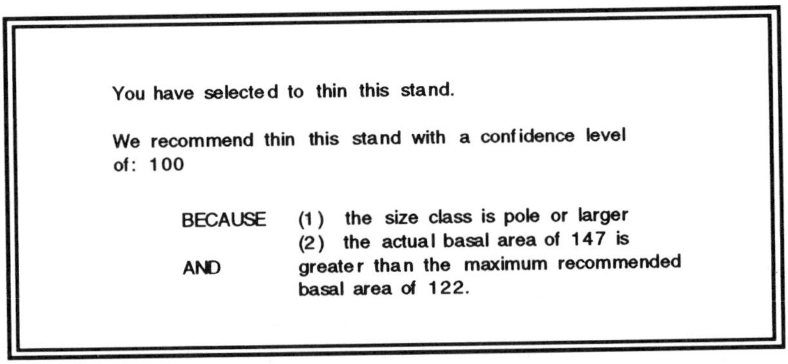

(b)

Figure 4–6. Multiple answers are not uncommon for users of RP-FMAS (a). When the user selects a preferred treatment, RP-FMAS simulates the effect of that treatment by modifying its internal representation of the stand in terms of numbers of trees and their sizes. A second display (b) provides the user with rationale for the decision that they selected.

solution. A forest management operational structure and a pest management operational structure are both presented.

4.3.3.6 RP-FMAS Operational Structure. Once the RP-FMAS application has been studied to determine its viability and utility, and its solution space has been identified, then the next task is to define the major modules and their functions (Figure 4–7). In other words, we need to specify the *operational structure*, i.e. describe the overall structure and function of the proposed knowl-

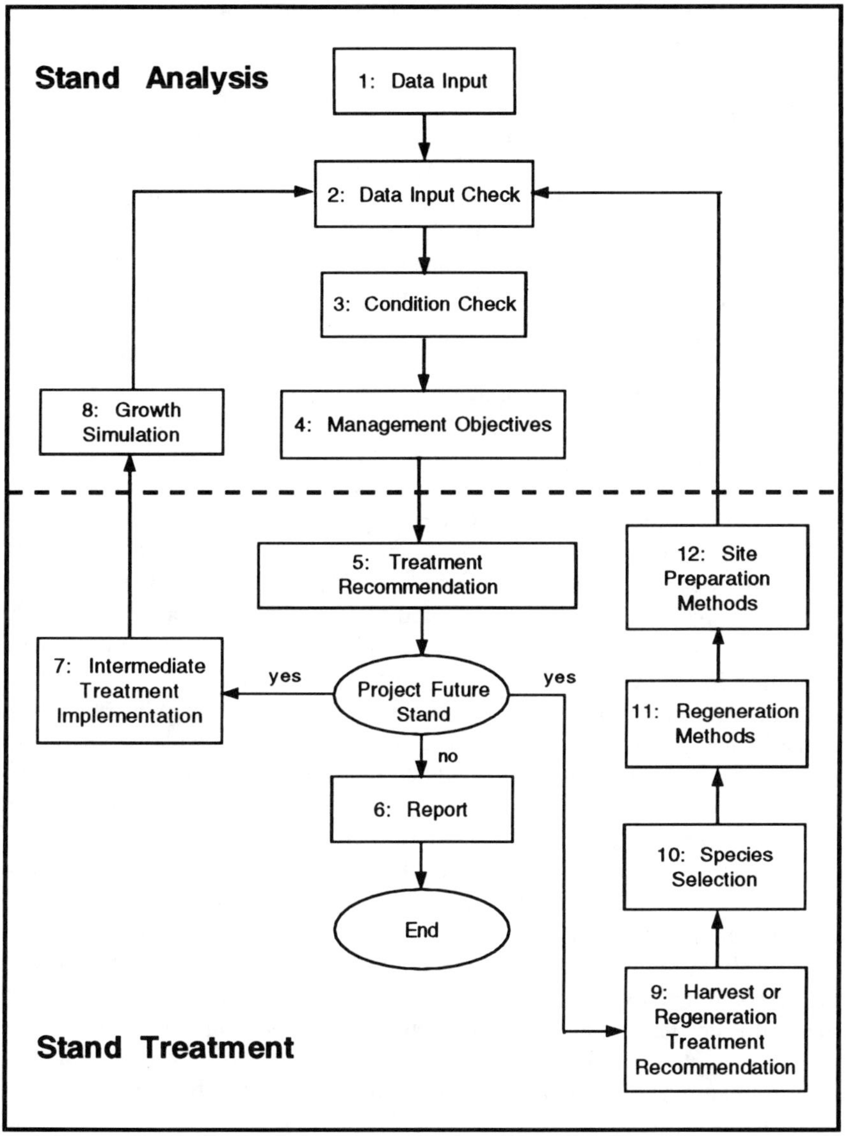

Figure 4–7. The operational structure depicted here describes one possible view of red pine management. This view consists of two major components, stand analysis and stand treatment, depicted in the upper and lower portions of the figure, respectively.

edge-based system. The operational structure is another major organizing tool leading to a good system design. A operational structure makes it fairly easy to see where and how most of the sub-components that make up a forest management knowledge-based system fit, and how they interact for solving problems.

In RP-FMAS (Figure 4–7), the beginning module, number 1, obtains data describing a subject stand. The next module, number 2, analyzes the input data for missing information. Missing data must be estimated through other means. Once the subject stand has been fully described, module number 3 analyzes the health of the stand to determine if management decisions will be impacted by significant insect, disease, or physical damage. The management objectives of the owner must be determined, module number 4, and then module number 5 can analyze the subject stand in view of the known background information to produce a list of reasonable treatment recommendations.

If an intermediate treatment is selected, i.e. thinning, weeding, or thinning and pruning, etc., then module 7 changes the description of the subject stand accordingly and passes control to module 8 which simulates future growth of the stand. The future stand description is then passed back to module 2, whereupon the cycle is restarted. For the purposes of module 2, it is irrelevant whether the stand description is entered directly by the user in module 1 or whether it is generated by the growth and yield program called by module 8.

If a harvest or regeneration treatment is selected, then module 9 is activated to simulate implementation of these treatments. At this time, the user may choose to retain the former species or convert the site to another species by exercising the species selection module, number 10. Given a choice of species, the regeneration methods module, number 11, is activated followed by the site preparation module, number 12. Finally, the simulated newly established stand description is fed back to module 2 for further processing.

The advisory system can be terminated each time a treatment is selected. Upon termination, a report of the entire session is presented to the user in a text file. The user can make annotations in this report file and save it for future use.

Obviously, such an operational structure is not created instantaneously through some creative inspiration. It evolves and matures with each cycle through the design-implement-evaluation process. Our experience has shown that this conceptual model is continually evolving and will never be "finished" any more than our understanding of this problem will be "finished".

4.3.3.7 PREDICT Operational Structure. In PREDICT (Schmoldt 1988), there are two possible modes of operation—either diagnosis of a stand pest problem or assessment of existing hazards for pest injury (Figure 4–8). After the user decides on what type of pest management question needs to be answered, then restrictions may be placed on the extent of the analysis. Restrictions may be stand location, i.e., the county in which the stand is located, or

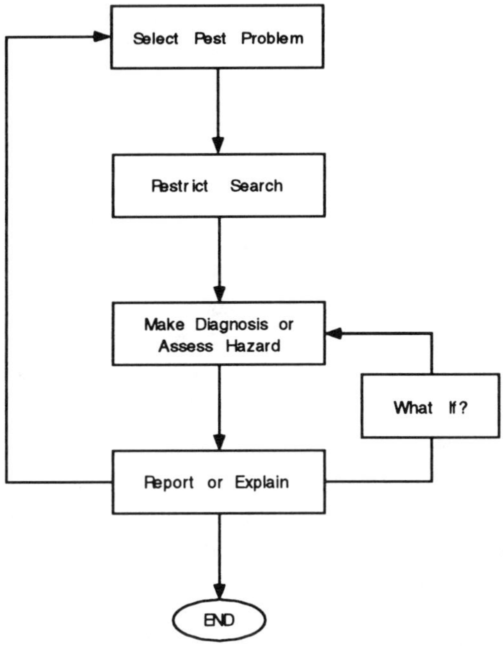

Figure 4–8. The operational structure of PREDICT depicts the distinct stages of a consultation.

consideration of certain pests or groups of pests only. County location is very important information because PREDICT has knowledge about likely pest problems for each county. Therefore, only those pest problems that are known to occur in a particular county are considered. Also, those pests that occur most frequently in a particular county are more likely to be the cause of observed damage (in the case of diagnosis) or are more likely to be a future problem (in the case of hazard assessment). Then, the user has the option to restrict the list of pests that PREDICT considers. Certain categories of pests, e.g., root/root collar pests, or defoliators, or even specific pests may be selected. By limiting the number of pests that PREDICT examines, the search space is reduced and the consultation proceeds much faster.

PREDICT then activates the rule set associated with each pest that has been selected, and performs either diagnosis or hazard assessment, depending on which was selected initially. In diagnostic mode, each pest that is within the restriction set of the analysis is considered as a potential cause of stand damage. In hazard assessment mode, each pest selected is assigned one of four hazard ratings independent of any other pest's hazard. A brief report displays a list of pests and the likelihood that each might be the cause of stand damage (diagnostic mode) or a hazard rating for all pests selected (hazard assessment mode).

At this point the user can proceed to a consultation examination step or cycle back to another diagnosis or hazard assessment. In consultation examination, the user can inspect the line of reasoning that produced the report or can change his/her answers to questions and then view the resulting changes in the report. The ability to change input information and see the resulting effects provides the user with very helpful "what-if" capability for viewing the impact of different items of information.

This operational structure does several things: (1) it allows the user to focus on those pests that seem most pertinent to the problem at hand, (2) it seems to parallel to some extent the strategy that pest specialists use to attack a diagnostic problem or assess hazard, and (3) it greatly restricts the number of information questions posed to a user. Effects (2) and (3) were two of the operational objectives for the system. The remaining effect resulted from the logical structure used to implement the other two objectives.

4.3.3.8 Characteristics of the Data and Knowledge. In the next two sections, we use the term *data* as inclusive of both *data* and *information,* as described in chapter 1. Data and knowledge used in problem solving can exhibit many different characteristics. Some characteristics of data are:

1. *timing* – when does a system have access to the data?;
2. *completeness* – all necessary data may not exist in all cases;
3. *certainty* – we may not be confident that all facts are true;
4. *form* – the types of relationships that may be represented by the data; and
5. *ambiguity*-- is there always agreement on the meaning of data items?

In some instances data may be volunteered as needed during problem solving, in other circumstances, all data items may be available prior to the search for solutions. In still other situations, some data may be available now and additional data may need to be collected, perhaps over a few days or a few weeks. Data availability was discussed in relation to selecting forward or backward control strategies in Chapter 2.

Data can be incomplete in the sense that not all informative observations may have been made. For example, just because some disease symptom has not been observed does not necessarily mean that it is not present. Our abilities to observe are often limited by time, effort, and skill. Therefore, some highly useful pieces of data may be missing. A system should be able to provide useful answers in spite of data gaps.

Sometimes, not all data items are known with absolute certainty to be true or false. For example, if asked for the average basal area for a stand of trees, a value of 120 sq. ft. per acre might be supplied by a user. The user may be guessing, in which case 120 would be highly uncertain. Or, a random sample of field measurements may indicate that 120 is a good approximation, therefore 120 is fairly certain (this would contain then probabilistic uncertainty). In many cases it is important that a system is made aware of any uncertainty in data items and that it is able to use it properly.

The data may be represented in several different ways and may have several different representational attributes, e.g. symbolic, numeric, relational, temporal, or spatial. The first two are the most common; however, representing temporal or spatial data and knowledge is much more difficult and has become an active area of research in AI (e.g., Ladkin 1986, Tsang 1987, and Vilain and Kautz 1986—temporal; Davis *et al* 1988 and Retz-Schmidt 1988—spatial). Data bases, for example, contain large amounts of data in temporal, numerical, and relational formats; work on intelligent data bases attempts to make that data more viable and functional (q.v. Parsaye et al. 1989).

When a system asks the user about a data item, there should be complete understanding by the user as to what is requested. Ambiguity may be inherent in the observation process. e.g., at what point is a soil containing sand labeled as a sandy soil? That is, human and machine sensors are not infallible, sometimes they provide us with conflicting messages (note: this is different from uncertainty, where a fact may have some believed truth value). Ambiguity may also result from a lack of definitional clarity of terms (e.g., how red is red?). Vagueness and ambiguity can present very difficult problems for ensuring that data are treated in a consistent and consentaneous fashion.

Data characteristics, like those mentioned above, help determine the sources of data, i.e. where we can expect to find such data and, reflexively, data characteristics are also partially determined by the sources available, i.e. typical data sources may have fixed characteristics that we must accept and deal with.

As in the case of data, there are several characteristics of knowledge that carry importance for "intelligent" systems. Some of these are:

1. *currency* – how up-to-date is the knowledge?;
2. *completeness* – all necessary knowledge may not exist in a particular domain;
3. *certainty* – the knowledge may be heuristic and, hence, may only be someone's best guess;
4. *functionality* – to what purpose the knowledge can be used; and
5. *commonality* – do most "knowledgeable" people agree on the validity of the knowledge?

Knowledge is dynamic; it continually changes as we learn more about the world. Once it has been made explicit within a system, maintaining its currency becomes extremely important. What are considered established scientific facts change from year to year, and appropriate ways to deal with those facts change also. Knowledge engineers must be aware of how tenuous knowledge can be, and must develop systems that can be adapted to innovations and revisions.

Because of its dynamic nature, knowledge also tends to be incomplete and uncertain. As noted in Chapter 2, most KBSs don't have a sense for their own limitations. KBSs don't possess any self-realization for what they know and what they do not know. Consequently, the knowledge engineer must continually add to the store of knowledge; however, until systems are able to learn on their own, this process must necessarily lag behind the creation of knowledge. Many systems do, however, have some methods for representing vagueness and uncertainty and using it in their reasoning (Chapter 3).

We have previously seen some functional capabilities of knowledge, e.g., procedural, control, explanatory, domain, common sense, analytical, etc. Careful treatment of any application requires the inclusion of at least several of these different functional characteristics.

When knowledge can generally be considered common to the members of some "knowledgeable" community, then any difficulties with accessing and using it are minimal. On the other hand, knowledge may depend on the source and vary considerably. When well-accepted community knowledge does not exist then either consensus should be reached prior to system implementation or differing opinions should be handled within the system or, perhaps, the system should not even be built at the current time.

In developing the RP-FMAS, we often encountered problems with incomplete and uncertain knowledge and data. The preferred solution would have been to use inexact reasoning methods to differentiate more or less known and certain knowledge and data from uncertain knowledge and data. We chose not to do this in the current version of RP-FMAS because: (1) the experts had extreme difficulty expressing knowledge uncertainty and its impact on the system using the currently available methods, (2) the user community generally prefers a single best answer and, hence, would be uncomfortable with an inexact answer that might required further interpretation; and (3) the inclusion of inexact reasoning in any meaningful manner will require a significant amount of future research in the forest management domain and has been relegated to a future version.

When dealing with incompleteness and uncertainty in PREDICT, it was apparent that not all information pertinent to a pest diagnosis solution would necessarily be available to the system (volunteered by the user). In the unrealistic situation where all important information is available, many diagnoses become relatively trivial tasks. If a system were designed to assume all data are always available, a lack of information would make it impossible to diagnose any pest. But yet, managers and landowners must still make pest decisions using the available knowledge, hence it was decided that PREDICT should allow for this eventuality. Therefore, an evidence accumulation approach was employed.

In this method most rules in the knowledge base contain a certainty level that describes how strongly the premise conditions infer the pest mentioned in the conclusion. This certainty level reflects the degree of support that the premises' evidence conveys for that particular pest. Other rules for a particular pest provide additional evidence and, hence, support for the pest. Support from individual rules for each pest are combined using MYCIN-like calculations (see Chapter 3). The result is a gradual increase in confidence for a particular diagnosis given accumulating evidence. Even with poor or contradictory data from the user, PREDICT will often provide some answer, although usually not a very conclusive one under these circumstances. While experts were uneasy about assigning evidence support values to each rule, with some coaxing and plenty of discussion it was possible to arrive at some reasonable values. Because many of these values are eventually modified in the refinement stage (Schmoldt 1989), agonizing over specific values, in absolute terms, during this early stage is not necessary. Relative value assignments are more important.

Another layer of complexity in pest diagnosis results from the ambiguity associated with symptom descriptions. As an example, a user may have to provide information about the sandy nature of the soil in a stand. It may be difficult to declare how "sandy" a "sandy loam" or a "loamy sand" is, that is, to what degree can the soil be described as sandy. As a result, PREDICT users are allowed to augment most answers to questions with some subjective numerical score indicating confidence in their answer. Confidence values for premises of a rule are combined with the certainty level of that rule to arrive at a total score of evidential support implied by that rule. Again, the reader is referred to the discussion of MYCIN-like uncertainty factors in Chapter 3.

4.3.3.9 Sources of the Knowledge and Data. Access to data for any particular problem instance may not only be limited to the data available to, and supplied by, the user. In addition to his or her specific input for a problem, there may also be applicable input from spreadsheets or data bases, from other application programs, from documented sources, or from equipment with real-time connections. These can be just as important and more or less as reliable as user data. Oftentimes, an inconsistency exists here, however. While the user may be able to append belief or plausibility values to his or her input, data received from other sources often does not receive any similar qualifying adjustment.

In the immediately preceding section, we peripherally alluded to sources and accessibility of knowledge during the discussion of commonality. Knowledge may be located in one individual, several individuals, books, manuals, other computer programs, etc. The location of this knowledge will determine to some extent how accessible it might be. Often, the "private" knowledge of individuals can be more difficult to acquire than that resident in more "public" sources. Documents display much of their knowledge in a manner that lends itself to easy examination. Multiplicity inherent in the community knowledge of several individuals may sometimes be more easily accessed than the personal knowledge of a single individual; accessibility of knowledge from a group can be complicated, however, by consensus difficulties. Each knowledge source has its own peculiarities and these vary from application to application.

This generalization of an association between accessibility and the notion of public/private can be somewhat misleading though. While hard-copy materials, like books, journals, and manuals, contain much explicit knowledge, they also contain knowledge structures and relationships that may be less obvious. Table of contents outlines and illustrative figures are good examples of the knowledge structures in hard-copy materials. Much application software also has hidden knowledge structure. Extracting these ideas from such "public" sources may be difficult and require the "private" knowledge of an expert.

Multiple sources of knowledge imply conflicts. These not only arise between individuals, but also occur between software, documents, and individuals. Anytime multiple sources of knowledge are included in system planning and design, conflicts should be expected and planned for.

The RP-FMAS is based principally on about 90 years of research performed by scientists of the USDA Forest Service's North Central Forest Experiment Station and the universities in the Upper Great Lakes region. The work of numerous researchers has been published, evaluated, and organized to produce a large amount of "public" knowledge. For example, Figure 4–9 is a dichotomous decision key for Red Pine Forest Management decision support published by Benzie (1977). The experts on our development team have learned this public knowledge and have also reformulated it based on their many years of field experience. Their "private" knowledge is invaluable because it contributes judgment and understanding so that we can use the public knowledge more wisely. Advisory system developers should strive to capture and clone this personal judgment and understanding as much as possible, in addition to the more public and formally organized knowledge of the domain.

Many pieces of knowledge about pest diagnosis in red pine can be obtained from the literature. However, the manner in which this knowledge is applied to stand problems is not documented. It resides only with those individuals that have absorbed the available information and that have then formulated tactics and strategies for its application. It is precisely this type of knowledge that we sought to elicit and to incorporate into the diagnostic capabilities of PREDICT.

4.3.4 Evaluating the Application

As noted in chapter 1, many authors have sought to identify likely candidate KBS applications by listing types of problems that seem to be amenable to the KBS approach. One disadvantage of making these lists and then comparing any particular application to ones on the lists, is that each application may differ

1.	Red pine stand	2
1.	Other stands or the area is nonstocked	14
	2. Even-aged stand	3
	2. Uneven-aged stand	12
3.	Stand has minimum or higher stocking	4
3.	Stand is below minimum stocking	14
	4. Stand is not mature	5
	4. Stand is mature	12
5.	Average tree dbh is less than 2 inches	6
5.	Average tree dbh is less 2 inches or more	8
	6. 400 or more trees per acre free to grow	7
	6. Less than 400 trees per acre free to grow -- RELEASE	7
7.	Less than 2000 trees per acre	17
7.	2000 or more trees per acre -- WEED	17
	8. Average tree dbh is less than 5 inches	9
	8. Average tree dbh is 5 inches or more	10
9.	Basal area is less than 160 square feet per acre	17
9.	Basal area is 160 square feet or more per acre -- CLEAN	17
	10. Basal area is less than 140 square feet per acre	11
	10. Basal area is 140 square feet or more per acre -- THIN	11
11.	Crop trees do not need pruning	17
11.	Crop trees need pruning -- PRUNE	17
	12. Old growth stands are needed for timber or other resources	13
	12. Old growth stands are not needed	18
13.	Temporary need until other stands mature -- EXTEND ROTATION	17
13.	Continuing need for mature trees on area	USE SELECTION
	14. Site is suitable for red pine	15
	14. Site is not suitable for red pine	MANAGE FOR OTHER SPECIES
15.	No merchantable stand on area	16
15.	Merchantable stand on area -- USE CLEARCUT	19
	16. Will have merchantable stand in 20 years or less	17
	16. Will not have merchantable stand in 20 years	19
17.	Low risk of injury or loss	WAIT
17.	High risk of injury or loss	CONTROL IF FEASIBLE
	18. Continuous tree cover needed -- USE STRIP SHELTERWOOD	19
	18. Continuous tree cover not needed -- USE CLEARCUT	19
19.	Adequate mineral soil seedbeds free of slash and competition	20
19.	Inadequate seedbeds -- PREPARE SITE	20
	20. Easy seeding chance	DIRECT SEED
	20. Poor seeding chance	PLANT

Figure 4–9. A dichotomous decision support tool for red pine forest management illustrates a publicly available source of organized knowledge for RP-FMAS.

substantially in important respects from other similar applications. Those differences may not be captured by the simplicity of the succeed/fail list comparison. Consequently, broad, blanketing lists may overlook "good" KBS applications and similarly may fail to identify "bad" ones.

Laufmann et al. (1990) detail a systematic procedure for examining any potential KBS application. Their method uses a scoring worksheet that asks specific questions covering many of the issues presented in the preceding sections. Their evaluation is separated into two separate analyses, a top-level analysis and a detailed analysis. At both levels the following four questions are addressed through more specific questioning:

1. What is the definition of success?

	Level One					Level Two				
	+ +	+	•	-	- -	+ +	+	•	-	- -
Goals										
Appropriateness										
Resources										
Nontechnical										

Figure 4–10. A scoring summary like this can be used to evaluate a potential KBS application across four categories: goals, appropriateness, resources, and nontechnical. Detailed analysis involved in each category can be found in Laufmann et al. (1990).

2. Is the KBS approach appropriate for the task?
3. Are the necessary resources available?
4. Are there any nontechnical impediments to developing a KBS solution?

The top-level analysis is designed to be performed by managers with consultation from a knowledge engineer or technical manager. If the first analysis uncovers no strongly negative results then a detailed analysis should be performed by experienced knowledge engineers and other technical people. In the interest of space, we do not reproduce a scoring worksheet here for either RP-FMAS or PREDICT. Interested readers should refer to Laufmann et al. (1990) for more detail. Answers to each question on their scoring sheet are taken from the following symbol set:

+ +	strongly favors a KBS approach
+	favors a KBS approach
·	neither favors or disfavors a KBS approach
–	disfavors a KBS approach
—	strongly disfavors a KBS approach

Then, answers to individual questions in each of the four categories in each analysis are summarized in a grid such as Figure 4–10. The total numbers of responses of each type (+ +, +, etc.) are summed and recorded.

Laufmann and his colleagues provide the following guidelines as a way to interpret these results.

1. If there are even a few strongly negative indicators, regardless of the number or strength of any positive indicators, the candidate task is a poor choice. Reject the task *as stated*.
2. When the scoring indicates mostly neutral indicators, with at most a few strongly positive indicators, the candidate task is probably not a good opportunity. Neutral

indicators identify areas where adequate information is lacking, indicating a high degree of uncertainty (that is, risk) regarding important criteria. Reject the task as stated, and work to better define the task.

3. Many positive indicators, together with a few to several mildly negative indicators, signals a relatively good opportunity.

One obvious strength of the Laufmann method is that it not only indicates when an application is not well suited to a KBS approach, but it also identifies where the description of that application fails. Unlike the use of lists, here there is no categorical elimination, rejection is based on careful scrutiny. In light of the possible interpretations stated above, it is conceivable that a rejected application could be salvaged based upon better information or by developing a revised plan.

4.3.5 Requirements Analysis

Given the results of the above analysis, it now can be possible to identify system requirements. Requirements should describe what the final system will be expected to do from the perspectives of all participants. Requirements serve as guidelines for selecting best approaches and techniques during actual system construction. Development efforts will become more efficient given a "best" set of requirements (Walters and Nielsen 1988). Best, in the sense that they closely approximate all expectations of the operational system.

Each participant group will generally have their own specific requirements. Some requirements may be left implicitly unstated and others may be demanded. Under certain circumstances, some requirements of a participant group may be known a priori. For example, management concerns usually focus on policy issues, like guaranteeing that some responsible individual makes the final decision and not a system, or forcing a system to conform to the current organizational information system structure, or providing an acceptable benefit-cost ratio. Conflicting requirements between groups are not unusual. While management may desire that an individual make all final decisions, user groups are often attracted to the KBS idea because they can let a computer make decisions for them—they are then able to get their job done much more quickly. Priority values may need to be assigned to each requirement; later, this will aid system developers in resolving conflicts as they arise. Weighting each group's requirements could be based upon the perceived influence each group will have on the final product and how substantially each group might be affected.

Requirements will be particular to each groups perceived needs. They will, however, fit into various categories that may be used by system developers to plan construction activities. Some of the categories that could be addressed are (Walker and Nielsen 1988):

1. *development resources* – time, money, equipment, and supplies;

2. *expertise availability* – frequency and duration of experts' time;
3. *functional capabilities* – things that the system will do for the user and what it will not do for the user;
4. *operational environment* – where and when it will be used, decision time frame, number of users, and integration with current computing resources;
5. *user interface* – what the system looks like to the user;
6. *information and knowledge sources* – the sources and reliability of the data and knowledge;
7. *system outputs* – decisions created, equipment controlled, and changes made to information sinks; and
8. *maintenance specifics* – *who* will perform *what* maintenance or modifications, and *how often*.

As noted above, there will be conflicts in this list—not everyone can get everything they want. In addition to noting a priority for each requirement in each category, any that are critical to the project should be identified. This may help to further resolve conflicts, e.g., where two non-critical requirements must be compared against a single critical one.

The development team recognized that the first versions of the RP-FMAS should be regarded as primarily research tools—research in designing and implementing a comprehensive theory of forest management using AI methods. After several cycles through the design-implement-evaluate loop, RP-FMAS can now be described as an application system for forest managers. Development resources available were essentially 4 person-years of part-time effort for 2 researchers, the knowledge engineer and the domain expert. Salaries were covered by research funds but little money was available for hardware or software purchases to support this project. Availability of the researcher's time was variable and largely up to the researchers themselves. RP-FMAS was designed to (1) create a comprehensive and testable hypothesis (the knowledge model) of the forest management process; and (2) to provide treatment advice to forest managers. It was designed to be used on a micro-computer by non-expert individuals for analyzing problems which would otherwise require the time of a forest management expert. Maintenance and improvement of the system will be performed cooperatively by the Lake States Forest Management Cooperative, an association of interested researchers and managers.

As with RP-FMAS, the primary thrust of PREDICT was to investigate research questions. It was believed, however, that the resulting software product could also be regarded as a useful decision aide to foresters. This pest management project did not benefit from a large base of supplies, money, or equipment. Shortcomings in those areas were compensated by greater expenditures of time by the knowledge engineer. There were two principal subject matter experts available and another four that were periodically consulted during the project.

As expected, the time constraints on the experts were many and it was very difficult to schedule quality time slices. PREDICT was designed to provide pest diagnosis; a hazard assessment component was added later. Additionally, it was hoped that a treatment recommendation subsystem could be added eventually. The ubiquitous personal computer was selected as the operating environment because the development software operated in this environment and PC's were readily available to end users. The user interface was limited to the facilities provided by the development software. There was great concern about the reliability of user input, consequently, care was exercised to make PREDICT tolerant to incomplete and uncertain data. Subject matter experts used for system construction were considered to be the best available and, hence, their knowledge was considered to be very reliable. Adjunct experts were often consulted to corroborate the knowledge of the principal experts. The developer, in discussions with subject matter experts, felt that system output should provide for the user a list a possible damaging agents along with some estimate for the likelihood of each. Maintenance issue were not directly addressed, with the exception that it was clearly understood that a completed system would be made available to users along with some documentation.

The preceding two paragraphs, and similar paragraphs throughout this chapter, present the way that our thinking proceeded on some of these planning questions within the development of a real-world application. At the time that RP-FMAS and PREDICT were developed, however, our problem definition efforts were not this structured. Rather, the highly organized layout of planning issues presented here is a result of hindsight thinking on our part and of additional work on building other KBSs. Nevertheless, we feel that the detail given in the examples is a minimum effort that should be expended by anyone going through this process.

4.4 SUMMARY

This chapter covers quite a few topics related to planning an application and to understanding the system development cycle. An attempt is made to emphasize the parallel nature of the entire development process. Our presentation of the material in this and subsequent chapters is necessarily serial, but in reality many issues must often be confronted coincidentally by the development team. A number of development approaches are described and can be seen to lie at different points along a continuum, from exploratory programming to software engineering. A three-stage approach is proposed as an alternative to somewhat traditional knowledge engineering methods.

To completely specify what a final system is expected to do, the development team should produce a problem definition. This defines characteristics of the subject matter area, specifies who are the players involved in system development, produces a world view of the application, details an operational structure, and identifies potential solutions and necessary knowledge and data. A careful analysis of these issues can illuminate whether a KBS approach is a good way to address a particular application problem.

Many of the specifications outlined in a requirements analysis can be made more concrete by the creation of a proof of concept system. Such a concept system can be coded fairly quickly using one of two Prolog programming approaches presented in Chapters 8–9.

CHAPTER 5

Knowledge Acquisition

We can scarcely imagine a problem absolutely new, unlike and unrelated to any formerly solved problem; but if such a problem could exist, it would be insoluble. In fact, when solving a problem, we should always profit from previously solved problems, using their result or their method, or the experience acquired in solving them.

Ploya

In the preceding chapter we examined the steps that might be used to survey a potential knowledge system application. Among other things, this included developing a problem description, conceptualizing a world view, and producing an application evaluation. From this point on we shall assume that sufficient support exists for continuation of the KBS development process.

Once a world view has been defined and the knowledge sources identified, it is time to acquire the necessary knowledge about the subject area in order to eventually build a **knowledge model**. A knowledge model includes the knowledge and rationale that must be applied to the problem in the domain. It also contains subject-area specific methods for arriving at solutions to specific problems. Knowledge models are presented in the following chapter. First, we will describe some of the ways that useful problem-solving knowledge can be collected and organized. A knowledge model follows from this **knowledge acquisition** phase and the model is then implemented using the PROLOG techniques presented in later chapters (Figure 5–1).

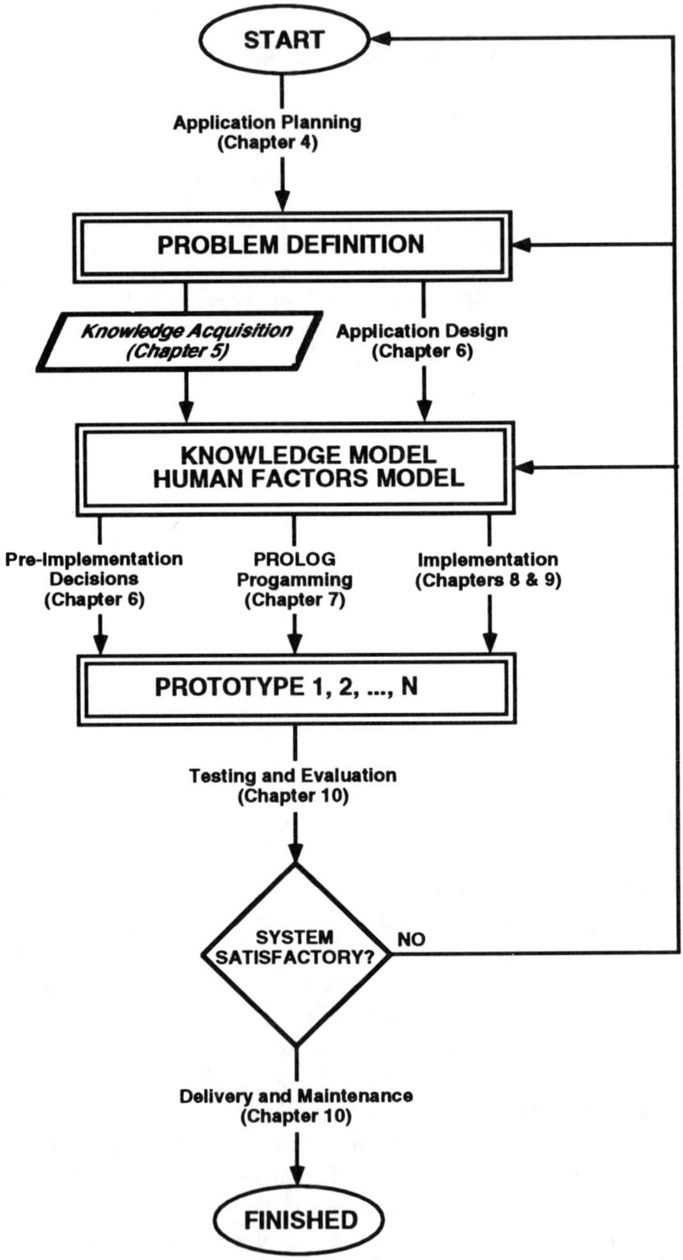

Figure 5–1. Knowledge acquisition (presented in this chapter) and application design (presented in the next) allow one to develop a comprehensive understanding of the application. This understanding is expressed via a knowledge model and human factors model.

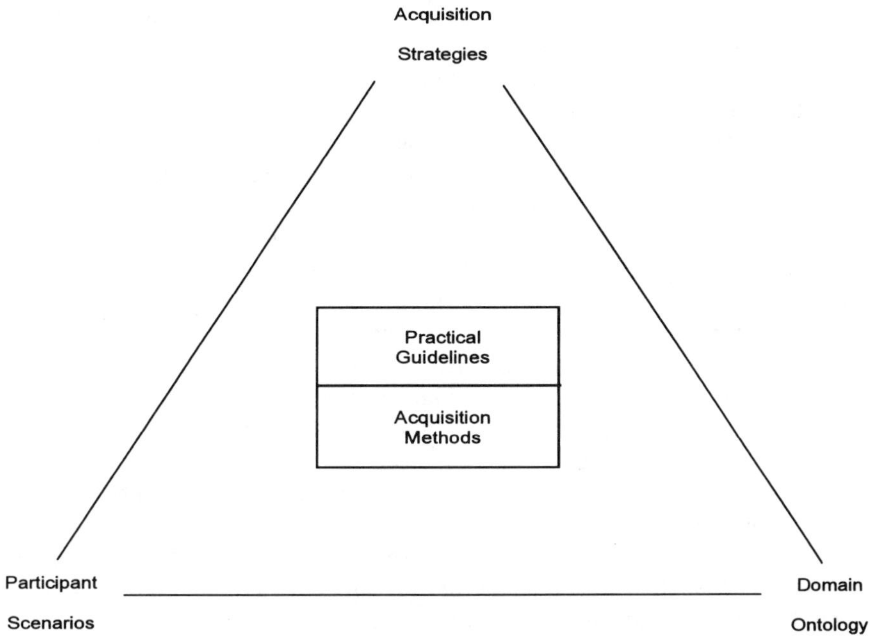

Figure 5–2. Knowledge acquisition can be viewed from several different and related perspectives. Each of those views, depicted as the corners of a triangle, relies on implementation techniques consisting of practical guidelines and acquisition methods.

A major obstacle to the development of knowledge systems is the expensive work of acquiring human expertise and encoding it into a computer representation of that knowledge. Ideally, of course, computers should learn from experience just like people do. That is, a human specialist teaches a computer what is correct and the computer adjusts its understanding and behavior in order to be smarter (Michie and Johnston 1985). We want to be able to show a computer examples of things, be they statements of fact, pictures, sample actions, etc. and have it, by a process of discovery, formulate ideas that connect those examples, finding patterns amid apparent chaos. Only in this way can the bottleneck of encoding knowledge into knowledge representation structures be overcome (Michie and Johnston 1985). Unfortunately, it will likely be some time before our "intelligent" creations are capable of learning in that manner. For the present we must be content to work closely, either directly or indirectly, with those individuals whose expertise we desire to capture.

Our coverage of knowledge acquisition in the following pages includes the different ideas presented in Figure 5–2. The corners of the triangle represent three different aspects of the knowledge acquisition process. Regardless of which aspect we are concerned with at any particular time, acquisition guidelines and methods (at the center of the triangle) can be applied. We preface the topics of Figure 5–2 by a brief overview of knowledge acquisition and by a short section on surface and deep knowledge.

Following this introductory material, we discuss several different knowledge acquisition *strategies*. These strategies constitute general overall procedures for conducting a knowledge acquisition exercise. Second, we describe how to create a **domain ontology**, i.e. a theory of reality for the KBS application domain. An ontology describes the "things' in the domain that we need to reason about, how they relate to one another, and how they can be organized for problem solving. Creating an ontology involves the analysis of three types of knowledge: factor, structure, and strategy knowledge. Third, brief introductions to a number of particular knowledge acquisition *methods* are presented. Fourth, we highlight three different *scenarios*, consisting of knowledge engineer(s) and expert(s), in which these methods might be applied. And finally, we enumerate some general *guidelines* and tips that knowledge engineers have found useful in the knowledge acquisition task. The knowledge acquisition guidelines presented are generally applicable regardless of the strategy, ontology, or participant scenario being considered. However, the reader should keep in mind that often the knowledge acquisition methods that are actually used are determined to some extent by strategy, domain ontology, and participant scenarios.

5.1 KNOWLEDGE ACQUISITION IN OVERVIEW

There are four, primary sources of knowledge: literature, human specialists, existing models (e.g. mathematical models, existing problem-solving procedures), and examples (Sell 1985). **Knowledge acquisition** refers to the process of collecting, organizing, and formalizing the information, concepts, and strategies that pertain to some subject area of interest. Figure 5–3 illustrates how the knowledge acquisition process fits into knowledge system development. The goal of the knowledge engineer is to combine the knowledge, that is scattered throughout these four sources, to produce a useful knowledge system. To do this, he or she must complete a knowledge acquisition task to create a knowledge model and must also perform a knowledge translation task (implementation) to encode that model into an executable computer program.

Methods of collecting, organizing, and formalizing knowledge vary substantially depending on the knowledge source. For example, knowledge in literature has already been collected and organized, but one must still formalize it so that it can be implemented in a KBS. Human expert knowledge, on the other hand, is woven from human experience and training into internal mental models of how the world works and therefore requires all three tasks of collection, organization, and formalization. Knowledge that exists in expert models is

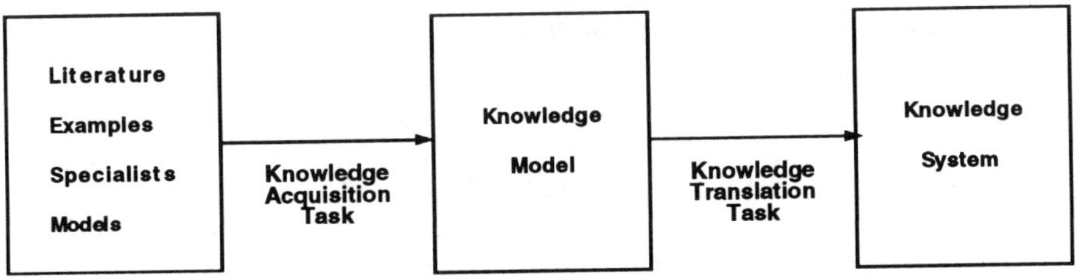

Figure 5–3. Knowledge system development contains two subtasks. The first creates a knowledge model through a knowledge acquisition process, and the second produces a KBS through a knowledge translation/implementation process.

already collected, organized, and formalized, but it may be necessary to re-formalize it or to integrate it with other knowledge. By carefully developing a domain ontology it becomes possible to combine knowledge from several different sources.

When knowledge must be extracted from human experts, the acquisition process is often referred to as **knowledge elicitation**. The job of knowledge collection from human sources can be very difficult due to the inexplicit nature of an individual's knowledge. That is, most people find it very difficult to explain how they know what they know, and how they use what they know. Consequently, knowledge collection, or elicitation, from experts has received a great deal of emphasis historically and is the primary focus of the following sections on knowledge acquisition methods.

Acquiring knowledge in a scientific domain is easier than most other domains because laws, principles, and experiences have been rigorously developed and reported in the scientific literature. In many scientific domains, algorithms have been developed and can be used in conjunction with the knowledge gleaned from books and scientific experts. On the other hand, in scientifically weaker domains, the sources of knowledge may be limited to human specialists, who must be interrogated about their knowledge, because little has been reported in writing (Sell 1985). Because knowledge-based system techniques allow us to encode human experience and judgment in the computer, many systems are based solely on human sources of problem-solving knowledge.

The development of knowledge within an individual follows a fairly typical pattern in the western world. Formal university education stresses abstractions and general theories and tries to build a conceptual, domain-independent framework for knowledge in the minds of students. This background teaches students how to properly "chunk" information thus organizing it and creating knowledge, i.e. organized information. Once this basic foundation has been laid, the student emerges and begins to accumulate domain-dependent information with which to fill his or her mental structures. This domain dependent information comes from experience and from mentors (Harmon and King 1985). After approximately 10 to 15 years of work in a specialty area, human experts of superior problem-solving abilities begin to emerge.

5.2 SURFACE VERSUS DEEP KNOWLEDGE

Experts perform well because they have a large amount of compiled, domain-specific knowledge stored in long-term memory (Harmon and King 1985). Heuristics and domain-specific theories upon which experts rely for problem solving are called **surface knowledge**. On the other hand, the first principles and general theories that an expert uses when his or her heuristic knowledge fails to adequately address a problem, is called **deep knowledge**. For example, when an entomologist enters a stand of trees, that suffers from some insect injury, the entomologist immediately looks for certain insect symptoms and signs that are characteristic of specific insects. The location of the stand, the type of site on which it is growing, the size and species of trees, etc., all tell the entomologist to ignore particular insects and focus on others. This sort of diagnostic capability has evolved as a shortcut substitute for a textbook examination that would be exhaustive and tedious. These "shortcuts" are surface knowledge that is readily available and works well for most cases encountered. On the other hand, for particularly difficult cases, the entomologist may need to resort to a more fundamental understanding of insect/tree/environment interactions, i.e. deep knowledge.

KBSs have traditionally included only surface knowledge (Cullen and Bryman 1988), although authors, such as Torasso and Console (1989), have developed and promoted systems that are able to operate at both the surface and deep levels. These have been referred to as **second-generation knowledge systems.**

A major difficulty with extracting knowledge from human experts becomes readily apparent whenever one asks advise from another individual. Often, the queried individual recalls a past experience that closely relates to the question posed to them; this is, in essence, a pattern matching process; it's also referred to as case-based reasoning. That is, their response to a question involves anecdotal recall. This seems quite natural because their expertise has been acquired over time through a series of anecdotal experience. This process of accumulating personal experiences may or may not have allowed them to distill the important knowledge structures of their specialty area. They may only have a collection of individual experiences that are loosely tied together conceptually. Consequently, a knowledge engineer can become quite befuddled by the rationale that a specialist's experience-based, problem-solving shortcuts often lack. As an expert is queried further about the application of past experiences to the current problem, he or she may experience great difficulty in producing the pertinent principles and ideas that substantiate such a shortcut. Although when pressed, an expert can often manufacture a more principled answer. It can seem like a tooth-pulling exercise, however, because much of an expert's knowledge is not acquired or stored as deep knowledge.

5.3 KNOWLEDGE ACQUISITION STRATEGIES

Prior to describing the creation of a domain ontology and presenting some specific knowledge acquisition methods, we would like to introduce several general strategies that might be used to govern an entire acquisition process. To clarify what we mean by this, acquisition methods are particular techniques, for example a particular type of interview, that can be used to extract knowledge. Strategies, on the other hand, describe the manner in which those methods are applied to arrive at a domain ontology, or a knowledge model. By their nature certain acquisition methods may only be applicable to certain strategies. For example, various machine learning methods would not be used in a manual acquisition strategy. The categories in Table 5–1 represent an extension to Cullen and Bryman's (1988) organization.

The most general classification places knowledge acquisition methods into one of two groups, either manual or automated approaches. In manual methods, an individual, e.g. a knowledge engineer, "interviews" an expert to capture his or her knowledge. Automated techniques, on the other hand, use computer programs either to find patterns in examples (induction by example) covering the range of problems to be solved, or to interview human experts by means of elicitation techniques similar to what a knowledge engineer might employ manually (automated interviews). The reader should not construe that necessarily only one or the other approach is used exclusively throughout the acquisition process. For example, it might be useful to begin with an automated method and then refine the knowledge elicited in that manner with additional manual techniques.

We can further subdivide manual methods into fast prototyping, evolutionary, ad hoc, and expert driven. *Fast prototyping* attempts to build a preliminary

Table 5–1. The general approach taken by a development team during knowledge acquisition can often be categorized into one of these strategies.

Manual Methods	Automated Methods
Fast Prototyping	Machine Learning
Evolutionary	Automated Interviews
Ad Hoc	
Expert Driven	

prototype system as quickly as possible and then uses the prototype as the primary focus for further knowledge acquisition. It has the advantage of producing a working system quickly and maintaining the system (the final product) in center stage throughout. *Evolutionary* acquisition employs a variety of methods and extends the acquisition process throughout the entire system development cycle. This is consistent with the strategy we have suggested previously, but certainly does not preclude using a prototype for acquisition, also. *Ad hoc* refers more to the lack of any formal or systematic knowledge acquisition methods than any particular strategy per se. An *expert driven* strategy is essentially one of the participant scenarios we describe later in which the developer and domain expert are the same individual. Because dialogue is entirely internalized in the developer-expert there exists no objective counterpart to help avoid biases and omissions. Which type of strategy or strategies a knowledge engineer employs will probably depend upon human and financial resources, time availability, project complexity, and familiarity with, and confidence in, available acquisition methods.

5.4 CREATING A DOMAIN ONTOLOGY

Before proceeding to specific knowledge acquisition methods, it is important to understand what types of knowledge we would like to capture. We have briefly alluded to several types of domain knowledge in the sections above. The three general types of knowledge are factor knowledge, structure knowledge, and strategy knowledge. Any knowledge acquisition effort should include an analysis of each of these types of knowledge for the particular subject area.

Identifying, labeling, and describing conceptual and physical objects in a domain represents the first step. These objects are decision *factors*, i.e. conditions, variables, and domain entities that an expert uses to describe problems in the subject area. *Factor knowledge analysis* creates an important vernacular from which it becomes possible to subsequently discuss knowledge structure and problem-solving strategies. *Structure knowledge analysis* identifies, labels, and describes the relationships between the factors identified in factor analysis. If we view knowledge acquisition and analysis as producing a language of decisionmaking within a particular domain, then factor analysis produces the lexical objects and structure analysis produces the syntax (how those factors are can be organized together). *Strategic knowledge analysis* focuses on specific combinations of factors and structures to indicate plausible avenues of search toward solutions. Strategy analysis adds meaning to our lexicons and syntax to produce a knowledge semantics, telling us when and how various concepts and structures should be used.

Let's examine more closely the *what* of knowledge analysis and identify what kinds of knowledge needs to be acquired. Because system development is an iterative process of building, revising, and rebuilding, knowledge analyses will be modified continually as new prototype versions are constructed.

5.4.1 Factor Knowledge Analysis

In Chapter 4 some general guidelines were proposed for characterizing data and identifying their sources. Now, as we move into the KBS application design stage the details of the problem-solving process must be analyzed and described. So, one of the key tasks to help understand a problem-solving process is to analyze the factors that are important for decision making. That is, what pieces of information are used to solve a problem? We are assuming here that what constitutes a valid solution has been stipulated previously.

First of all, any objects in the domain that will be part of the decision-making process must have a name. This may not be as trivial as it sounds. Some abstract concepts are difficult to enclose within a meaningful label (Benfur and Furbee 1989). Most experts encounter substantial difficulty with the task of creating lists of decision factors off the top of their heads. When presented with a partial list a priori, however, it becomes much easier for them to focus on important factors, refining those factors already present on the list and adding others to it. Hence, problem examples (actual or hypothetical), books, manuals, etc. are often the most helpful sources for enumerating decision factors and finding unique names for them. Schmoldt (1987) used this approach to enumerate symptoms, signs, and predisposing conditions of various red pine insects and diseases. At least initially, objects on a decision-factors list should be reduced to the most general terms possible. An attempt to introduce more detail during this step forces one to begin considering how the factors are used in decision making and how they will be represented as part of implementation. Some implementation issue may need to be addressed at this stage, but others that surface cannot be resolved yet given the current stage of model development.

Once a list has been created it is likely that it will contain some redundancies. As each object is labeled, it should also be given a definition to distinguish it from all other terms. While this is being done, it should be possible to identify those terms which are duplicates of others.

When a list of reasonably unique terms has been created, each term should be assigned a set of possible values that it may assume. For terms with numerical values, there may be some range of values that is reasonable; for more vague terms there may be discrete values such as high-medium-low or red-yellow-brown. In situations where the entities are more complex than single terms, i.e., each entity may possess several attributes, then each of these attributes may have values rather than the entity itself. As an example, we could describe facts about trees within the forest in a number of ways. Each different attribute of a tree may have a separate label, definition, and value; or alternatively, if there are many characteristics of a typical tree that the system should be able to reason about, it would be more elegant and efficient to regard each tree as distinct, but each with a standard set of descriptors or attributes. This begins to sound very much like a frame-based representational structure, hence, rather than just describing entities, we've begun to consider how they will be used. Consequently, issues of knowledge representation begin to surface. Just because the different topics of system development are presented sequentially does not imply that there are serial design decisions to be made. Often different aspects of design are intricately interwoven and ideally must be considered in parallel. Again, as we noted previously, it is difficult to do this well, hence the reliance on an iterative evolution of prototypes.

One final aspect of the knowledge-term vocabulary should be clarified during this listing process. Some terms constitute the basic data that describe a problem, for example, "tree diameters" or "forage species present". These objects are often referred to as *atomic* factors because they are elemental and observable. In general, all decision factors can be either observed directly (atomic) or derived from atomic factors. Some atomic factors can be observed or, if the necessary observation cannot be made, can also be derived from other atomic factors. For example, a tree can be observed as dead, or alternatively its death can be inferred from the observations of dead shoots, buds, and needles. When a factor is derived from other atomic factors, it is often known with much less certain than if it had been directly observed.

Factors, that represent collections of atomic facts in a conceptual manner, are often called *hypotheses*. Hypotheses represent higher level concepts that must be inferred and cannot be observed directly, e.g., "moisture stress", "stand stocking level", or "germination condition". Hypotheses and their formation are often very important in the problem-solving process. Hypotheses—sometimes also called *intermediate factors*—often serve to organize and consolidate information present in atomic factors and other lower-level hypotheses into a solution-building and reasoning process.

At this stage of factor analysis it is oftentimes difficult to define a possible range of values for some of the intermediate hypotheses. Possible values for intermediate hypotheses are often dependent on how particular factors combine to infer these hypotheses. Consequently, some factor knowledge analysis may need to be completed later after some knowledge structure has been imposed on these factors and hypotheses.

5.4.2 Structure Knowledge Analysis

A list of terms without any structure conveys little knowledge about a domain. Just as words that don't adhere to any grammatical rules result in poorly formed sentences, so too, knowledge terms without any structure make it difficult to represent complex ideas. The second important aspect of knowledge analysis is the understanding of relationships among decision factors. Structural knowledge expands the number and types of concepts that can be represented, because it is possible to combine separate factors in different ways. Then, knowledge begins to take on an aspect of depth, rather than the one-dimensional nature of knowledge terms.

Structure specification requires two properties: (1) a description of the relationship, i.e. type, and (2) the strength of the relationship, i.e. proximity between two terms. Relationship type is most important; strength becomes important when it is necessary to make finer judgments based upon a relationship. For example, two different insects that damage tree roots may both girdle large roots, which is very important information relating particular insects and tree symptoms. One of these insects, however, may girdle large roots more frequently or during certain times of the year. This additional information further describes the insect/symptom relationship by qualifying the association. The distinction between relationship type and relationship strength implies, that for KBS development, type structure should be defined first and, if necessary, discriminate terms and their usage in decisionmaking on the basis of strength of association.

Because relationships between several terms can sometimes indicate a new, additional concept, structural analysis may fill in any omissions remaining from the factor analysis. All knowledge pieces should then be present for an analysis of strategy knowledge, the next step.

5.4.3 Strategy Knowledge Analysis

The third, and probably most difficult, aspect of knowledge analysis is the elicitation of problem-solving (strategic) knowledge. Eliciting and applying strategies for the knowledge acquired in factor and structural analyses is at least as important as that knowledge itself. So, in some sense, strategic knowledge is knowledge about knowledge, or meta-knowledge.

Strategy analysis seems difficult because an expert often has a better understanding of what knowledge is used for problem solving rather than how it is used. It is often quite readily apparent what specific pieces of knowledge are applied to a problem. To understand how that knowledge is actually used requires that an expert introspect about internal thought processes or retrospect about previously solve problems—activities with which most people have great difficulty. Whereas a large portion of factor and structural knowledge resides in the public arena, most strategic knowledge is private to each individual specialist. Consequently, most acquisition methods for strategic knowledge employ indirect approaches to infer underlying strategies. These strategies surface from observed or proposed behavior possibly supplemented with explanatory discourse.

5.5 KNOWLEDGE ACQUISITION METHODS

Most of the knowledge acquisition techniques mentioned in the following pages aim to systematize an expert's knowledge into a form that is universal and, hence, less anecdotal. Anecdotal knowledge is often limited in application to situations that are very similar to the prior experiences that germinated that knowledge. On the other hand, universal knowledge is more general and can be applied to a broader range of problems, many of them might only be vaguely similar to an expert's original experiences. Because our purpose here is to impart a sense of the methods that are available and where they may be most appropriate, we urge interested readers to seek out the respective authorities listed in the bibliography regarding additional details about particular methods. Figure 5–4 depicts a taxonomy of the various methods presented in the following sections.

5.5.1 Unstructured Interviews

Unstructured interviews are characterized by a lack of organization. A knowledge engineer (interviewer) and expert sit down and the expert responds to questions posed by the interviewer. Questions follow no preset or designed format and topics are pursued in whatever order or at whatever length seems best to the knowledge engineer. Successive interviews tend to expand on topics covered in prior sessions. Unstructured interviews often become the primary method of elicitation used in an ad hoc strategy. They are also employed quite often to develop a quick prototype for a feasibility study.

These types of interviews may be useful during initial stages of knowledge acquisition. Their lack of structure permits a sort of free association dialogue that may illuminate many of the major issues that are important, but they also have some drawbacks. They are by nature unstructured and, hence, may be very inefficient at collecting knowledge because of redundancies and omissions. To work effectively, the interviewer must be a very skilled communicator and have an ability to identify key issues through incisive questioning. The latter talent implies some unconscious systematic methodology by the interviewer, which means that he or she may in essence be conducting a structured interview. Except as an ice-breaker technique, unstructured interviews cannot be strongly recommended as a valuable or effective elicitation method.

5.5.2 Structured Interviews

Within the general realm of structured interviews there are a large number of methods which may be utilized (Gordon 1989). A fundamental intent of these methods is to provide the expert with guidance for his or her responses,

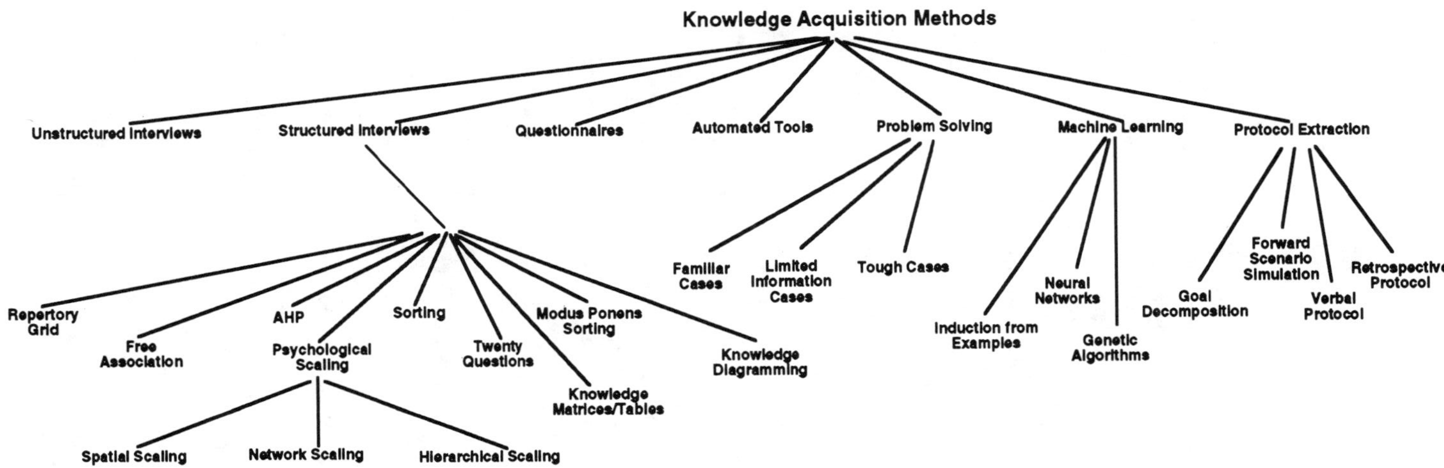

Figure 5–4. A taxonomy of knowledge acquisition methods identifies the major acquisition categories and also some particular techniques within each category.

thereby increasing interview efficiency. Explicit guidance allows an expert to focus on the subject matter rather than on how responses should be formatted. The following paragraphs briefly describe a number of these techniques.

- Anderson (1983) proposed a model of human memory called *spreading activation theory*. Basically this theory proposes that whenever one thinks of a particular concept, all concepts that are closely associated with that initial concept can very easily be recalled (activated). On the basis of this spreading activation idea, Mitchell (1987) suggests a *free association* knowledge acquisition technique. Here, various key concepts in the subject area are posed to an expert one at a time and he or she is asked to respond with other concepts that are in any way related to each of them. It then becomes possible for the knowledge engineer to construct a matrix or graphical road map of terms/concepts in the subject area. An example of such a terminology graph appears in Figure 5–5. Here, three defoliators of red pine are linked with terms that are related to these insects. The knowledge engineer might ask an expert to respond with all pest diagnosis terms that are related to these three insects. This method not only elicits those concepts that are part of the general theory of a subject area, but also those things that are idiosyncratic and, hence, private to an expert's knowledge (Mitchell 1987).

- Results from the use of free association could be subsequently enhanced by one of several *psychological scaling* techniques. Cooke and McDonald (1987) discuss a number of scaling methods. Proponents of scaling methods claim that in addition to associations between concepts in memory there also exist strengths of association. These strengths have some relation to cognitive proximity. So, the stronger an association is, the closer, or more related, two concepts are. Whereas free association can provide a graph or matrix of all concepts and whether they are related, scaling methods go one step further and require that the relationships be assigned some value from a numerical scale to indicate relatedness. This association matrix (or distance matrix) can then be transformed into a spatial representation (using multidimensional scaling, Shepard 1962), a hierarchical representation (using cluster analysis, Johnson 1967), or a network representation (Cooke and McDonald 1987) of an expert's knowledge about a subject. The graph in Figure 5–5 can be converted into a matrix with numerical values that represent closeness (Figure 5–6). This representational transformation accentuates any factor analysis information and elucidates knowledge structure, also (see 5.4.2 Structure Knowledge Analysis).

- In addition to psychology research, a number of useful elicitation techniques have also been developed in anthropology (Benfur and Furbee 1989). To understand the language and thought patterns of vastly dissimilar cultures anthropologists have been forced to develop numerous methods to describe the acquisition of personal/cultural knowledge—knowledge that is quite difficult for indigenous peoples to impart to others from outside their particular culture. Such cultural roadblocks are analogous to the situation of a domain expert and his or her subject matter specialty, which can also be viewed as a culture of sorts. Many of these methods are detailed in a book by Brokenshaw et al. (1980).

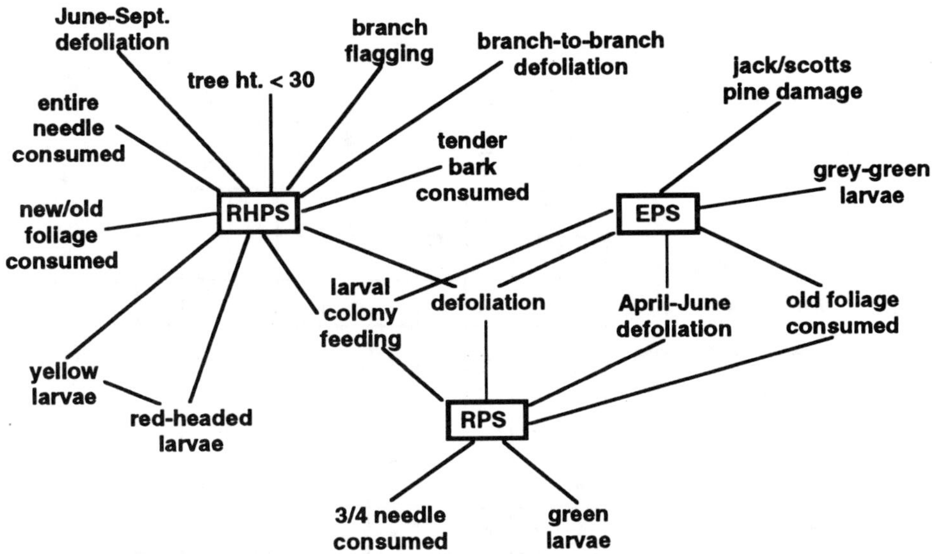

Figure 5–5. Free association by a subject matter expert can identify many of the important concepts within his or her knowledge base. Associations between these concepts can then be represented graphically or in a tabular manner using a matrix. RPS, EPS, and RHPS refer to red pine sawfly, European pine sawfly, and red-headed pine sawfly, respectively.

One technique, referred to as *sorting*, uses a stack of cards each with a term/concept from the domain written on it. An expert is instructed to sort the cards into different piles using whatever criteria seem appropriate. Following a sort, the expert is asked to provide a verbal description of the sort criteria used. When the cards contain different possible solutions only and no intermediate or atomic factors, this method works very similarly to the repertory grid technique (see below). By organizing concepts into "similar" stacks using different sorting criteria on each pass, the expert is essentially performing a psychological scaling exercise identical to that described above (except perhaps more interactive in this case). With each pass the expert is scaling the concepts along a different dimension.

• Schweickert et al. (1987) compared sorting with two other elicitation methods, *twenty questions* and *modus ponens sorting*. Twenty questions is essentially the game familiar to most readers. The knowledge engineer collects several task scenarios prior to the interview. Then, during the interview, an expert is instructed to ask questions about a particular task much like would transpire in an actual problem-solving situation. The expert also supplies rationale for asking each question. Our defoliator example from above is used to illustrate the twenty questions technique in Figure 5–7.

In modus ponens sorting (named as such by the authors), referred to as "structured interview" by Schweickert et al. (1987) an expert is asked to list variables/factors

	green larvae	yellow larvae	red-headed	new/old foliage	old foliage	entire needle	branch flagging	grey-green larvae	bark consumed	3/4 needle consumed
green larvae	0									
yellow larvae		0								
red-headed			0							
new/old foliage				0						
old foliage					0					
entire needle						0				
branch flagging	8	3	3	2	8	2	0			
grey-green larvae	2	6	8	9	4	3	8	0		
bark consumed							1	9	0	
3/4 needle consumed							9	7		0

Figure 5–6. By creating a matrix of pairwise relatedness values, one can extend the graphical representation in Figure 5–5 produced by free association. Ten concepts were selected from the free-association graph and values (0-9) have been entered for two rows of the matrix. Because the matrix is symmetrical, it can be displayed as upper or lower triangular. Distances between concepts can be used to calculate clusters of similar concepts.

important to making a decision in the subject area and also to list all possible outcomes (final decisions). Then, an expert must connect, in the form of if-then rules, variables to each other, outcomes to each other, and variables to outcomes. As you can see, this is a sorting method based on if-then relationships, hence our use of the term *modus ponens* sorting. If the number of variables pertinent to each solution is small, then it may be possible for the knowledge engineer to construct all "reasonable" combinations of variables for each solution and present these to an expert for critiquing; this method was used by Schmoldt (1987). When Schweickert et al. (1987) compared sorting, twenty questions, and modus ponens sorting for producing if-then rules, they found modus ponens sorting and twenty question superior to sorting. Their sorting method, however, was not formulated to explicitly create if-then rules as part of the interview process, hence, it exhibited a poor showing in their comparison.

• Graphing domain concepts and the relationships between those concepts, as revealed by an expert, can be a useful visual aid. This *knowledge diagram* creates a

Where is the stand located? [*Certain insects are limited to particular geographic areas.*]: **Portage county**

How large are the trees? [*Some insects only affect trees of a certain height.*]: **25 ft.**

What type of damage has occurred? [*Gross symptoms can indicate certain classes of pests.*]: **Defoliation**

Are new needles or old needles or both affected? [*Different defoliators consume either old needles or both new and old.*]: **New**

When did defoliation occur? [*The timing of defoliation can be very diagnostic.*]: **June**

Are larvae present, if so what color are they? [*A good larval description can identify a particular defoliator absolutely.*]: **No**

Figure 5–7. When example test cases are available for knowledge acquisition, a subject matter expert can be asked to solve particular cases by having the expert ask probing questions and provide rationale for each such question. This is often referred to as the twenty questions technique, similar to the game by the same name.

sort of road map of an individual's cognitive structures. Then in subsequent interviews, the interviewer and expert have a record of where they have been and what topics may need to be expanded further. Figure 5–5 can be expanded into a knowledge diagram by adding connecting phrases to the links and by expanding and grouping various concepts (Figure 5–8). For example, the relationship between "larval colonies" and "defoliation" could include the connector <cause>, as in "larval colonies" <cause> "defoliation." Also, the different types of needle consumption, e.g., "entire needle", "3/4ths of needle", "branch to branch", could be grouped under the concept, "needle consumption." Certain graphical scaling technique, mentioned above, create this sort of graphical representation.

The attentive reader will notice a strong association between the knowledge diagram and a semantic network knowledge representation. In fact, one important advantage of this acquisition method is that it places knowledge directly into the form of a knowledge representation method. Although, this representation may not turn out to be the one that is actually used for implementation later.

Knowledge diagramming has also been used with question probes by Gordon and Gill (1989) and Graesser and Clark (1985), and also by Schmoldt and Bradshaw (1988) as part of multi-expert elicitation. In the latter application, the authors used knowledge diagramming to acquire wildfire prevention knowledge for development

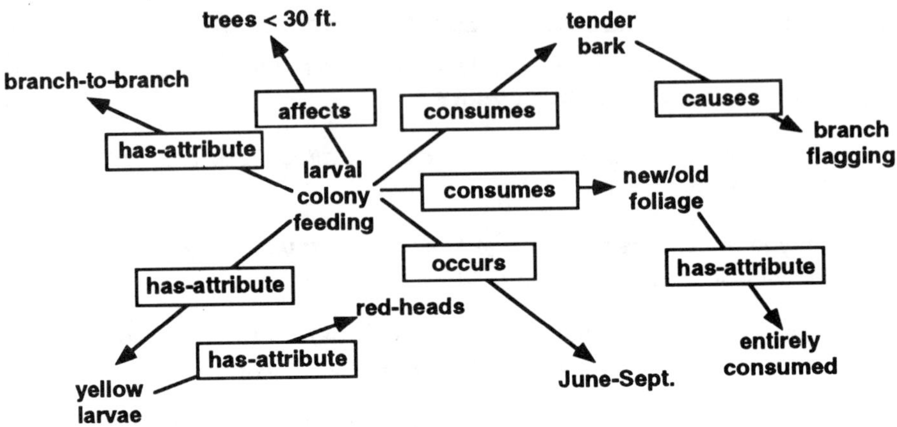

Figure 5–8. A portion of the free-association graph of Figure 5–5 appears here with different concepts connected by labeled links. This graph provides more structure and more detail than the free-association graph.

of an intelligent tutoring system. Actual diagram creation can be highly recommended even if the knowledge diagram is not utilized in successive interview sessions. Diagramming can also be used as part of system verification (see Figure 10-3); this becomes particularly valuable if diagramming is used throughout the development process, elicitation → model → verification (see Figure 5-3). The activity of creating a diagram helps the knowledge engineer visualize and better understand the domain's knowledge structure.

• The *repertory grid* technique originated with Kelly (1955) and his theory of personal constructs. This theory posits that each of us operates somewhat like a "personal scientist", i.e. we attempt to organize, predict, and control our own world by categorizing and classifying our experiences. This is similar to a scientist who develops and tests theories about the physical world. A repertory grid is a clinical technique that Kelly developed to identify and analyze personal constructs, i.e. mental models. Boose (1985, 1986) automated and applied this technique to the acquisition of expert knowledge, a very natural extension. Basically Boose's process attempts to fill in a matrix, where each column corresponds to an element that is to be discriminated (i.e. final solutions), and each row represents a personal construct (decision factor) that differs across several of the column elements. Each construct has a diametric description, e.g. short/tall or good/bad. Then each column element (solution) can be assigned a value for each construct (decision factor) that indicates the association between it and each diametric construct.

Figure 5–9 contains part of repertory grid for our defoliator example. Each element receives a score in the range 1–5 depending on how much of the construct it possesses. Many of the constructs for this example present in Figures 5–5 and 5–7 do not necessarily have a natural range of values, but rather have discrete values, e.g., the location of the stand or the description of the larvae. These types of

Elements			Constructs
RHPS	RPS	EPS	
2	5	4	new/old
2	4	4	spring/summer
5	4	2	% needle consumption
3	1	1	no flagging/flagging

Figure 5–9. The repertory grid technique can effectively discriminate between a set of elements (solutions) based on various bipolar constructs. All the solutions are scored for each construct on some arbitrary scale, e.g., 1–5.

constructs are not easily incorporated into the repertory grid and must be added to the knowledge representation later.

The repertory grid has the nice advantage of forcing an expert to create a tabular representation of his or her internal concepts about a subject area. This grid can then be used for structural analysis or actual rule construction (Boose 1986). But, Boose (1986) notes that it is limited in application to declarative types of knowledge; procedural, strategic, and causal knowledge are difficult to represent with this technique.

A technique that is very similar to the repertory grid is the use of *knowledge matrices* or *knowledge tables*. These are presented in some detail in chapter 10 with respect to knowledge system verification; readers are referred to that chapter for more specifics. Basically, knowledge tables allows the knowledge engineer to explicitly describe the associations between a particular hypothesis and all the factors upon which it depends. In the table, various combinations of values for the factors are associated with particular values of the hypothesis. These associations can then be translated directly into if-then rules, if desired.

• Another aide to the analysis of a decision process is the *analytic hierarchy process* (AHP) developed by Saaty (1980). It allows persons with decisionmaking expertise to structure a complex problem in the form of a hierarchy. The process requires an ability to enumerate all possible decisions, i.e. alternative solutions, a priori. Then, criteria are established that are used to evaluate those decisions. Likewise, there may also be criteria that can be used in evaluating each of the previous criteria. This leads to the formation of a hierarchy (Figure 5–10).

At each level, pair-wise comparisons are made regarding the relative likelihood, relative preference, or the relative importance of each criteria versus each of the other criteria at the same level. For our example, erosion hazard rating, establishment likelihood, political/social impacts, and downstream values would be compared in a pairwise manner. Similarly, runoff potential and soil cover would be compared, as would on-site seed viability and historic success, etc. In the AHP, these comparisons can then be converted into numbers which represent each criterion's contribution to the overall decision. At each of the leaf nodes, a score is made by an expert for each

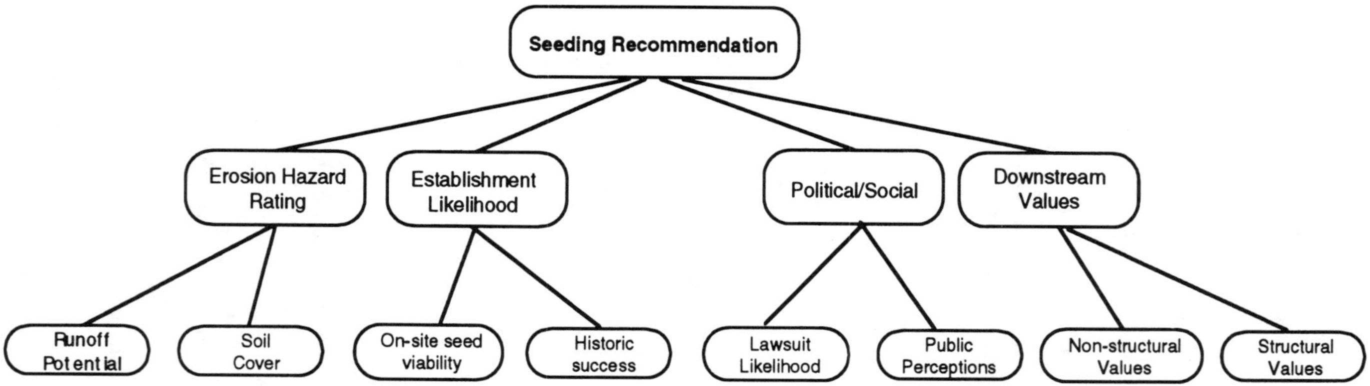

Figure 5–10. The analytic hierarchy process (AHP) dissects a decision process into a hierarchy of criteria, where each criterion may have subordinate criteria. At each level of the hierarchy pairwise comparisons are used to generate a priority value for each of the criteria at each level. A finite and known set of decisions can then be compared by scoring each leaf node decision with respect to preceding criteria and then multiplying those values by successive priority values between the leaf node and the root node.

possible decision with respect to the criteria immediately above that node. Suppose that, in our example, the two possible decisions are "seed" or "do not seed." Then, both of these decisions are scored according to their impact on runoff potential and soil cover (two lower level criteria). The score for each is then multiplied by the priority value for each of these criteria, runoff potential and soil cover. Each of these two scores is then multiplied by the priority value for the next higher level node, i.e. erosion hazard rating. A similar process proceeds from each set of leaf nodes up to the root node, where the scores for each possible decision are summed across the top-level criteria. The process of combining and propagating scores up through the hierarchy to the root node results in an overall likelihood, preference, or importance, score for both of our decisions, *seed* or *don't seed.*

For knowledge acquisition purposes, it would probably not be useful to duplicate exactly this behavioral model of decision making with a KBS. The AHP produces a decision structure that is valid *at one point in time* and *for a particular instance.* This general analytic process, however, does provide a great deal of knowledge about how an expert might structure and think about a class of problems. The primary advantage of this analytic method is that it provides a systematic and detailed description of an expert's decisionmaking criteria and each criterion's relative contribution in the decision process.

Several drawbacks to this approach are immediately apparent. First, not all decisionmaking situations can be structured in a hierarchical manner necessarily. Second, the question also remains of what to do when certain criteria are not known. Third, priority values in the AHP are static; however, in many real-world situations the relative contribution of different factors may vary over time or with different users and their specific problems.

5.5.3 Questionnaires

Occasionally, it may be difficult to physically meet with an expert or it may be necessary to extract very detailed and specific knowledge about some topic. *Questionnaires* can be quite useful in both circumstances (Olson and Rueter 1987, Schmoldt and Peterson 1991). Also, when several experts are to be interviewed, the time and effort of the knowledge engineer can be reduced if questionnaires can be used in place of face-to-face meetings. This situation was encountered with a KBS development effort for seeding recommendations (Figure 5–10). The content of questionnaires may be very specific and require only short answers or they may be more general and intended to elicit longer prose. Multiple choice or short answer questions can often help an expert focus on those specific details that a questionnaire is designed to address. It may also be desirable to allow an expert the flexibility to add comments for explanation or emphasis. Schmoldt (1987) successfully used short answer questionnaires to verify and augment lists of factors related to pest diagnosis; these factors were categorized as either predisposing conditions, symptoms, or signs. Using categories of lists helps an expert narrow his or her attention to a specific task with specific types of output. Schmoldt and Peterson (1991) used tables and short answer questionnaires to elicit knowledge about air pollution impacts in wilderness areas from a large group of scientists and land managers. These questionnaires were completed in a workshop setting where: (1) they provided a focus for discussion and (2) their presence ensured that participants were aware of tasks to be completed in the time allotted.

5.5.4 Protocol Extraction

Encouraging an expert to explicitly detail how either typical or specific problems are solved can elucidate many specifics of his or her decision-making process. Explication of an expert's problem-solving *protocol*, identifies factors that are important and also indicates strategies of how and when these factors are applied. A number of protocol extraction methods have been utilized.

- *Goal decomposition* represents one of the most basic methods of formalizing strategies. An expert is asked to enumerate the steps to be followed (subgoals) as a problem is solved. A particular problem scenario may be used as an example, or general types of problems may be discussed more in an abstract manner. These subgoals can then be used for subsequent goal decomposition until the subgoals mentioned are fairly simple and readily accomplished tasks. This approach may be used in combination with knowledge diagramming, or other methods presented above.
- Another method, *forward scenario simulation* (Grover 1983), is almost identical to goal decomposition but prompts an expert for additional information, such as decision factors and explanations, in addition to subgoals. It is referred to as a simulation because an expert does not actually solve a problem scenario, but only describes how it might be done (Gordon 1989). Forward scenario simulation represents a very general protocol method because it elicits both decision factors and problem-solving strategies.
- In *verbal protocol*, an expert is asked to solve a particular problem in the domain and to verbalize his or her rationale at the same time. Hoffman (1987) suggests the use of three different types of tasks for verbal protocol: (1) typical and familiar tasks, (2) limited information tasks, and (3) rare or tough cases. Gordon (1989) mentions a variant of typical verbal protocol, *retrospective protocol*, in which the actual problem solving and explanation portions of the protocol are performed separately. That is, an expert first solves a particular problem and then reflects on the methods and rationale used for problem solving. The technique in the following section also requires that an expert solve example cases, but without any accompanying explanatory rationale.

5.5.5 Observed Problem Solving

An expert may work differently when he or she is not required to justify problem-solving steps. To avoid overly self-conscious and, hence, unnatural decision-making situations, an expert may be asked to solve problems without providing explanations. *Observed problem solving* may occur either in its natural environment, i.e. on the job, or in an artificial situation. Different types of tasks, i.e., familiar cases, limited information cases, and tough cases, may also be used with this method. Because problem-solving steps are not made explicit by the expert when using this knowledge acquisition method, the knowledge engineer must infer implicit strategies that are employed to solve various types of problems (Gordon 1989).

5.5.6 Machine Learning

Most of the extraction methods described in the previous sections seek to collect knowledge from experts. But, as we noted at the beginning of this knowledge acquisition section, some knowledge also exists in examples. That is, there is much we can infer from prior situations and then apply to new or similar circumstances. *Machine induction* methods attempt to discern patterns present in collections of decision factors and corresponding solutions (e.g. Langley and Carbonell 1984, Michalski et al. 1986). Many of these programs attempt to induce rules from examples using an algorithm similar to the ID3 algorithm of Quinlan (1983b). This is a bottom-up approach that creates general and universal associations from specific problems and their attributes. The result is often a discriminating hierarchy similar to a decision tree. Like empirical relationships derived using traditional statistical approaches, however, decision rules resulting from machine induction are very sensitive to, and only applicable to, the range of problems from which they were developed. This implies that examples used for learning must be selected very carefully (Hart 1986).

Another approach to machine learning uses *neural network* structures to represent the association between decision factors and possible solutions. The first description of the operation of a neural network was probably given by Hebb (1949). They were studied extensively by Minsky and Papert (1969) and received a thorough treatment in Rumelhart and McClelland (1986) and McClelland and Rumelhart (1986). A neural network consists of one or more layers of simple processing units (Figure 5–11). Each processing unit is connected to one or more processing units or input variables or output solutions. The output of any processing unit is determined by the inputs of its connections and the activation function that is used to transfer its input values to an output. A network is then exposed to a variety of input and output combinations which constitute a training set. If network output does not agree with expected output from the training set pair, a learning algorithm modifies the strength values associated with processing unit connections. Different network architectures, different learning algorithms, and different activation functions cause networks to exhibit vastly different properties.

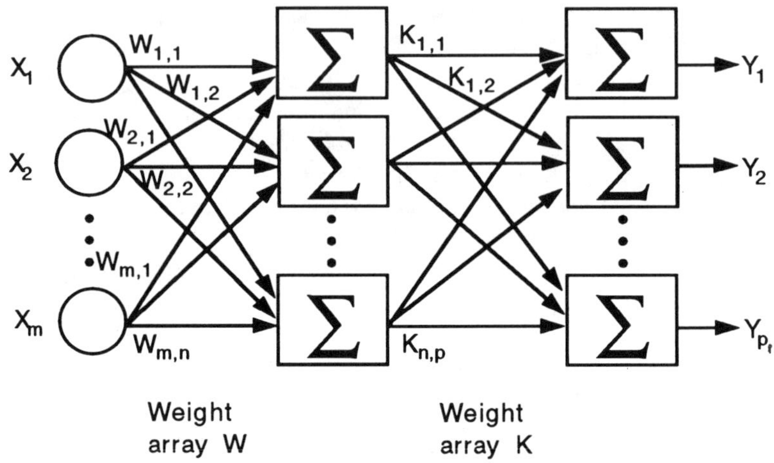

Figure 5–11. Neural networks can be useful for knowledge acquisition because they are able to "learn" patterns in examples and reproduce the performance and breadth of that training set when presented with new unfamiliar examples. Their "knowledge" of the training set is implicit in node connections and connection weights, there is no explicit representation of knowledge that can be examined.

These two machine learning approaches possess several similar characteristics. Both are dependent on the quality and quantity of examples used for system construction. Each requires that decision factors are selected a priori, this may necessitate the use of other knowledge acquisition methods initially. Certain decision factors may later be eliminated, however, as they are found to be redundant or uninformative to the correct operation of the system. Structural knowledge, represented as relationships between concepts in the domain, remains hidden in the discrimination tree or in processing unit connections. Strategy knowledge can possibly be inferred from a discrimination tree but has no intelligible representation in a neural network. Machine learning can often produce very applicative systems, but they are opaque with respect to providing explicit accounts of domain knowledge.

A third machine learning approach is *genetic algorithms* (Goldberg 1989). This technique borrows ideas from traditional biological evolution theory. In a population of individuals, those with the "best" score, along some measure of usefulness or fitness, survive and reproduce. Offspring are better adapted because they contain the best attributes of their parents and, hence, new individuals of the evolutionary system can arise, which possess new combinations of attributes that were not previously present in any other individuals. In this sense, the genetic system has produced, or "learned", a new and unique individual.

When applied to machine learning, genetic algorithms are generally constructed as classifier systems. That is, a genetic-based machine learning system attempts to create a set of classifiers that are useful for describing a particular data set. A genetic algorithm begins with a set (population) of candidate classifiers. In most implementations of genetic algorithms for machine learning, classifiers are represented by strings of 0's and 1's. Those classifiers that are useful, i.e. successfully describe some aspect of the data set, receive a payoff that determines how well they perform in their environment in the future. Members of this classifier population reproduce in the next cycle proportional to their "usefulness" score. After reproduction, patterns of 0's and 1's in members of the new population cross over, i.e. two population members contribute a portion of their pattern to produce a new member (classifier) of the population. In this way, new classifiers are created from the best classifiers of the previous population. A random mutation step is also applied to ensure that populations do not stagnate (a random 0 or 1 is changed with a very low frequency). Over several generations, populations produce individuals (classifiers) that become more "useful" in describing a data set than individuals of previous populations. Genetic-based machine learning can, therefore, learn new classifiers that were not present in the original set (often a randomly generated set) of classifiers.

5.5.7 Automated Tools

Various computer programs have been developed to perform many of the interview tasks described above, thereby relieving a knowledge engineer of some of those chores. These automated knowledge acquisition tools, such as ETS (Boose 1986), lead an expert through various on-line exercises to identify decision factors, knowledge structures, and problem-solving strategies. Some of these tools even create an operational system directly from interviews. Many rely on graphical aides to help an expert visualize his or her knowledge structures. Several such tools are available commercially, but most have been developed and are used in-house or represent academic research systems. Gordon (1989) provides a brief survey of several of the tools available.

5.6 DOMAIN ONTOLOGY AND ACQUISITION METHODS

From the descriptions of acquisition methods above we can extrapolate and suggest some methods that may be particularly useful for each of the three knowledge analysis steps presented previously. Table 5–2 summarizes the application of particular acquisition methods to the three components of a domain ontology. Several techniques, such as free association, modus ponens

Table 5–2. Several different knowledge acquisition methods can be applied to each of the three components of knowledge analysis.

Factor Analysis	Structure Analysis	Strategy Analysis
Free Association	Sorting	Problem Solving
Modus Ponens Sorting	Modus Ponens Sorting	Protocol Extraction
Repertory Grid	Repertory Grid	AHP
AHP	AHP	Twenty Questions
Questionnaires	Questionnaires	Goal Decomposition
Automated Tools	Automated Tools	Machine Learning
Twenty Questions	Knowledge Diagramming	Unstructured Interview
Unstructured Interview	Unstructured Interview	Modus Ponens Sorting
Machine Learning	Twenty Questions	Automated Tools
Protocol Extraction	Psychological Scaling	Knowledge Diagramming
	Protocol Extraction	

sorting, repertory grid, analytic hierarchy process, questionnaires, and automated tools, specifically request decision factor information. Others, such as twenty questions, unstructured interviews, protocol extraction methods, and machine learning methods, often extract decision factors as part of their acquisition process. It deserves repeating that literature and examples are also very useful for identifying factors and should be considered as a first source. Because a knowledge engineer will want to review the available literature for his or her own background education in the subject area, initial acquisition via literature can occur simultaneously with preparation for interview sessions.

Most of the acquisition methods that seem to elicit structural knowledge do so on the basis of relationship type. Sorting, modus ponens sorting, knowledge diagramming, repertory grid, questionnaires, and automated tools usually attempt to explicitly identify and label relationship types. Using a hierarchical structure and priority values, the analytic hierarchy process elicits both the type and the strength of relationship knowledge. Unstructured interviews, protocol extraction methods, and twenty questions illuminate structure less directly and often less completely. Psychological scaling methods, however, highlight the proximity between terms, rather than actual types of relationships that exist.

A number of acquisition methods utilize a problem-solving scenario. To facilitate problem solving, an expert is asked to consider either a generic example problem or a specific problem instance. An expert may then solve the problem and provide explanation as in twenty questions and protocol extraction. Retrospective protocol splits the activities of problem solving and exposition into two separate tasks. Problem solving and forward scenario simulation eliminate exposition altogether, where in the former an expert solves a problem and in the latter only problem-solving steps are elicited. Goal decomposition also extracts various components of solving a problem without actually asking an expert to solve one. In the analytic hierarchy process, criteria and their priorities explicitly specified in a hierarchical structure precisely define a decisionmaking strategy; in some sense the AHP *is* the strategy. Induction from examples creates an empirical strategy from previously solved problems. It is also possible for strategic knowledge to crystallize during modus ponens sorting, unstructured interviews, knowledge diagramming, and automated tool usage, but strategic knowledge results from verbalization rather than from performance.

From Table 5.2 it is apparent that many knowledge acquisition methods can be applied to all three components of an ontological analysis. Although, certain methods are more effective for particular analyses, the knowledge engineer can carefully apply one of several methods and, in so doing, can extract all three forms of knowledge to some extent.

5.7 KNOWLEDGE ACQUISITION SCENARIOS

In a preceding section we presented the types of knowledge that need to be extracted for viable and useful knowledge systems. Then, ontological analysis components were associated with various knowledge acquisition methods to help guide the reader in applying methods where appropriate. It is also possible to tie acquisition techniques to various system development scenarios in which the persona and cardinality of human expertise may change from system to system. The following three subsections cover the three most common cases where: (1) the knowledge engineer and expert are the same individual, (2) the knowledge engineer(s) works with one expert, and (3) the knowledge engineer(s) works with several experts.

5.7.1 Knowledge Engineer = Expert

Neither of the authors have been in this situation when developing a knowledge system. So, we can only note, from simple inspection, which acquisition methods would not be appropriate due to their reliance on the presence of a second individual, often in an interviewer role. Twenty questions, unstructured interviews, questionnaires, and problem solving all seem to require someone to formulate questions or engage in dialogue or interpret expert's actions/responses. Therefore, these would be of little use in the knowledge engineer = expert scenario. Of the remaining methods, certainly machine learning and automated tool usage are affected the least by the absence of a "true" knowledge engineer. Repertory grid, psychological scaling, sorting, and modus ponens sorting would also be little impacted. The remaining methods might be a little awkward at times or suffer somewhat from a lack of external prompts, but could probably be applied to this scenario.

We would comment, however, that it might be instructive for prospective knowledge engineers to develop a simple system for self education in which the knowledge engineer acts as his or her own expert. A role-playing scenario like this may provide valuable practice for future acquisition circumstances by giving the engineer some feel for internal processes that an expert may typically encounter.

5.7.2 Knowledge Engineer(s) With Expert

This scenario represents the most common situation encountered during system development. Being the most typical, all acquisition methods are applicable. Which ones are selected will depend upon the expert's available time, the current stage of system development, the knowledge engineer's familiarity with various techniques, and how comfortable an expert feels with particular methods. Each acquisition method possesses certain strengths and weaknesses, many of which we have enumerated above. It is advisable to exploit different methods and their characteristics rather than to rely on a single technique alone.

5.7.3 Knowledge Engineer(s) With Multiple Experts

When multiple experts are available it is advisable to utilize as many experts as possible. Opinions from several individuals can be useful for application selection, verifying community knowledge, and taking advantage of differences in specialization (Mittal and Dym 1985). Difficulties associated with pursuing multiple experts include: additional time consumption, obtaining agreement between experts, and providing a forum, document, or medium which records and facilitates dialogue between experts. Despite these difficulties, however, benefits can be substantial (Mittal and Dym 1985).

In the case of a large, complex application a knowledge engineer may find that multiple experts are necessary. When system components are fairly disjoint and when separate experts are available, this situation may be regarded as multiple and independent instances of knowledge engineer(s) working with one expert (e.g. Schmoldt 1987). Hence, the methods discussed above may be used.

A number of useful group interaction techniques have been developed in the field of decision analysis that may be applied to the multiple expert problem. Probably the most familiar, *brainstorming*, simply provides for face-to-face discussion between individuals with the intent of idea generation. The *nominal group technique* (Van de Ven and Delbecq 1971) has a similar group meeting scenario, but with slightly more structure. Each participant offers an idea in turn and group discussion follows each idea. When no more new ideas can be offered, the session ends. The *Crawford slip method* (Crawford and Demidovich 1981) is a variant on the nominal group method in which ideas are written on individual slips of paper; these are then discussed with each idea having anonymous authorship.

Another group technique that aims to maintain anonymity for participants' ideas and opinions and to avoid confrontation is the *Delphi technique* (Dalkey and Helmer 1963). Standard implementation of the Delphi employs questionnaires to which each member of a group anonymously responds. Questionnaires are repeatedly administered to members of the group for revision, intermixed with feedback of questionnaire summaries until some consensus has been reached. Group interaction is minimized to avoid voice dominance by position or persuasiveness, and to reduce the group pressure to conform (von Winterfeldt and Edwards 1986).

There are three critical components of any multiple expert acquisition method. These are: (1) some common medium for knowledge expression by all members of the expert group, (2) a procedure for obtaining group consensus, if it exists, and (3) allowance for, and representation of, unresolved differences of opinion. Schmoldt and Bradshaw (1988) utilized knowledge diagrams and glossaries of terminology as a knowledge representation medium. Boose (1986) used the repertory grid in a similar fashion. Schmoldt and Peterson (1991) found questionnaires to be a useful structure for representing and focusing discussions of knowledge content and knowledge organization in a workshop setting. In Schmoldt and Bradshaw's formulation, a *knowledge packet*, consisting of a knowledge diagram and a glossary, was routed successively to each expert until no more changes were needed. Boose used repertory grids, that were completed by experts, to facilitate structured negotiations. Boose (1986) suggested having a suite of repertory grids, imbedded in an expert system, available for a user to select from, each one would contain knowledge specific to an expert or group of experts.

In many ways this last idea seems very appealing because it mirrors the way that we currently operate when seeking expert opinion. We often seek opinions from several individuals and then weigh their contributions with our subjective confidence in the opinion giver(s). We don't currently have a mechanism for automatically consolidating individual knowledge bases created by separate individuals. If we were able to do this, we could create a "consultant" knowledge base for any subject area. Each subject area would have a panel of accepted expert knowledge bases. Then, the user could select to receive the separate opinions from any subset of the panel of expert systems, or the user could, instead, ask for a consolidated consensus opinion from any subset of the panel. For such an organization of knowledge to be efficiently realized it would be necessary to develop some automatic procedure for combining expertise from separate knowledge bases. However, the availability of many combinations of knowledge bases could provide a decision maker with large amounts of expertise and also the flexibility to apply it in whatever manner seemed most appropriate. It also means that there would exist no clear cut decision because recommendations would be dependent on the group of expert systems selected for consultation and on whether the systems were consulted individually or as a group. Consultations with human experts is currently done in an analogous manner, unfortunately, many people express a sense of uneasiness with a lack of clear cut decisions.

5.8 KNOWLEDGE ACQUISITION GUIDELINES

There are some useful tips and caveats which a knowledge engineer should be aware of before beginning discussions with experts. Many of these can be found in Davies and Hakiel (1988) and Surko (1989). Some of these suggestions help to prepare for an interview and others relate to particular social interactions that a knowledge engineer must continually be alert to.

- First of all, the proper experts should be sought. It is essential that experts possess both exceptional knowledge and skills in their field and also profess substantial interest in the development project. This ensures that the human knowledge source is both valuable and cooperative.

- It is also extremely important that a knowledge engineer be well prepared in advance for knowledge acquisition sessions. That is, he or she should read as much as possible about the domain of interest as general preparation. The knowledge engineer is not expected to become an expert, but preparatory homework allows the engineer to discuss the subject area intelligently. Then, for each acquisition session, questions and topics should be well organized. Organization minimizes wasted time and enables the interviewer to interact intelligently with an expert in the subject area. Experts are likely to become frustrated by having to tutor the interviewer while at the same time explaining difficult ideas.

- Outline the project development process. This can be useful for the knowledge engineer, but more importantly, provide the expert with a road map of how the process will unfold. It is best to do this during the first meeting with the subject matter specialists. This enables everyone to view the big picture and ease into the interviewing process gradually.

- For the expert, the entire knowledge acquisition process can seem more like a clinical examination than a well-intentioned development activity. Continual probing and prodding and critical examination of professional abilities can be very irritating. Remember, experts more commonly have their opinions and expertise taken at face value rather than having them scrutinized and criticized. So, at some point during interviews experts may become very protective and somewhat testy. It is up to the knowledge engineer to ensure that this potential antagonism is avoided. Experts are solicited because they are experts, and a fair amount of prestige and pride goes with this. Treat them as valued colleagues.

• Spend no more than an hour and a half during any session. Most people tire from this type of intensive cognitive activity after that amount of time. Limiting interviews to relatively short time intervals will help ensure that the time spent is most meaningful and does not lead to ruffled feathers or bruised egos.

• Cultivating a friendship with the expert can help make interview sessions proceed smoothly. When knowledge acquisition sessions have a tone of social relaxation, experts will be more receptive. They may be able to think more clearly without straining and there may be a reduced possibility that feelings will become frayed during difficult moments.

• Two knowledge engineers can work well as team interviewers. Schmoldt and Bradshaw (1988) found that it was very useful for one interviewer to ask questions while a second takes notes. While the first interviewer is engaging the expert in conversation, the second can be pre-analyzing his or her responses. In this way, fine points and details, which might otherwise remain unnoticed until later, can be fleshed out and thoroughly discussed.

• Knowledge acquired from one particular session should be processed and absorbed before proceeding into the next session and new material. Often, analysis of one session's output will lead to further questions, which may then be covered at the beginning of the next interview. This provides for a very systematic movement between topics and allows everyone to focus on only one small aspect of the domain at any point in time.

• Keep the expert included in the progress of the project. This gives him or her a sense of ownership and helps ensure continued interest. Identify some milestones in the development process and tie those into the expert's contributions. It is then apparent that the expert's efforts are making a valuable impact on the project. Interest can also be cultivated by giving periodic rewards, such as junior authorship on appropriate manuscripts, letters of commendation or appreciation from superiors in the organization, small gifts acknowledging their efforts in the project, or naming the system after the expert. Any appreciative gestures made toward the expert will engender a positive and enthusiastic feeling for the development effort.

• In general, it is often difficult to verbalize all aspects of a complex subject area. Visual aides, however, can greatly facilitate the conveyance of ideas. Pictures, graphs, tables, figures, and drawings can often clarify subtle points and major ideas. One visual aide, that is discussed more below, is the use of a prototype of the system being developed. A prototype system provides a good focal point for fine-tuning knowledge and demonstrating project advancement. As an expert critiques a prototype he or she understands better how the knowledge is being used, can correct errors, and makes suggestions for specific improvements.

• Like visual aides, example problems, similar to the ones that will be solved by the final system, can be useful for directing discussions. A number of the knowledge acquisition methods, presented previously, depend upon examples ranging over various degrees of difficulty, e.g. eliciting protocols or problem-solving methods. Oftentimes, an expert's heuristic, compiled knowledge will be readily apparent, but further inquiry can elicit deeper principles upon which experts sometimes rely.

- Ask questions to ensure that both you and the expert agree on specific points. "Why" questions are often the most difficult for an expert and may make him or her uncomfortable, but these questions ultimately reveal deeper, more fundamental knowledge structures. Recapitulate ideas in your own words and ask for verification from the expert; this guarantees that your understanding of it represents an accurate paraphrase. Suggest ideas that are alternative to or conflicting with the expert's, and ask why yours are not viable. Forcing the expert to reason against erroneous knowledge helps elucidate valid aspects of his or her knowledge. These particular interview guidelines are most useful during less structured sessions when the expert has not been asked to perform specific tasks.

- Much knowledge acquisition work occurs before and after interview sessions through the analysis of previous sessions. Therefore, it is imperative to have an accurate record of what transpired during a session. Video recording or audio recording are the best mechanisms for limiting any loss of information gathered during discussions. Hand written notes are also useful, but relying solely on note-taking can interfere with dialogue during an interview. It can also protract the time allotted for an interview, consequently less is accomplished in any given amount of time. Questionnaires can also serve as a reliable account of recorded knowledge, but they record only the specific knowledge that they were designed to elicit. Various combinations of these media are a knowledge engineer's best guarantee against lost information and wasted elicitation opportunities.

5.9 EXPERT BIASES AND ELICITING UNCERTAIN KNOWLEDGE

Often the knowledge that we wish to acquire for a KBS has some uncertainty associated with it. This creates special difficulties when we elicit concepts, relationships, and strategies using the acquisition methods above. Experts have trouble estimating uncertain quantities because there exits no natural scale that is directly related to their experiences (Saaty 1980). Therefore, when their knowledge and experiences are translated into commonly used scales (e.g. probability values or confidence factors), biases often appear.

The essence of this problem, then, is how to determine the likelihood (a.k.a. probability, degree of truth, certainty, or confidence factor) for a fact or rule (Parsaye and Chignell 1988). Unfortunately, experts usually do not associate numerical values with the knowledge they possess. Nevertheless, when asked to do so by a knowledge-based system developer, they can usually make some estimates. But these estimates are likely to be influenced by a number of biases. Three of the most common classes of biases that experts can fall prey to are *named representativeness*, *availability*, and *anchoring*. Many others have been identified and described (Cleaves et al. 1987, Saveland et al. 1988).

The representativeness bias leads experts to assign a certainty for the occurrence of an object based on how typical the object is for a particular class rather than on how likely it is that the object will occur. The essential features of an object or event A are compared to a stereotype B (Saveland et al. 1988). If A is judged similar to B, then the likelihood of A being a member of set B is judged high. This is perfectly reasonable and usually a useful heuristic. However, it is important to realize that the representativeness bias can lead to errors in judgment. A common error occurs when a small sample is erroneously trusted.

> "For example, it is fairly common to flip a coin four times and get three heads, but the chance of getting a proportion so out of line with the real odds of 50:50 is small after 100 flips. Although most people are aware of the problems with small samples, they often ignore them... When a person takes a sample from memory, it is necessarily small based simply on human limitations of recall [and limited human experiences]. For example, a fire specialist may see a dry May and try to recall other years that had a dry May. The sample will be small (Saveland et al. 1988)".

If the small sample is truly representative of the entire population, then all is well. If not, a bias occurs and even expert judgment will be wrong.

Another frequent representativeness bias is due to ignoring base rates and their resulting overconfidence. The match between a single event and the expert's personal sample of the population to which that event belongs determines his or her degree of confidence (Saveland et al. 1988). For example, people are prone to express great confidence that a person is a forester when given a description of the person's personality that matches the stereotype for foresters. This high level of confidence is found even if the description is scanty, unreliable, or outdated, and even if the base rate of foresters in the general population is very low (Tversky and Kahneman 1974). When the significance of the consequences to the expert increases, so does the tendency of the expert to be overconfident (Saveland et al. 1988). These findings are not meant to imply that expert judgment should not be trusted, but merely to point out that experts are as fallible as the rest of us even when working in their own field of expertise.

An availability bias may lead experts to assign a higher certainty to knowledge that is easy to remember. Typically, frequently occurring events are easier to recall, but not always. Sometimes, events which have been personally experienced, recently experienced events, and events which people are prone to pay particular attention to (i.e. firefighters and large fires) often lead to biased certainty values (Saveland et al. 1988).

The anchoring bias refers to a tendency for people to assess certainties for events based upon some natural, observable starting point and then to adjust judgments of subsequent probabilities away from the initial estimate. The adjustments tend to be conservative and thus the initial estimates tend to act like an anchor (Saveland et al. 1988).

It is difficult to make experts aware of these biases and take them into account. People in general, including domain experts, make frequent mistakes combining events with different likelihoods (or belief values) especially if the reasoning chain involves more than a few events (Ross 1988). Again, as noted above, this is due in part to the scale mismatch between an individual's internal truth (or belief) scale and the scale we force their thinking processes into.

Several methods have been developed to help experts quantify uncertain knowledge. First, simply making the expert aware of the various biases may help (Saveland et al. 1988). Second, the knowledge engineer can employ specialized techniques to minimize biases, for example, to use visual props such as probability wheels or graphical confidence scales, one for each fact or rule which can then be visually compared by the expert (Spetzler and Staël von Holstein 1975). This technique reduces the need for the expert to wrestle with numerical intricacies of any particular certainty calculation method selected. Third, Parsaye and Chignell (1988) discuss several ranking schemes useful in helping experts assess probabilities. Fourth, to reduce overconfidence, the knowledge engineer can ask the expert to think of reasons why the certainty value may be wrong. Fifth, anchoring biases can be minimized by asking the expert for extreme values before asking for the most likely value. Sixth, providing the expert with independently obtained base rate data for an event in question appears to be effective in minimizing the representativeness bias (Saveland et al. 1988). And finally, Selvidge (1975) presents guidelines for assessing the probability of rare events.

The main lesson that can help guide us is to make changing the certainty factors in the knowledge-base system as easy as possible in order to encourage experimentation with many different combinations until the "right" one is found. It usually takes several iterations to find uncertainty levels that will yield the desired results when combined. Schmoldt (1989) presents a method for iterative refinement of belief measures for rules of classification KBSs. It uses categories of belief for rule consequents and the information content of rule premises to suggest revisions to rule belief values. Experience has shown that knowledge-based systems using uncertainty will not work satisfactorily the first time around. This leaves the development team with three choices (Parsaye and Chignell 1988):

1. change the uncertainties assigned to facts and rules
2. change the rules to add or delete premise conditions
3. change the paradigm being used to combine uncertainties

Some combination of all three above must usually be used to achieve satisfactory system performance.

5.10 SUMMARY AND ADDITIONAL READINGS

To acquire the necessary knowledge for system development is a difficult task because of the abstract and complex nature of knowledge, especially private, human knowledge. Performing a good job during this phase of system development is essential, however, to produce an adequate knowledge model. This model will later be implemented as computer code in the knowledge system. A poor job of knowledge acquisition will produce a faulty knowledge model, which will, in turn, result in a faulty knowledge system.

Knowledge acquisition can be viewed from several different perspectives: strategies used throughout the acquisition process, analysis of knowledge as an entity in itself, and different personnel scenarios for acquisition. Acquisition methods have been developed in many different disciplines, such as decision theory, psychology, management science, computer science, and anthropology. Each method has its own strengths and weaknesses. Consequently, it is best to apply several different methods to obtain a complete picture of any subject area and to find effective methods for reasoning about that knowledge.

Hart (1986) provides a more extensive presentation of knowledge acquisition than we have space for here. There is also a journal, *Knowledge Acquisition*, that contains the latest research in this area. Also, the *International Journal of Human-Machine Studies* publishes many articles on knowledge acquisition techniques and systems. In general, however, the level of activity devoted to understanding and capturing human knowledge has lagged behind efforts in hardware and software implementation.

CHAPTER

6 Designing the Application

Each of us carries around in his or her head a model of the world, ...It is the only way we have of relating to other people and to our larger surroundings. We draw upon these models whenever we plan for the future in any way.

R. C. North

In chapter 4, we discussed planning an application project and began with the *problem definition*. Red Pine Forest and Pest Management examples were introduced and defined. Once an application has been defined and knowledge sources identified, problem-solving knowledge must be acquired and organized. In the previous chapter, we reviewed some relevant *knowledge acquisition* concepts and methods to do this.

The current chapter focuses on *application design*, which along with knowledge acquisition forms the second step leading toward a completed initial prototype (Figure 6–1). Application design entails developing a **knowledge model** (ontology) and a **human factors model**. Next, pre-implementation topics, such as the selection of *knowledge representation*, *search methods*, and *knowledge-based system architecture* must be addressed for the initial prototype. Finally, the selection of an AI programming environment is explained.

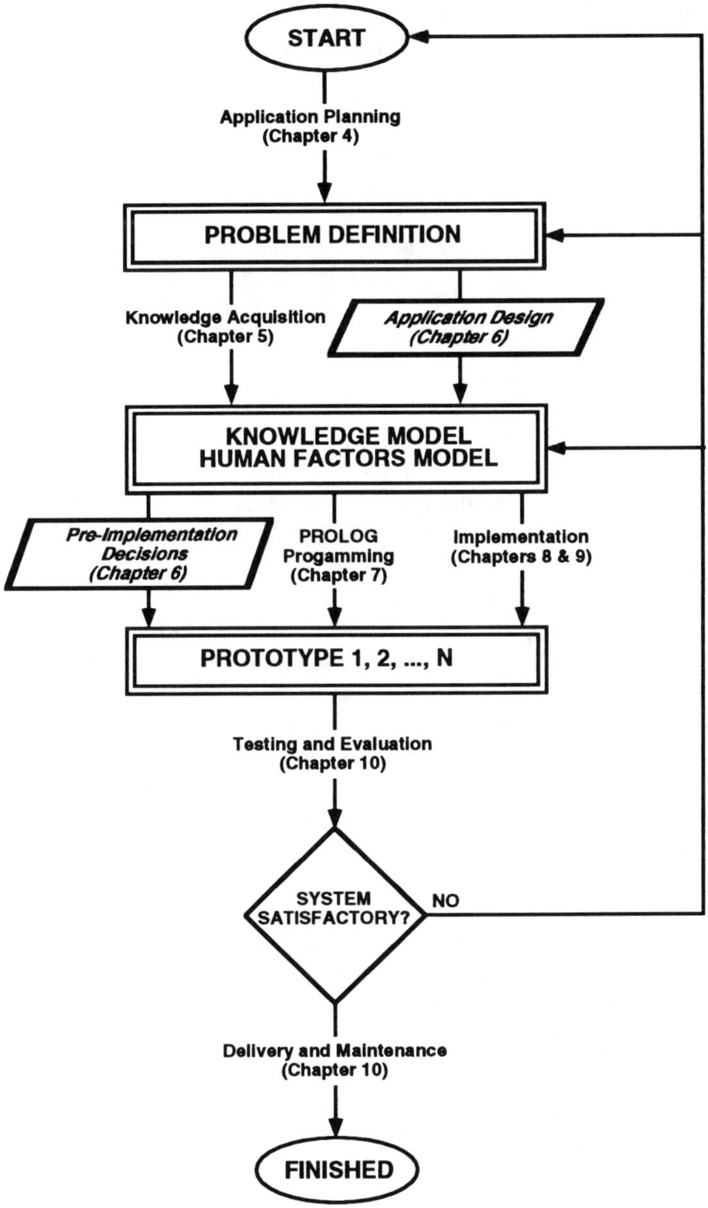

Figure 6–1. A knowledge model contains the domain knowledge collected and organized in the knowledge acquisition phase of system development. The knowledge translation task merges human factors issues, problem space models, and implementation considerations with the knowledge model to support the coding phase of development.

6.1 APPLICATION DESIGN = KNOWLEDGE MODEL + HUMAN FACTORS MODEL

Designing a knowledge-based application involves two basic components. These are: a conceptual model about the subject matter of interest (a knowledge model), i.e. a theory of reality in the subject area, and a human-machine interaction, or human factors, model.

Reports of knowledge-based systems in natural resources management began to appear in the literature in significant numbers in the mid-1980's: 12 in 1983; 17 in 1984; 38 in 1985, 52 in 1986; and 45 in 1987 (Davis and Clark 1989). The primary focus of these reports has been on describing how a KBS functions rather than on presenting the details of the knowledge base itself. Descriptions of the organized and synthesized knowledge in the knowledge base, i.e. the knowledge model, has been variable, ranging from brief (Rice et al 1989, White and Morse 1987) to reasonably complete (Buech and others 1990, Davis and Laut 1989). Appreciation is growing that the description of the knowledge base as an organized and evaluated *ontology* for an area of human understanding may be the most important and lasting result of a KBS project.

When designing a human factors model, the central question is how the system should interact with users and how it presents to a user what it knows. To do this, there must be some model of the user's expectations and technical competence and the existing culture of the organization and user group that the software is expect to operate in. A user's role in the decisionmaking process may range from bystander and observer on one hand, to being the central object and focus of system activities on the other, e.g., KBS monitoring of a manufacturing process would have the user as an observer, whereas, tutoring is a user-centered activity. The user interface typically consumes up to 60% of the computer code in a system (Ross and Kirk 1988 cited by Rewerts and others 1990). So it seems reasonable to devote a proportional amount of time on human factors engineering during the design phase.

There are a growing number of Ph.D. level programs in universities around the world with the title of "software" or "instructional" designer or human factors engineering. This field of expertise covers the structure and content of user interfaces in corporate software. As we have noted previously, a highly effective problem solver, but one that lacks a good interface, may be awkward to work with and, consequently, fall into disuse.

6.1.1 The Knowledge Model

A knowledge model is a thinking aid essential to the unambiguous understanding and communication of a KBS (Shlaer and Mellor 1988). A knowledge model consists of highly structured text, diagrams, tables, graphs, etc. describing the organization and interaction of the knowledge base components precisely and concisely. Graphical aids can be used to communicate an integrated overview of the set of components (modules) in the knowledge base. In discussing communication and comprehension of ideas, Halpern (1989) presents several *graphical organizers* that are analogous to our concept of a knowledge model. All these organizers involve building a meaningful representation of knowledge that matches the one used by the person conveying the knowledge. This can be said to have been accomplished successfully when the knowledge model exactly reflects the knowledge structure in the originator's mind. Halpern's (1989) graphic organizers are: (1) *linear arrays*, e.g. a causal chain of reasoning or a sequence of events, (2) *hierarchies* (we have seen many examples of these, e.g., dichotomous keys, analytic hierarchy process), (3) *networks*, e.g. the semantic networks of Chapter 2 or the knowledge diagrams or terminology graphs of Chapter 5, (4) *matrices* or *tables*, e.g. the repertory grid of Chapter 5 or the tabular knowledge model of Buech and others (1990), and (5) *flow charts* (this is the representation chosen for the RP-FMAS knowledge model, below). Because many subject areas can be quite complex, several different graphic organizers can be useful within a single knowledge model to comprehensively capture the underlying knowledge structure.

The function of a knowledge model is to organize and present the design of the knowledge base so clearly and completely that it can be used as an implementation blueprint. The knowledge model identifies all modules, presents the organization of the modules, discusses their attributes, and defines their interactions. Once a knowledge model has been formulated and presented, the implementation of that model in a shell or an AI language, such as LISP or PROLOG, is a relatively straight-forward task. Experience shows that more effort goes into developing the knowledge model than into its implementation.

The most effective way to illustrate what a knowledge model should look like is to present one. The Red Pine Forest Management Advisory System (RP-FMAS) was introduced and defined in section 4.3. The first step toward a knowledge model is to define the general logic structure of the system and identify the major logical modules and their functions. This was accomplished as part of the problem definition discussion in Chapter 4. Given this definition of the 12 major modules (Figure 6–2) and the accompanying description of the operational structure of the system, the next task is to describe each module in more detail. An abbreviated discussion of the purpose, logic, and references used for each of the 12 modules is presented below along with a figure identifying relevant subordinate modules and a description of a typical cycle through the system.

The level of detail presented here for each of the modules would certainly not allow anyone to implement the system. More specific details would consume considerable space in the text and would be of interest to only a few readers. The full details of the knowledge model, that would allow an individual to actually implement it, are available in a hypertext electronic document which can be found by referring to the Frontmatter.

Introductory screens begin the program and explain its purpose to the user. Then module 1 requests the user to enter standard stand description data.

Module 1: Data Input
Purpose: prompt the user for a description of the subject stand.

Logic: This module is very simple and has no subordinate tasks. It offers a data input screen which allows the user to enter the "standard" stand description variables (*age*, diameter at breast height (*DBH*), *basal area*, *trees per acre*, *site index*, *dominant height*, *stand area*, and *volume per acre*) at the start of a consultation. These are required values. The user enters as many of these numeric values as are known.

References: None.

The variables require numeric values, which are frequently available and provide the greatest accuracy if known. Failure to fill in some or any of the values needed, will cause module #2, Data Input Checking, to generate estimates or to ask the user for qualitative values.

Module 2: Data Input Checking (Figure 6–3)

Purpose: review the user input for missing values, provide estimates for missing values as needed, and compute the values for secondary variables, i.e., those that can be calculated from the primary variables entered by the user.

Logic: The program never knows how much or how little data the user can provide the system. It is the responsibility of this module to intelligently analyze the input data set, estimate missing values where possible, or ask the user for qualitative input rather than the previously requested quantitative input. For example, if *site index* is not provided by the user, but *age*, *basal area* and *volume in cubic feet* or *volume in cords* are given, the system can estimate *stand height* as a function of those three and then estimate *site index* with the values for *height* and *age*. Module 2 has the effect of making user input much more flexible because the user is free to enter none or any combination of values at hand.

Once a full set of primary variables is available, the derived variables, *stand size class*, *maximum recommended basal area*, *minimum recommended basal area*, *trees per acre*, and *stocking level* are calculated. The primary and derived variable values are saved so that they are available to the other modules in the system.

References: Benzie 1977, Buckman 1961, Buckman 1962, Gevorkiantz 1957, Lundgren 1981, Lundgren and Dolid 1970, Wambach 1967.

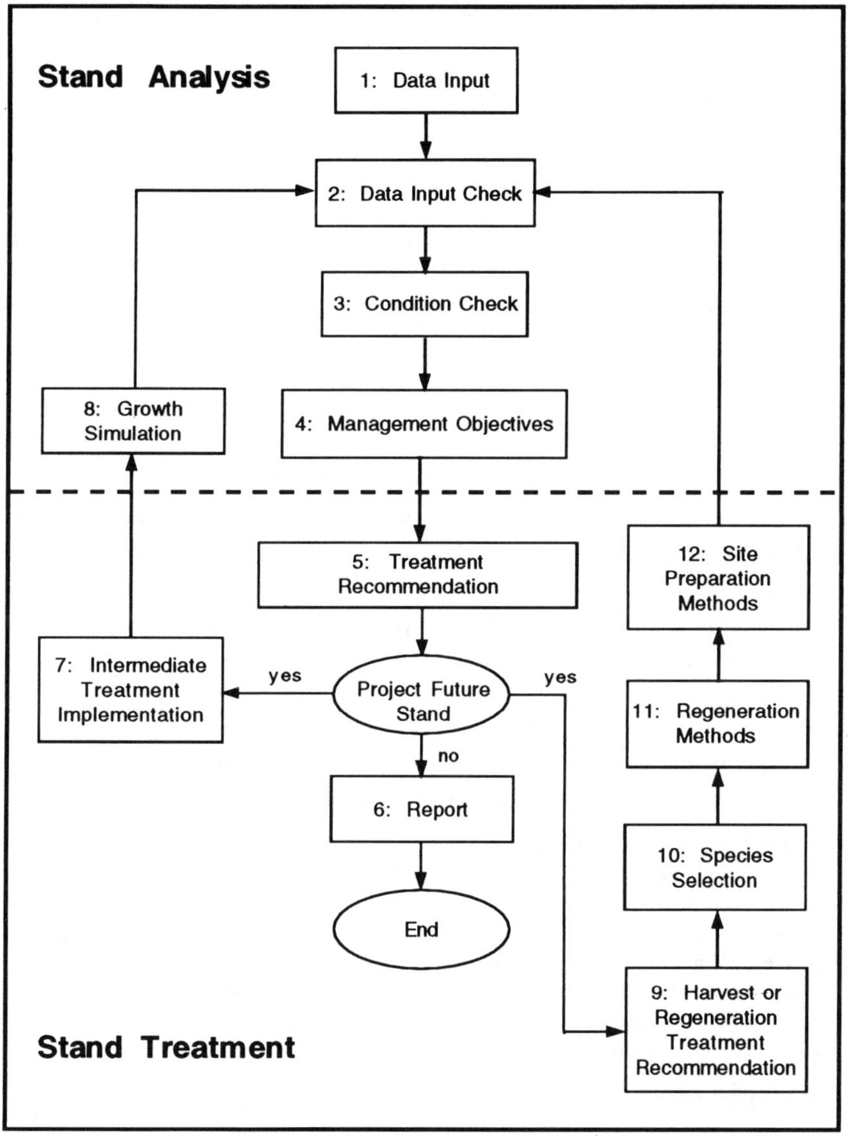

Figure 6–2. A graphical display depicting the operational structure of knowledge model components for RP-FMAS clearly shows the relationships between modules in a flow chart format.

```
2.0   Checking complete
 |
 |--- 2.1   Check age
 |--- 2.2   Check site index
 |--- 2.3   Check dominant height
 |--- 2.4   Check mean dbh
 |--- 2.5   Check trees per acre
 |     |
 |     L--- 2.5.1 Reproduction density
 |
 |--- 2.6   Check basal area
 |     |
 |     L--- 2.6.1 Density
 |
 |--- 2.7   Check stand volume
 |--- 2.8   Calculate size
 L--- 2.9   Calculate stocking percent
```

Figure 6–3. Organizational diagram of module #2 shows the required submodules relating to checks of the input data.

The program continues to stand condition analysis, module #3, which determines the health of the red pine stand.

Module 3: Condition Analysis

Purpose: To analyze the health of the subject stand. Serious insect, disease, or physical damage to the stand will influence the treatment recommendations.

Logic: This knowledge module has been defined at the most general level and fit properly into the overall organization of RP-FMAS. Appropriate submodules have not yet been developed. This illustrates one of the major advantages of the top-down design strategy for program development. It is possible to develop modules to any degree of complexity required and have them function properly in the knowledge-based system. Any module can always be decomposed further into sub-modules as the need becomes evident and the resources become available. At the same time total system design remains comprehensive and operational.

References: none.

The management objective Module, #4, asks the user to select the timber product objective for the stand being evaluated so the *DBH* goal for trees at the end of the rotation can be set and the rotation age can be estimated. The problem of determining what product objective (pulpwood, small sawtimber, large sawtimber, poles and piling, multiple products, etc.) to choose for a given stand is not part of RP-FMAS at this time and must be determined externally by the user.

Module 4: Management objective analysis (Figure 6–4)

Purpose: to determine the management objectives for the subject stand and calculate several variables dependent on those objectives.

Logic: The previous data input Modules, #1 to #3, have all been concerned with describing the physical attributes of the subject stand. Selecting the *management objective*, an administrative variable, is an important decision over which the user has a great deal of control. The selected objectives become the focus of RP-FMAS and numerous consequences result from them. It is not possible to practice sound forest management without having some sort of management objective. One of the major weaknesses of the 1989 version of RP-FMAS is the lack of a module to assist the user in making this critical decision. We do test whether the *site index* of the stand is high enough to support the product objective selected. The user is warned if the actual *site index* of the stand is less than the minimum recommended for the chosen product. By design, we offer the user the freedom to ignore this warning and pursue a product objective not suitable for the site. The user is thus able to convince himself or herself that such a choice will result in sub-standard productivity.

Only timber production objectives are currently defined in RP-FMAS. The system has, however, been designed with the flexibility to add wildlife and recreation objectives at some point in the future.

Given a *timber product objective* value, *product quality objective*, *average stand DBH goal at harvest*, and *thinning basal area multiplier* can all be assigned values. For example, given a *timber product objective* value of "small sawtimber", *product quality objective* would be set to "minimum number and size of knots", *average stand DBH goal at harvest* would be 15 inches, and *thinning basal area multiplier* would have a value of 0.7. This stand would not be considered mature until *average stand DBH goal at harvest* is greater than or equal to 15 inches. This stand will be considered overstocked and in need of thinning whenever the actual or simulated *basal area* value equals or exceeds 90% of *maximum recommended stand basal area.* In general, the smaller the physical dimensions of the product selected, the greater the suggested density of the stand up to the point beyond which extreme competition results in overall stand volume growth reductions. This maximum recommended density is based on the stocking guide for red pine reported in the manager's handbook.

Finally, an algorithmic procedure is called to calculate the expected rotation age for the subject stand. That is, a growth and yield procedure estimates the number of years the subject stand will take to reach the *DBH* goal required by the product objective.

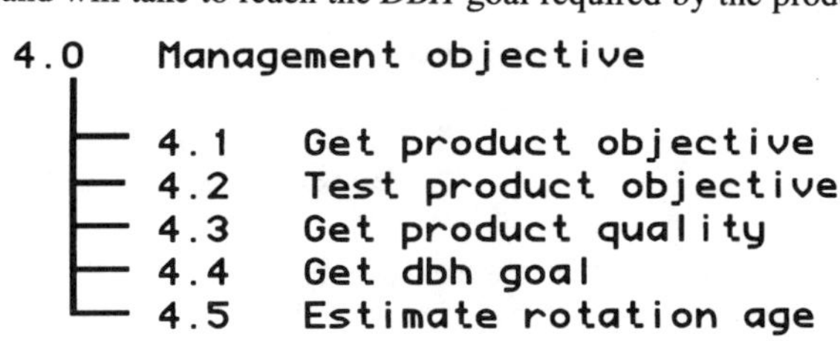

Figure 6–4. Organizational diagram of module #4 shows the required submodules that help define the management objective for the stand and associated objective variables.

References: Benzie 1977, Buckman and others 1983, Buckman 1962, Kallio and Benzie 1980.

Module #5 displays a status screen indicating what is known about the stand at this point. The user is then shown a list of treatment recommendations for the current stand, and asked to select one to implement. After making this selection, the user can request to see the rationale that might lead to this particular treatment recommendation. For an example of these two displays, please refer back to Figure 4-6.

Module 5: Treatment recommendation analysis (Figure 6–5)

Purpose: to present the user with a list of possible treatments recommended for the subject stand and accept the choice.

Logic: The user is first presented with a screen showing the current values of the variables describing the subject stand. A knowledge base is consulted in order to generate a list of possible treatments along with suggested certainty factors. The certainty factors are meant to convey to the user the degree of support each treatment has on a scale of 0–100. The certainty factors are merely crude indices of the merit of each treatment for the specific conditions of the subject stand. We are not entirely satisfied using these generally accepted, numeric certainty factors. In future versions, we will seriously consider using qualitative descriptors such as: highly recommended, moderately recommended, or not recommended.
The user is asked to select one of the listed treatments whereupon the system displays a screen explaining the chain of reasoning that might support the chosen treatment. We recognize that it would be better to make these explanation screens available to the user *before* a treatment is chosen. We plan to improve this in the next version.

References: Benzie 1977, Malpas 1987.

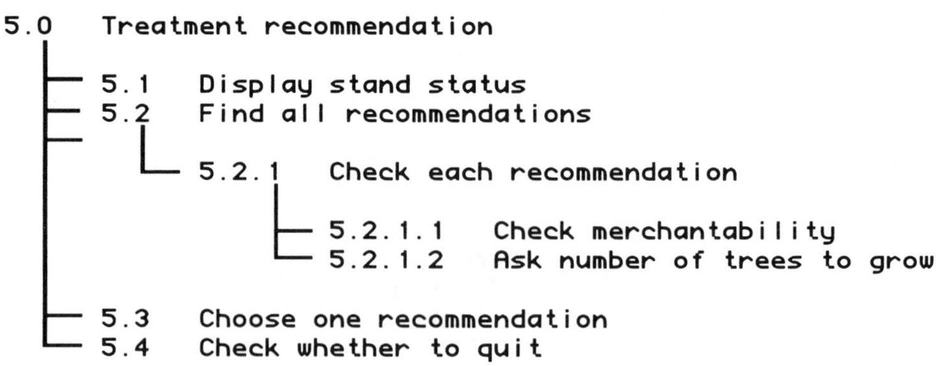

Figure 6–5. Organizational diagram of module #5 shows the required submodules to analyze the stand treatment recommendations.

Whenever the user has selected a treatment, the advisory system provides an exit choice. If the user chooses to end the consultation, module 6 is activated, followed by program termination screens.

Module 6: Report

Purpose: to present the user with a text file containing a history of the just completed session along with an explanation of which rules fired and why they fired.

Logic: The user can browse back and forth as much as desired, can edit this file to make annotations, and can print or save this file for future reference. This so-called "why" explanation is provided in clear, understandable sentences using common forestry terminology.

References: none.

Depending upon the treatment chosen, either module #7, intermediate treatment, or #9, harvest treatment, is activated next. One possible intermediate treatment is to "do nothing" with the stand at the present time. In this case, module #7 proceeds directly to module #8 for growth of the stand.

Module 7: Intermediate treatment implementation (Figure 6–6)

Purpose: to change the stand description thus simulating the implementation of the selected intermediate treatment.

Logic: This module contains the logic to simulate implementation of treatments. For example, if the recommendation is to "clean" the stand, then *basal area* is reduced by 1/2 and the other affected variables, such as *trees per acre*, are also recalculated. If thinning is selected, then the only action taken is to suggest a *residual basal area* value to the user. The user can either accept this value or choose to enter a new *residual basal area* value. A growth and yield simulation program, called from

```
7.0   Intermediate treatments
 |
 |--- 7.1   Implement recommendation
 |     |
 |     |--- 7.1.1   Recalculate trees per acre
 |     |--- 7.1.2   Recalculate basal area
 |     |--- 7.1.3   Calculate residual basal area
 |
 |--- 7.2   Check residual basal area
       |
       |--- 7.2.1   Ask residual basal area
```

Figure 6–6. Organizational diagram of module #7 shows the required submodules to deal with intermediate treatment of the stand.

module #8, simulates thinning the stand "back" to the *residual basal area* value set by the user here. Thinning is an integral part of the growth and yield model used. It seemed best to let stand changes due to thinning be handled there instead of in Module #7. A weakness of the present version is the inflexibility of the thinning procedure. It is only possible to simulate a thinning method such as row thinning where *average stand DBH* does not change, even though *basal area* and *trees per acre* decrease. Crown thinning or thinning from below cannot be simulated. We hope to correct this deficiency in the next version.

References: Benzie 1977, Buckman and Wambach 1966.

The growth simulation Module, #8, is responsible for simulating the growth of the subject stand. A growth simulation description screen communicates the growth changes to the user while a stocking figure charts the dynamics of stand *basal area* and *trees per acre.*

Module 8: Growth simulation (Figure 6–7)

Purpose: to prepare the control variables needed to guide the growth and yield simulation module.

Logic: Before the growth and yield module can be activated, the advisory system must obtain from the user one of 5 control options to activate in order to stop growth. The user can elect to stop growth based on the following criteria:

1. grow until the year when the next thinning is due;
2. grow to rotation age without thinning;
3. grow until the mean annual increment culminates;
4. grow until a given DBH is exceeded;
5. grow for a given number of years.

If the user instructs the system to grow until the next time a thinning should occur, the system uses *thinning basal area multiplier*, determined in module #4, along with *maximum recommended basal area*, from module #2, to suggest a *basal area threshold* value, which when exceeded, will cause the growth and yield model to stop

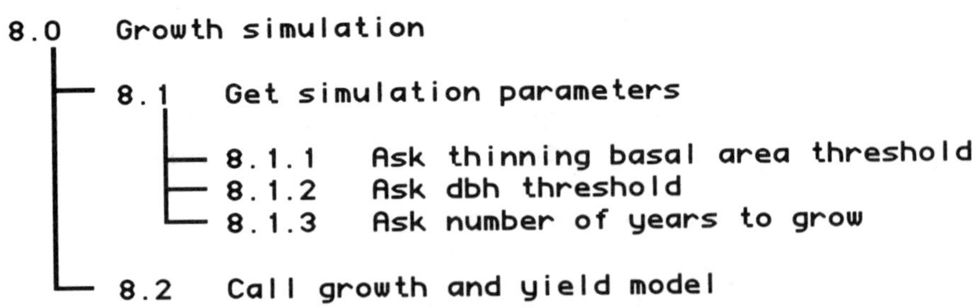

Figure 6–7. Organizational diagram of module #8 shows the required submodules to establish simulation parameters and run simulation.

processing. The user can change this suggested *basal area threshold* value if desired. When all is ready, the control variables and the current stand description variables are passed to the growth and yield program. We modified the red pine stand model RPAL (Ramm 1990), which itself is a variant of REDPINE (Lundgren and Belcher 1982), to simulate growth. If thinning is needed, it is performed first and then simulation continues until the selected stopping control rule is activated.

References: Lundgren and Belcher 1982, Ramm 1990.

The stand description is altered to reflect the new stand and module #2 is again called continuing with the consultation as previously described. Note that module #2 does not need to distinguish which module, #1, #8 or #12, immediately preceded it. This type of module independence is absolutely essential for easy maintenance of large systems. The advisory system cycles repeatedly through modules #7, intermediate stand treatments, and #8, growth simulation, until the user either chooses to end the consultation or harvest the stand. When harvesting or regeneration is chosen by the user as a treatment, the harvesting-stand-establishment cycle is initiated. If the treatment selected by the user is to regenerate the stand without harvesting then module #9 is skipped and module #10 is activated.

Module 9: Harvesting treatment implementation (Figure 6–8)

Purpose: to help the user select a silvicultural system, a harvesting system, and to simulate the harvest.

Logic: If harvesting is selected, a *silvicultural system* value must be selected. The only option in the current version is "clearcutting". In later versions, "selection" will be added as a choice for red pine management. To set the harvesting system to use, the user must select one of three choices:

1. full tree – bole and crown is removed;
2. tree-length – complete bole is removed;
3. log-length – only a part of the bole is removed.

We do not currently have a module to help the user decide which harvesting system should be used.

The implementation of the harvest takes the form of removing all stand description variables relating to the former stand from system memory. The *covertype* variable

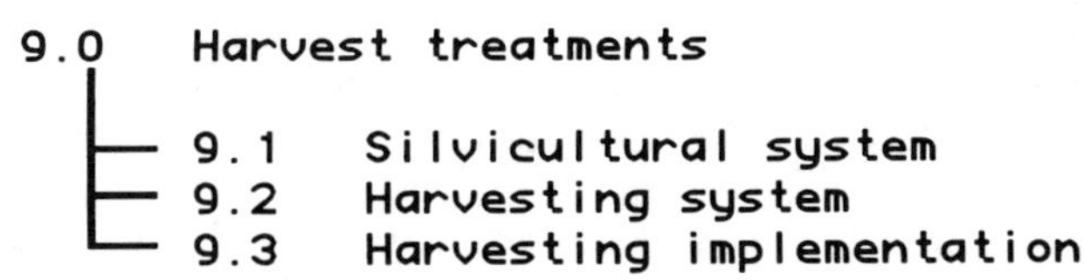

Figure 6–8. Organizational diagram of module #9 shows the required submodules implementing a harvest of the stand.

is set to "none" and *former covertype* and *former site index* are preserved as is *stand area*.

References: Zasada and Benzie 1970.

The species selection, module #10, inquires whether the former species will be used for the next rotation or whether a new covertype will be established.

Module 10: Species selection (Figure 6–9)

Purpose: to evaluate species-site compatibilities and recommend a list of preferred species for user consideration.

Logic: Once a harvest has been simulated or regeneration without harvesting has been selected, the user may choose to simply re-establish the former species on the site or, alternatively, to conduct a consultation to analyze species-site compatibilities. Questions concerning drainage, soil texture, and nutrient availability are asked in order to establish the site type. A list of 18 species/species groups with attached certainty values is then presented to the user in decreasing order of certainty. The user must select one out of this list as the species to re-establish on the subject site. If the selected species is red pine, the consultation continues; if not, the user is advised that the current version does not support establishment of the selected species and the consultation's termination sequence is begun.

References: Buol and others 1973, Perala, Pregitzer, & Johnson personal communication.

If red pine is the species selected for the next stand, the user is advised by the regeneration method, module #11, that planting is the only option currently

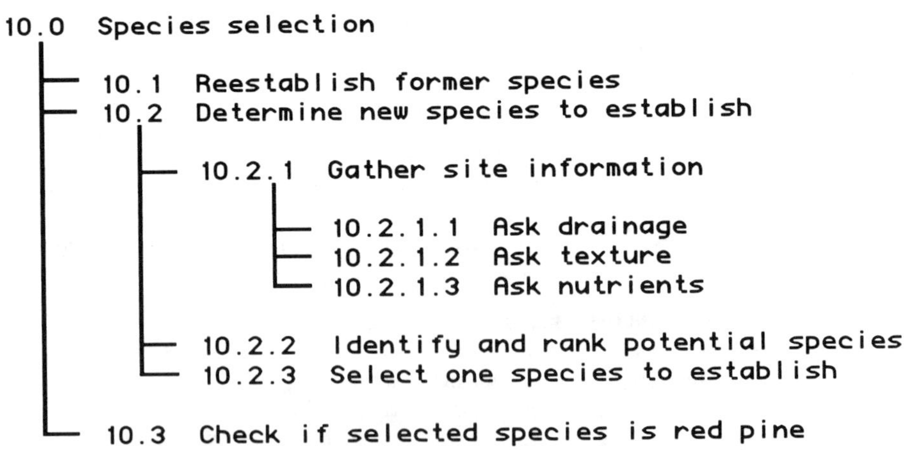

Figure 6–9. Organizational diagram of module #10 shows the required submodules for analyzing species-site compatibilities and for recommending regeneration species.

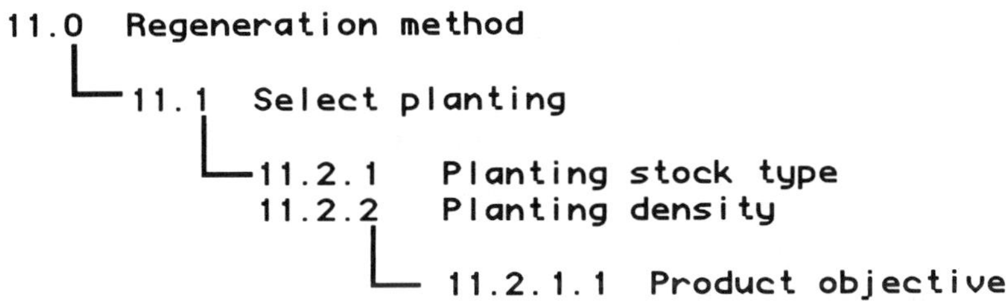

Figure 6–10. Organizational diagram of module #11 shows the required submodules to select a regeneration method.

recommended. The advisory system also asks the user to select a new *timber product objective* value for the next rotation.

Module 11: Regeneration method (Figure 6–10)

Purpose: to assist the user in selecting an appropriate regeneration method.

Logic: Planting is the only regeneration method currently allowed for red pine. A future improvement to the system will add seeding and natural regeneration options. The two major decisions to make in planting red pine are: (1) the stock type and (2) the planting density. The system does not yet assist in selecting stock type. Planting density depends upon the product objective for the next rotation. At the extremes, a pulpwood objective would result in a recommended planting density of 1200 trees per acre while a large sawtimber objective would produce a recommendation of 600 trees per acre. The other product objectives fall somewhere in between.

References: Bassett 1984, Benzie 1977.

Finally, the system helps the user select a site preparation method, module #12, based on drainage, litter conditions, vegetative competition, and other factors.

Module 12: Site preparation method (Figure 6–11)

Purpose: to assist the user in the selection of a site preparation method for the subject site.

Logic: Having determined a species and regeneration method, the last step in the establishment of a new stand is to prepare the site for regeneration. Any relevant information is asked, if it is not already known, and the user is presented with a list of eligible site preparation methods. One method must be selected. The final task for this module to perform is to change the stand description to reflect new values for *stand age*, *trees per acre*, and *timber product objective.*

References: none.

- 12.0 Site preparation method
  - 12.1 Find covertype class
  - 12.2 Find site preparation methods
    - 12.1.1 Harvest system
    - 12.1.2 Slash
    - 12.1.3 Vegetative competition
    - 12.1.4 Drainage
    - 12.1.5 Former covertype class
    - 12.1.6 Litter
  - 12.3 Choose one site preparation method

Figure 6–11. Organizational diagram of module #12 shows the required submodules to choose a site preparation method.

The new stand description is generated and the RP-FMAS cycles back to the data input analysis, module #2, as if the user had entered a 1 year old stand in the first place. These cycles continue until the user elects to terminate the consultation.

Because the pest management application in PREDICT is much less complex than forest management, the knowledge model is relatively straightforward. Hence, a detailed description is not provided here, only a general overview. The knowledge model of PREDICT attempts to model the process that pest specialists' use to solve insect and disease problems.

When pest specialists are asked to provide advice about diagnosing a pest problem or identifying potential pest hazards, there are a number of steps that they go through. These strategy-knowledge steps were gleaned from conversations with experts and by distilling the important aspects of their experiences during knowledge acquisitions sessions. To diagnose a pest attack, experts do several things: (1) determine the geographic location of the stand, (2) determine various site characteristics that can be predisposing for particular pests, (3) based on (1) and (2) generate a prioritized list of likely pests, (4) try to eliminate certain pests on this prioritized list by tree injury symptoms, and (5) ask questions about symptoms and signs that could imply one or more of the remaining pests. For hazard assessment of a stand, experts perform the first two steps listed above, i.e., they determine the geographic location of the stand and particular site factors that may predispose the stand to damage by particular pests.

It was expected that the system would need to solve these problems in consultation mode, as in a phone call, relying on the user to provide whatever data was available. To avoid asking the user all possible pieces of information about the stand, meta-knowledge was used to prune the list of likely pests that need to be considered for a particular instance. This meta-knowledge implements the strategy steps listed above. Then, PREDICT activates the knowledge associated with each pest that has been selected for inclusion in the analysis by the strategy knowledge. For both diagnosis and hazard assessment, knowledge was eventually coded in the form of If-Then rules. In diagnostic mode, each rule set contains three types of rules as depicted in Figure 2-3. These rule sets are evaluated in order to: (1) eliminate a pest from further consideration or if that fails, (2) infer a pest with absolute certainty or if that fails, (3) accumulate evidence for that pest. Without absolute visual verification of a pest (either insect or disease) and its symptoms, a specialist is only able to indicate those pests that *might* have caused the damage.

Under some circumstances, a specialist can also suggest the relative likelihood of each suggested pest. Therefore, it was important to incorporate inexact reasoning and allow for uncertain data. In watching pest experts diagnose problems and by talking extensively with them, this logical structure seemed to mirror their practical methods. For hazard assessment mode, each pest selected is assigned a hazard rating of *none*, *slight*, *moderate*, or *high*. These assessments are made based on: (1) location of the stand and (2) predisposing conditions present in the stand. No attempt was made to relate these qualitative labels to likelihood of infestation or to damage, because extensive historical data were lacking and annual variation in predisposing conditions is so large that likelihood values would be meaningless.

Once this knowledge model was developed, creating each rule set proceeded fairly smoothly. It remained, of course, to acquire the knowledge and to deal with some of the implementation details, such as inexact reasoning. But, by following the knowledge model blueprint, there was never any question as to how things fit together and whether the system would work.

6.1.2 Human Factors Model

The traditional label used for this topic is "man-machine" interface. We have chosen a more gender-neutral term. Regardless of the label, this topic includes all aspects of interaction between user and computer. The computer embodies hardware (screens, pointing devices, keyboards, etc.), software (operating system, knowledge system, etc.), and some conceptual framework for how the user is expected to interact with the hardware and software. These factors constitute some very important design issues and often require extensive amounts of time to implement.

The role of the computer and the role of the human working with it should be clearly specified in the developing system. When the human acts as a monitor of a system's activities, there exists the danger of the out-of-the-loop problem (Gordon 1989). The human in a bystander role becomes detached from the decision process. Then, in a critical situation, when human interruption of a machine process may be necessary, the human may not be able to react properly or promptly because his or her skills have diminished from lack of use. Human performance degradation may not, however, be a problem when the application task is relatively tedious or requires only common sense to react properly.

For most systems there will be some interaction between computer and user. Then both *physical* and *intellectual* interaction must be considered. Physical interaction refers to visual and tactile input and output, usually referred to as the user interface. Intellectual interaction includes the level of understanding about the subject area that must be expected of the user and the level of complexity at which the computer must relate subject matter ideas and concepts to the user. That is, how intelligent must the system be? There must be some reasonable match, anthropomorphically speaking, between the understanding of the user and that of the computer.

The classical knowledge-based system dialogue paradigm, where the system asks a single question and waits for the user to make a response is not always best. We found in RP-FMAS that asking the user the standard set of stand-level data described above in Module 1 item by item, one by one, quickly exhausts the user's patience. It is much more efficient and user-friendly to design an input screen where the user can quickly fill in the known data (Figure 6–12). Of course, the system never knows in advance what combination of data will be entered, if anything at all. This situation requires Module 2, above, to check the data input and query the user further if necessary, this time asking for qualitative input rather than the more precise quantitative input in Module 1. Such a scheme creates more work for system developers but the result is a smarter and more user-friendly program.

Some applications may require graphical components in the interface. A picture can often convey much more meaning than text, especially when the subject area relates to physical environments, such as a bridge or a geographic area or a manufacturing plant. In RP-FMAS, a graphical illustration of the red pine stocking guide (Figure 6–13) succinctly summarizes the progress of the stand as its growth is projected and keeps the user aware of the condition of the stand over time. With the sophistication of object-oriented interfaces, distinct components of a graphic representation may themselves be accessed (often by some pointing device) to perform some action or provide greater detail than the larger graphic (this capability is often referred to as "hot graphics"). Interface building tools are now becoming readily available to aid the programmer in creating attractive and highly functional user interfaces that can include text, graphics, sound, and video.

On the intellectual side of the human factors model, some intelligence must

Please enter as much data as you can. Missing data will be estimated where possible by the system. You may be asked to provide estimates if the system cannot. The better the input data is, the better the prescription recommendations.

PARAMETER	**VALUE**
AGE (yrs)	16
AVG. TREE DBH (in)	3.5
BASAL AREA (ft2/acre)	0
TREES PER ACRE (#)	800
SITE INDEX (ft)	65
DOMINANT HEIGHT (ft)	0
VOLUME (ft2/acre)	0
VOLUME (cords/ac)	0
VOLUME (MBF/ac int'l 1/4)	0

Figure 6–12. The initial date entry screen in RP-FMAS asks for the user's best estimates of several parameters and advises them on the effect of their estimates.

be expected of the system. Just how much a system must understand a user's input and how well-developed its knowledge of the subject area must be are two things that require close examination by the development team. Can the user input English-like sentences with a flexible syntax or must input be very rigorously prescribed? Does the system exhibit any sense of conversational context (ellipsis) or any common sense knowledge? Making erroneous assump-

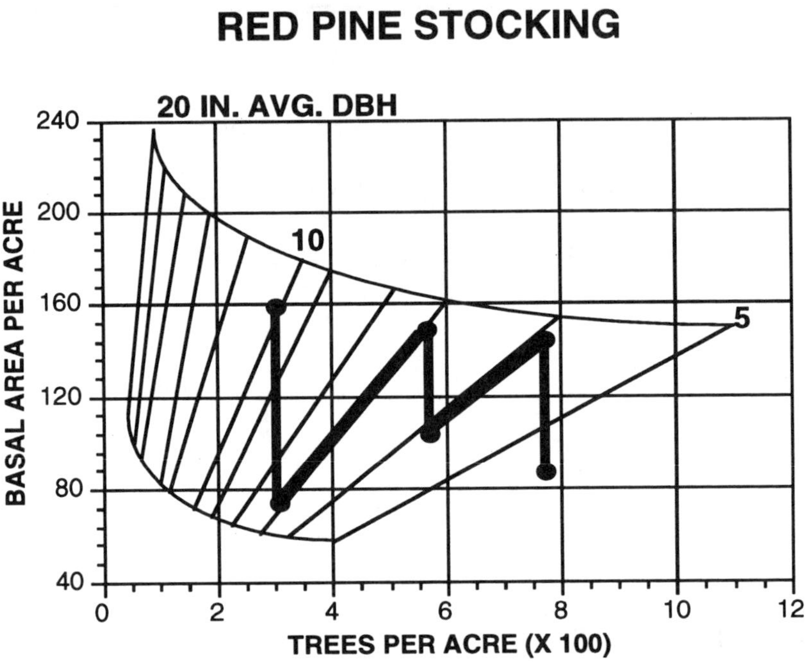

Figure 6–13. The graphical nature of the stocking chart in RP-FMAS enables the user to visually track the progress of a stand through thinnings and growth periods. Stocking curves graphically relate stem density, basal area, and average tree diameter within recommended settings. Bold lines indicate the cycle of growth and thinning for the user's current stand beginning at the upper left with a thinning.

tions about intellectual interaction needs can create large roadblocks for actual deployment of a system.

It is quite common for a human factors model to be designed and implemented generically because most AI programs have very similar needs (Rewerts and others 1990). There must usually be some kind of *ask* facility to manage user input and a *show* facility to present output to the user either on screen or in a file. A standardized human factors model is a necessary and very important part of the overall system. In chapters 8 and 9, we implement two different software interfaces in PROLOG based on the same human factors model of the user group.

6.2 PRE-IMPLEMENTATION TOPICS

The astute reader will have noted that no mention has yet been made of software, i.e. languages and shells needed to implement the system. We emphatically counsel that system planning (chapter 4), problem definition (chapter 4), knowledge acquisition and organization (chapter 5), and design of the knowledge and human factors models (chapter 6) be made without consideration of specific software and hardware platforms that may be used to implement the system. Planning and design issues should be regarded as independent of implementation considerations. The danger is that selection of implementation software and hardware brings with it constraints and conceptual tendencies that have nothing to do with the application problems at hand. It is too easy for developers to try to fit a "round" problem into a "square" implementation system, where the fit is awkward.

Having come this far in the development process, we can finally turn our attention to some implementation considerations such as: selecting knowledge representation scheme(s); selecting inference and control methods; selecting a system architecture; and selecting an AI programming method.

Most knowledge-based systems require three levels of logical organization. It has been recognized by a variety of researchers that successful implementation of any but the simplest knowledge-based systems requires these three separate logical levels to achieve satisfactory performance (Bundy and Welham 1981; Sterling 1984; and Chomicki and Minsky 1985):

1. domain knowledge representation
2. reasoning and user interface control
3. system architecture

Single-level problem solving systems (level 1 only), essentially the early knowledge-based systems, have been found effective only in carefully selected domains where the other two levels are not necessary. In general though, single-level problem solvers have been judged naive and inadequate (Davis 1980, Sterling 1984).

We have visited these three levels before using slightly different terminology. The first level is quite straightforward; it is concerned with domain knowledge representation. The second level focuses on controlling the reasoning process, explanation management, and human factors issues. Within this overall tactical control level, there can also be strategic control at the inferencing level. We have seen this previously in the form of meta-knowledge and the strategic level of our domain ontology. The third level is concerned with system level organization and managing interactions between the major modules of the system. These levels will provide the organizational focus for the following sections as well as for implementation examples in chapters 8 and 9.

6.2.1 Selecting a Knowledge Representation Scheme

Knowledge representation method(s) must be selected based on the knowledge model, human factors issues, and an analysis of the inference and control methods needed. Because different knowledge representations are not mutually exclusive, it can often be advantageous to combine the useful aspects of different schemes. Table 6.1 lists various problem characteristics and suggests particular representation schemes that are often applicable. Most developed applications will possess aspects of several problem characteristics. This indicates that multiple representations could be used effectively.

Most requirements of system design can be accommodated by just about any of the representation schemes, however some of them may be excruciatingly awkward to implement under certain circumstances. So, the ease with which the desired knowledge representation method can be implemented in a particular programming environment will and should influence the choice of implementation software.

We ultimately selected *predicate logic* for the initial prototype (chapter 8) of the Red Pine Forest Management Advisory System. Our knowledge model

Table 6–1. Application characteristics indicate the use of various knowledge representation schemes, although not expressly or exclusively.

Characteristic	Knowledge Representation
Knowledge is largely heuristic and unstructured with few well-understood interrelationships.	Rules
There exist highly complex and interrelated pieces of information of differing types.	Frames, Objects, Networks
It's not important that the application provide the user with extensive decision-making rationale.	Predicate logic
Concepts and entities in the domain are largely independent and have well-defined ways of interacting.	Objects
Additional facts or values may be assumed when certain other facts are given.	Use hierarchical representation: frames, objects, networks
It will be necessary to reason about objects and particular instances of them.	Frames, objects
All possible solutions are known in advance.	Rules
Solutions must be assembled, modified, and refined.	Frames, objects, networks, predicate logic

revealed that our knowledge was relatively unstructured and largely heuristic indicating that it would lend itself well to this implementation approach. In addition, forest management experts had already organized their mental knowledge base using heuristic rules-of-thumb that could readily be translated into symbolic rules for computer processing. Frames and symbolic objects could also have been used to represent the knowledge but we saw no particular gain for the extra conceptual and implementational complexity.

6.2.2 Selecting Control Strategies

A wide variety of paradigms exist for modelling the reasoning process (see section 2.4). As with choosing a knowledge representation scheme above, the choice of a reasoning control strategy is to some degree arbitrary because many different strategies can be used to solve the same problem. *Backward chaining* is a simple and surprisingly effective strategy for solving many problems, particularly those where a large but finite number of goals can be formulated. RP-FMAS falls into this category; therefore, we selected backward chaining as the major control method.

It does not take much experience with rule-based systems to realize that purely backward chaining for reasoning control is not efficient and is often prone to asking questions of the user exhaustively for all possible goals. This plethora of question is often irritating for the user. In a rule-based system using backward chaining, the reasoning process should be augmented by using *meta-rules* and *procedural rules.*

Meta-rules are rules which direct the reasoning required to solve a problem rather than perform the reasoning directly. Because meta-level knowledge deals with knowing what you know, it makes possible a system that has the ability to examine, abstract, and direct its own application (Davis and Lenat 1982). To do this meta-rules often mention other rules in their premise or conclusion or produce assertions that direct the use of other rules. In contrast to meta-rules, those inference steps which perform domain reasoning exclusively are referred to as *domain-level* (Jackson 1986).

Meta-level inferences can be performed to decide at each step which set of inferences should be attempted next (Jackson 1986). As an example, meta-rules are commonly used to prevent domain-level rules from being evaluated (META-RULE 1), or to schedule the order (META-RULE 2) in which domain-level rules are evaluated (Bowerman and Glover 1988).

META-RULE 1
IF NOT branch flagging is present
THEN Don't investigate root collar weevil (*H. rhyzophagus*).

{Note: branch flagging is always present for this insect. Also, the information contained in this rule could be implemented either as an domain-level rule—using *modus tolens* inference to provide evidence against root collar weevil (see Inference in chapter 2), or as a meta-rule (as we have done here) to direct search away from even considering this insect.}

META-RULE 2

IF branch flagging is present

AND the county where the stand is located has a high incidence of root collar weevil infestation,

THEN Investigate root collar weevil before Pales weevil.

Olson (1989) suggests that "control facts" may also be used to manipulate rule evaluation. In this technique, a meta-rule could assert a fact such as, "no pest has been diagnosed as certain." This fact would then appear in the premises of all rules that need to be controlled by the meta-rule. A "control fact" does not really represent any knowledge about the domain, it is used exclusively to orchestrate the evaluation of domain-level rules. This is illustrated in META-RULE 3.

META-RULE 3

IF the certainty of all diagnoses is less than certain,

THEN no pest has been diagnosed as certain.

Domain-level rules may also contain a form of meta-level knowledge. They can do this by having carefully ordered premise clauses (e.g., META-RULE 4) to limit which facts will be investigated in which order. Here, the first two clauses address facts which can each easily be obtained from an initial problem description or from a single rule. If either of these should fail, there would be no need to determine the truth value of the fact "large roots have been girdled", which might not have been examined or which might be difficult to determine. If, on the other hand, the last premise clause were listed first in the premise, it would result in needless effort for the cases when either "branch flagging" or "sandy soil" turns out to be false. Therefore, clause ordering can provide a very useful and powerful means of controlling inference, although it is much more subtle than explicit rules. On the other hand, the subtlety of using rule order for meta-level knowledge is dangerous because it hides control information (careful commenting of the rule base can mitigate some of this danger, however). Inference control using meta-rules is domain dependent, that is, the rules must change with each new application.

META-RULE 4

IF sandy soil is present in the stand

AND branch flagging is present of the trees

AND large roots have been girdled,

THEN conclude root collar weevil damage.

One other aspect of control involves *procedural control.* That is, program control statements similar to those used in traditional procedural languages. Most often these perform some user interface activities, for example, reading from or writing to a file, calling an external program to execute, or displaying some information for the user. These are referred to as procedural because they are not part of inferencing or controlling the search for solutions. Rather, they execute some bookkeeping, file transfer, or screen management functions required by the program and are not related to any problem-solving knowledge. For example, PROCEDURAL RULE 1 illustrates obtaining an integer from the user and saving it into the working memory, clearly procedural activities.

PROCEDURAL RULE 1
IF age of stand is needed
THEN ask *AGE*
AND remember *AGE*

A final aspect of control is when to terminate the reasoning or consultation process. The system designer must set the criteria for termination. One solution state, all possible solution states, or some subset of solution states may be designated as the termination criteria. For example, PREDICT's (Schmoldt 1987, red pine pest diagnosis) termination criteria are: one particular agent has been inferred with 100% confidence, or else all agents selected by the user have been investigated. Generally, when all possible inferences have been made and no solution state has been reached, then a KBS may state that no solution has been identified, or the system designer may elect to have some alternative action performed in the absence of a solution.

6.2.3 Selecting a Knowledge-Based System Architecture

Often a knowledge-based system can become too complex, with too many different, yet related, logical branches, to be comfortably framed in a single module. Also, these different modules, or knowledge sources, may not have identical structure, e.g., as production rules, and may or may not be autonomous (Shoham 1994). A multiple-module knowledge-based system requires a control level which defines an overall system architecture (see section 2.5 for a detailed discussion of alternative architectures). The *blackboard* framework is a very general and flexible KBS architecture (Feigenbaum 1988) and is the one that we selected for RP-FMAS. We can re-organizing the 12 major modules, from the operational structure presented in Figure 6–2, into an implementation viewpoint using the blackboard framework. The resulting scheme appears in Figure 6–14. A PROLOG implementation of the blackboard framework architecture is detailed in chapters 8 and 9.

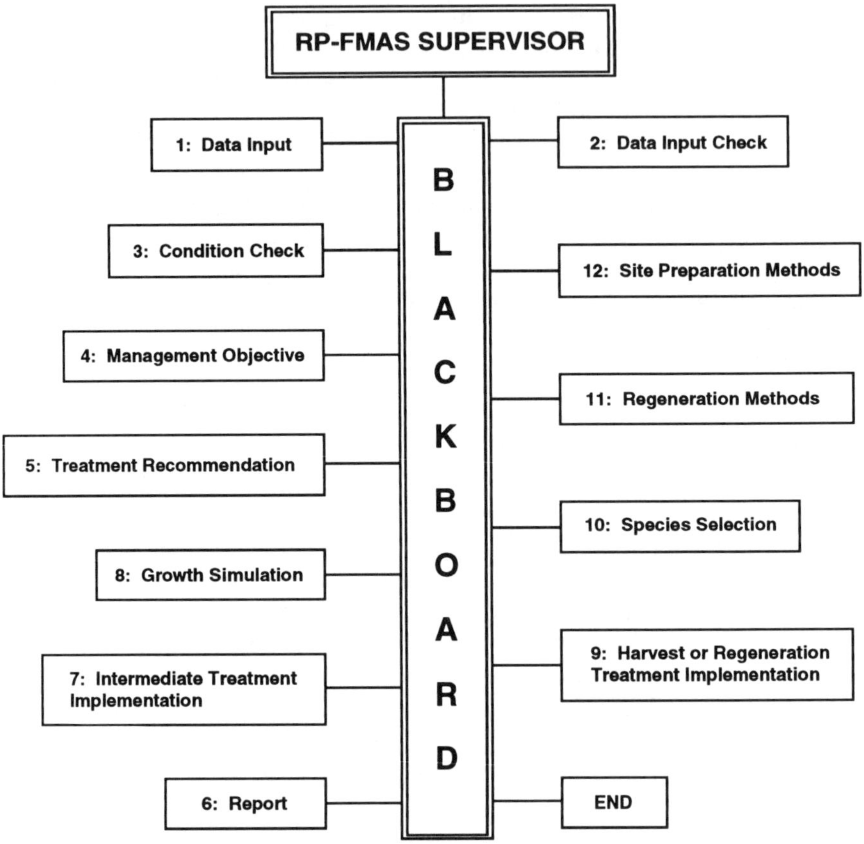

Figure 6–14. The implementation viewpoint of RP-FMAS using a blackboard architecture includes the previous modules as separate knowledge sources. Each module has access to the blackboard knowledge structure. Overall system control is orchestrated by the supervisor module.

6.2.4 Selecting an AI Programming System

Having chosen rules for knowledge representation, backward chaining for inference control, and the blackboard framework as system architecture, we are now ready to face the task of choosing the AI programming software methods to implement RP-FMAS. We face a bewildering number of options that are changing every year. Much confusion exists among developers and users over which software tools to select for a particular knowledge base development project. It is a difficult and yet important decision to make.

KBS can be developed in any language, including FORTRAN, Pascal, BASIC, C, Assembler, etc., the AI languages such as LISP, PROLOG, or one of many specialized KBS development shells. The difference between KBS and conventional programs (CP) is not the language used to build them but whether the software adheres to the concepts that make up the KBS model depicted in Figure 2-1 and discussed in chapters 2 and 3.

6.2.4.1 Conventional Programming Methods. A conventional program (CP) may be operationally defined as software (1) which distributes its problem solving knowledge throughout the computer code, (2) where program statements cannot be understood without reference to which other statements have been executed previously and on the results of these executions, and (3) which solves a problem that can be specified completely, i.e. what is required to be done and how to do it are completely known (Lazarev 1989). This type of programming results from both the type of entities that are manipulated, i.e., data items, and also the constructs built into, and available in, the language.

In the early years of computer programming, people wrote programs that did not differentiate between machine instructions that dealt with control of the solution process and those that specified the problem. Each time the data changed, the computer program had to be changed. As programmers gained experience writing computer programs, they realized that by separating the data *describing* the problem from the control statements describing *how to solve* the problem, then they could use the same program, unchanged, to solve many different problems of the same type. So, data files and data bases were commonly used as separate adjuncts to algorithms. Only the data needed to be changed for each instantiation. This led to increased efficiency by lowering software maintenance and development costs.

During the 1970's, researchers in AI realized that an identical result applied to the knowledge contained in computer programs. This led them to separate domain knowledge from inference and control procedures. This abstraction was a major break from traditional ways of writing program code and serves as one way to distinguish between KBS and CP system. CPs treat knowledge much like the early computer programs treated data. That is, they use knowledge "rules" which are distributed throughout the computer code and mixed inextricably with statements that control the solution process.

Given a "complete specification" of a problem, it is possible to write a CP algorithm. A complete specification is one in which the problem to be solve has been extensively researched and exact solutions are found by just "cranking through the data". A CP algorithm is a precisely defined, step-by-step procedure, that guarantees a correct conclusion will be reached when the correct data have been entered (Harmon and King 1985). The data given to the algorithm to process is free to change at any time (within limits set by the logic of the algorithm). If a CP algorithm can be developed for a problem, it should be used because it will be, by definition, more efficient than a KBS and guarantees a correct solution besides.

6.2.4.2 Knowledge-Based Methods. Unfortunately, completely specified problems constitute only a small minority of all the problems that humans must, and do, solve daily. We use knowledge along with general reasoning capabilities to successfully solve problems that, by and large, we do not understand enough to develop a CP algorithm.

About 20 years ago, researchers in artificial intelligence created the first KBS by discovering methods to make computer control of the solution process generic. The basic theory driving research in KBS is known as the *Knowledge Principle*:

> The power of artificial intelligence programs to perform at high levels of competence on problems of intellectual difficulty depends on the *quantity and quality of knowledge* these programs contain about their problem domain. The reasoning method plays a secondary role (Feigenbaum and others 1988).

By using a symbolic reasoning system to make inferences and forcing knowledge to be expressed in some knowledge representation language, researchers found they could treat knowledge just like data for conventional programs. At the present time, AI represents the leading edge of computer science moving in the direction of increasingly capable knowledge management systems (Bowerman and Glover 1988).

The Knowledge Principle allows us to develop completely general reasoning engines (Figure 2-1). These reasoning engines are, in fact, algorithms that define precisely how to use a knowledge representation scheme, such as rules or frames, to solve a problem. Whether the problem can be solved or not depends upon the makeup of the rules or frames in the knowledgebase. The algorithmic reasoning engine will process rules or frames the same way all the time. A generic reasoning engine can be used to solve many different kinds of problems by exchanging knowledge bases when necessary. In effect, AI researchers invented a totally new concept of what computer programming is (Waldrop 1987). There has been a switch from telling a computer what to do (the *procedural* method) to telling it what to know and let the machine perform the steps necessary to solve the problem (the *declarative* method). This is important because it is much easier to write, read, or modify a knowledge base that is basically a description of the problem that needs solving. Details of how to solve the problem have been extracted from the program code making it much easier to understand.

"Knowledge-based" systems, in contrast to algorithmic programs, are not-guaranteed to be correct. Table 6–2 summarizes many of the distinctions

Table 6–2. The distinction between conventional programs and knowledge-based system programs can be seen by comparing them on various attributes.

Attribute	Conventional Program	Knowledge-Based System
Problem-solving knowledge	In procedural code	Separate from code
Type of knowledge	Algorithmic	Heuristic
Problem-solving paradigm	Mathematics and data processing	Logic and heuristics
Computational units	Data (numbers, strings)	Symbols and belief values
Input/Output	Absolute	Uncertain
Applicability	Well-understood problems	Poorly-understood problems

between KBSs and CPs. KBSs solve problems that are not completely specified. The knowledge and data input are not precise and tidy but rather fundamentally messy and ad-hoc, like the real world. Although KBSs still exhibit brittleness, defined as the inability to deal with the unexpected, they are, nevertheless, more resilient than traditional algorithmic programs. Conclusions are often allowed to contain uncertainty, which provides a measure of their plausibility (Harmon and King 1985). Occasional incorrectness need not be a substantial drawback as long as the reasoning process used by the system can be examined (see 3.1 Explanation and Justification). For our purposes, we now eliminate the non-AI languages from further consideration and concentrate on comparing the AI languages and Shells to each other.

6.2.4.3 Comparing AI Programming Options. For those wishing to develop knowledge-based system applications, there are 7 development strategies to choose from (Table 6–3):

1. First are "low end" expert system shells which can provide a well supported working environment, so the user can concentrate on the problem and not the tools. Many people have their first AI experience learning and using a simple expert system shell. But it is a fairly typical experience that using a "low end" expert system shell to solve even moderately complex real-world problems brings the developers quickly to the boundaries of the capabilities of the shell. Features and capabilities are often needed that are simply not present in the shell and, lacking access to the proprietary source code, the developers are powerless to implement them. The distribution rights of finished systems may be restricted or charged a royalty. General purpose expert system shells are simply not suited for all applications (Helm and others 1985). Experience shows that in many circumstances, a knowledge-based system must be designed specifically for an application because required domain specific features are not generally found in many commercial products.

Table 6–3. An evaluation of knowledge-based system development environments based on four criteria: development speed, development freedom, development power, royalty-free delivery.

	Development			Royalty-free
	SPEED	POWER	FREEDOM	DELIVERY
Low-end Shells	+	−	−	−/+
Low-end Shells + Language	+	+	−	−/+
High-end Shells/Toolkits	−	+	+	−
LISP	−	+	+	−/+
PROLOG	−	+	+	+
LISP + ES Toolkit	+	+	+	−
PROLOG + ES Toolkit	+	+	+	+

[+]evaluates well on criteria

[−]evaluates poorly on criteria

[−/+]depends on particular product examined

Assumptions:

1. development by a "typical" programmer
2. delivery of the finished product for micro-computer hardware
3. OPS5, SMALLTALK, C++ not evaluated

Development SPEED = how much effort the "typical" programmer must expend in the development of a given expert system.

Development FREEDOM = how much freedom the developer has to configure a set of toolkit modules specifically for a particular domain problem, i.e. maximum choice of explanation methods, control methods, inexact reasoning methods, knowledge representation methods, etc.

Development POWER = how much conceptual and practical support the development system provides for unusually large and/or complex expert systems.

Royalty-free DELIVERY = no run-time charges need to be paid in order to distribute applications to users

2. To overcome the weaknesses of low end shells, several manufacturers of "low end" shells allow developers to write their own modules in a programming language, typically in PROLOG (Intelligence/Compiler by IntelligenceWare, Inc.; IBIS-Plus by Intelligence Manufacturing Company) or C (CLIPS by NASA/ Johnson Space Center). Using these extensible shells, the developers must view the proprietary shell as the global program control environment within which their own modules may exercise local control. However flexible this may be, the global control environment is still inaccessible to modification by the knowledge-based system development team. Nevertheless, this option can be attractive if the trade-off is favorable between increased program development speed and complexity realized by external coding and possibly messy semantics of a shell's knowledge representation language.

3. Another approach avoids the frustration of feature limitations by choosing a "high end" shell containing powerful and numerous features. However, the "high end" shell languages are at least as complex as PROLOG or LISP to learn to use. They must be complex because they contain features powerful enough to tackle various types of complex problems (Knaus and Blecker 1990). Flexible knowledge-representation formalisms are coupled with powerful reasoning techniques. Even if any single problem requires only a subset of these features, there is no choice offered to streamline the "high end" shell. This overhead translates into high software costs, high hardware costs, and, in general, a lack of easy portability. Another problem is that the current generation of "high end" shells severely limit access to control of the inferencing process (Cherubini and others 1989, Vrba and Herrera 1989). Finally, once a knowledge-based system is running there are portability problems. Typically, rather high distribution costs must be incurred due to the need to use proprietary software. Some "high end" systems offer conversion options to the "C" language, but this procedure is not necessarily easy or inexpensive either. Those "high end" shells that are tied to specialized runtime systems often make it difficult if not impossible to interface to external databases and spreadsheets where much of the real-world information is stored (Vrba and Herrera 1989).
4. The LISP programming language has traditionally enjoyed great popularity in the United States AI research community. It is flexible and has great power to express different knowledge representation formalisms as well as inferencing strategies. It is a relatively low-level language in that its relationship to PROLOG is analogous to the relationship that C has to Modula-2. A given programming task will often take longer in LISP than in PROLOG, but then the developer has more freedom of expression in LISP than in PROLOG. Another disadvantage of LISP is hardware related. Modern implementations of LISP for micro-computers are still relatively slow, cumbersome, memory-intensive, and most importantly, interpreter-bound rather than compiled. However, more and more implementations of LISP, e.g., Scheme and Allegro Common Lisp, provide run-time generators that produce executable code from LISP programs.
5. Expert system toolkits have been developed for LISP to reduce the development time and make LISP more attractive as a development environment (Tong and others 1987, Watanabe and others 1989). This recent development, along with more common true LISP compilers that can generate executable code, enhance LISP's position as both a development and delivery language for KBS applications.
6. PROLOG provides great flexibility and no distribution problems, but in the context of an immature language with relatively poor expert system development libraries and software toolkits. Toolkits (see the one developed in chapter 9) and libraries provide encapsulated, highly modular program segments that perform prespecified and "standardized" operations. The IMSL FORTRAN subroutine library for mathematical and statistical computing comes to mind as a well developed example of a software toolkit. Unfortunately, there is a large up-front investment not only in learning to use PROLOG well but also in developing entire knowledge-based systems from low-level predicate structures.

7. Expert system shells in PROLOG have been published by several authors and are generally available for anyone to use and modify (Merritt 1989, Malpas 1987, Schnupp 1987, Ross 1989, Covington and others 1988). These authors would be the first to admit that their shells are primarily intended as "teaching" systems with limited features rather than full development environments. At least one PROLOG expert system shell with source code is commercially available (Berkshire Software Company 1987). But it, too, is a rather limited shell. Given time and money, it is certainly possible to study the source code, disassemble the useful pieces, and then try to reassemble a new expert system shell in PROLOG to meet the needs of a particular application. In practice, however, few beginning to intermediate level PROLOG programmers find this a realistic option. Several sophisticated expert system development toolkits for PROLOG have been reported in the literature (Oliveira 1984, Lee 1986, Torasso and Console 1989). Unfortunately, these toolkits are limited to research labs and are unavailable for general use. Having made a theoretical comparison, let's next review what developers actually chose to do.

6.2.4.4 Programming Systems Chosen By KBS Developers. Rauscher and Hacker (1989) reported on the results of a world-wide survey of 74 AI projects in natural resource management. It is most instructive to examine the development software from several viewpoints. First, by major subdivisions:

AI language	46%
AI shell/toolkit	38%
Conventional language	16%

It can be seen that the AI languages (PROLOG, LISP, SMALLTALK) were used in about half of the projects reported. That means that many developers have decided to invest the necessary resources to learn a new programming methodology. The language of choice among this group was PROLOG:

PROLOG	77%
LISP	18%
SMALLTALK	5%

In the AI shell/toolkit category, EXSYS and INSIGHT were the most frequently used:

EXSYS	22%
INSIGHT	13%
CLIPS	9%
GOLDWORKS	6%
VP-EXPERT	6%
BRUCE SHELL	6%
OTHERS	36%

The range of shells/toolkits reported was very large, so 2 projects out of a total of 32 resulted in a 6% representation. Obviously, the sample size is small and the results are unstable. Only 14 projects used conventional computer languages as a main or supporting language for their AI applications:

C	43%
PASCAL	36%
FORTRAN	21%

The majority of these AI applications were developed for IBM PC-compatible microcomputers:

IBM PC-compatible	60%
SUN	11%
MACINTOSH	11%
OTHERS	18%

Mainframe and most mini-computers are less frequently targeted for AI software delivery systems by this group of developers. Because the number of AI programming environment options is steadily increasing, these results would likely be different if another survey was conducted today.

At least two other surveys of this nature have been conducted more recently. The Operational Research Society of the United Kingdom conducted a survey of the application of AI techniques among society members (Doukidis and Paul 1992), 265 of 1038 members responded. The survey results showed than 62% of the respondents used AI shells, 30% used PROLOG, and 18% used LISP (note: a respondent could use more than one tool). PROLOG was clearly the dominant European AI language although LISP was quite popular. Durkin (1993) summarized over 2500 expert systems reported in the literature across all fields of study. His results showed that about 45% of developers used some kind of hell, 25% used LISP, and 10% used PROLOG. While there are some differences between these 3 surveys, the common theme is the preference for shells and for LISP and PROLOG to maintain their popularity.

In summary, selecting a programming tool for implementing a knowledge-based system depends as much on personal preferences as it does on the characteristics of the available shells, environments, and tools. To a very large extent there is no absolute right or wrong choice. Previous experience and accomplished skill with one of the traditional languages, say FORTRAN or BASIC or C, may more than offset the advantages of AI-related tools, such as LISP or PROLOG or an AI shell. After all, useful knowledge-based systems have been developed using all of these tools.

Every developer of AI systems must eventually choose the hardware and software combination that best meets their needs. Regardless of the sometimes acrimonious arguments about which language or shell is "best", any solution is good as long as it produces a good product. For RP-FMAS, we chose the blackboard framework for strategic control to partly hedge our bets. With the blackboard framework coordinating the activities of 12 separate executable modules (Figure 6–14), it does not matter which language is used to construct these modules. In RP-FMAS, most are in PROLOG, our AI language of choice, with BASIC as our language for performing algorithmic/procedural tasks. We could just as easily have written some modules in C or PASCAL or FORTRAN or LISP.

For the implementation examples in Chapter 7, 8, and 9, we chose PROLOG. Ideally, we would like to have presented the same examples in PROLOG, LISP, and in at least one of the more popular shells. Unfortunately, this was not possible and we were forced to make a choice. So then the question might be asked "Why PROLOG?". Section 7.1 in the next chapter addresses this question.

6.3 PROTOTYPING

Sometimes prototypic implementations are constructed to test and supplement the feasibility of an idea. These proof-of-concept systems may be essential to procure necessary development resources—both interest and funding may be at stake. Where extensive resources are required for complete implementation, a cursory test version can indicate whether such expenditures are warranted. Under these circumstances there may not be sufficient time to devote to well-structured knowledge acquisition. It may only be given brief consideration due to limited time and little knowledge source availability.

We note below that a "flashy" user interface can sometimes be a more prominent selling point than sophisticated knowledge structures or inferencing methods. When introducing and marketing the idea of KBSs, an early proof-of-concept system can be invaluable. For demonstration purposes, it may be useful to attach a flashy interface to an empty knowledge base, which also means that an expert's valuable time is not required. A proof-of-concept system can readily be implemented using either one of the approaches to Prolog programming presented in Chapter 8 or 9.

6.3.1 Different Prototyping Styles and Their Uses

When system design and construction begins in earnest, several different classes of prototypes are possible. Each has advantages determined by the end use of the prototype, the size and complexity of the application, and the resources available. These prototyping styles have been described by several authors (Floyd 1987, Walters and Nielsen 1988); we will refer to them as plenary, horizontal, and vertical prototypes.

6.3.1.1 Plenary Prototypes. Occasionally it may be possible to prototype the entire application at once. This is often the case when the application covers a very narrow, carefully-defined problem and it involves little complexity. For example, in Figure 6–15a, the goal is to diagnose three possible defoliators of red pine. In this case, the prototype contains the complete or plenary knowledge about the types of problems to be solved. Because the entire application is within focus and is tightly restricted, the selection of knowledge representation and search methods becomes a relatively easy task. There are fewer choices to make because there is less complexity in the application that needs to be accommodated.

Also, an attempt should be made to use the same software tool both for development and delivery in this kind of prototype. This will help avoid any duplication of effort necessary for translating a prototype into a finished product. Size and complexity limitations inherent in a small problem area mean that the initial and advanced prototyping stages dissolve into a single stage. All the extensions that would normally occur in an advanced version are included in the initial prototype.

Although a single prototype stage has been described for the plenary prototype, it consists of numerous revisions to a first-pass prototype. Walters and Nielsen (1988) advise developing a broad but shallow version of the system at first. This will flesh out the major structural components (e.g., knowledge representation, control, and inference). System evaluation of each prototype should then indicate whether these designs and implementations work adequately. Later prototype versions can complete the lower level details of each subcomponent. The first-pass version and all iterations of it should perform several useful functions:

1. they should allow developers to assess how well the evolving system conforms to the specifications set out earlier;
2. they should facilitate dialogue with the expert(s) about their knowledge or, alternatively when no experts are used, they should provide a compendium of knowledge acquired to date; and
3. they should permit the identification of performance errors or user-interface problems.

Early prototype versions should focus more on the first item, completing specifications, while later versions should be more concerned with knowledge and human factors issues (2 and 3).

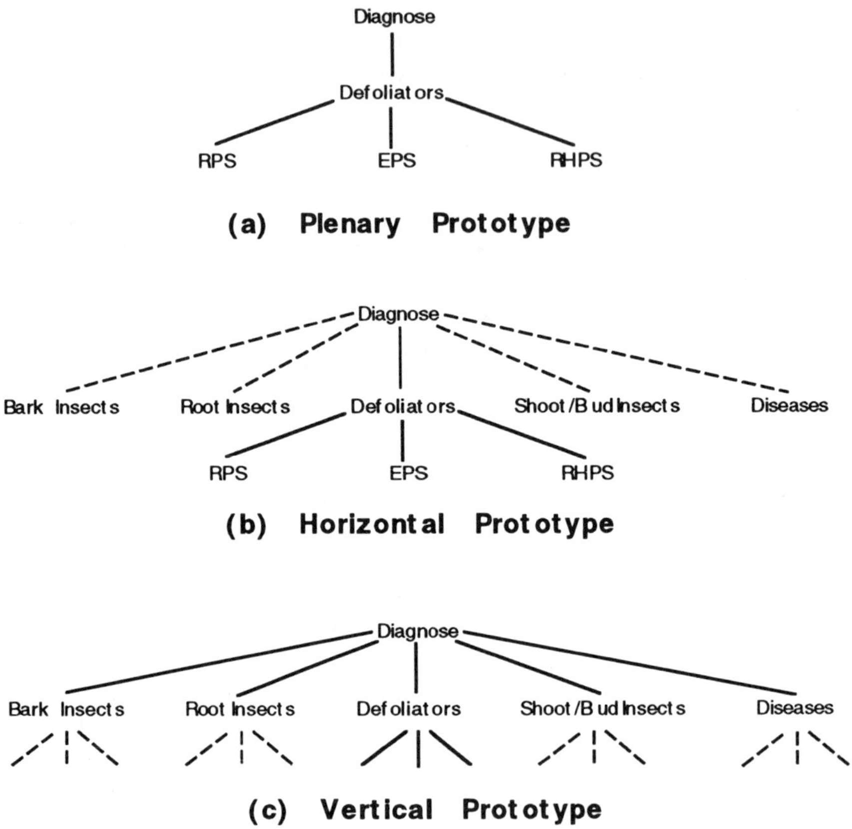

Figure 6–15. Three different prototyping styles, plenary, horizontal, and vertical, are depicted. Dashed lines indicate components that are not fully developed in an initial prototype. Those ancillary components can be developed separately (horizontal) or by other developers (vertical).

Besides being useful for small applications, plenary prototyping may also be useful in larger applications that consists of two or more relatively independent components. Each separate component would then be treated as autonomous and complete in itself. The vertical prototyping method, described below, can incorporate this plenary approach very easily.

6.3.1.2 Horizontal Prototypes. As applications become larger and more complex it becomes increasingly difficult to prototype an entire system at once. Yet, often it is desirable to maintain a broad view of the larger problem during system construction. Figure 6–15b depicts an expansion of the red pine insect diagnosis system above. Here, diseases, bark insects, root insects, and shoot/bud insects are diagrammed in addition to the defoliators. Only specific pests in the defoliator category are implemented initially, however. In circumstances such as this, it can be useful to develop a skeletal system. Here, all classes of red pine damage are addressed in overview at the outset, but initially only one or two are developed in their entirety. Because of the increase in application size and complexity, the two stages of prototyping described in Chapter 4 become necessary. An initial prototype stage would conclude when a skeletal framework and one or two components were completed.

The skeleton developed initially in Figure 6–15b provides a framework on which the remaining aspects of an application can later be attached. A skeletal version includes all the implementation methods (e.g., knowledge representation) thought to be necessary for all aspects of the problem. Hence, there must be at least some cursory consideration of those system components that are not built completely during the first pass. When implementing and running a skeletal system, default values may be used to represent those components that remain uncompleted. Consequently, during evaluations of horizontal prototypes only partial examples of real-world problems can be used. The ability to perform modular tests of completed components can be an advantage because large and complex test cases need not be collected nor their results analyzed. After an advanced prototype has been built, the three prototype functions noted above can be used to guide improvements toward a final version.

6.3.1.3 Vertical Prototypes. A third method of prototyping, the vertical prototype, can be useful for large, complex applications in which various subcomponents of the larger problem are independent. Again, we can view the diagnosis of red pine pest problems in this framework (Figure 6–15c). Each of the five components identified in this example can be developed relatively independently of the others. These independent and separate areas may then be combined into a final product later. Because autonomous subcomponents cannot, in reality, be completely independent and yet be part of the same application, some amount of interaction between them will exist. Any overlap should, however, be confined to the linkage mechanisms that allow the separate components to communicate and share information. There needs to be a supervisor program that performs this linkage and coordinates the operation of the separate knowledge systems (see section 2.4 Knowledge-Based System Architectures).

This vertical approach allows the developers to dissect a large problem into smaller, easier to manage units. These may then be distributed to any number of widely dispersed developers for completion, if desired. An advantage of this approach is that knowledge from geographically or ideologically or technologically dispersed groups can be included.

Each small, independent knowledge base may be developed using one of the two previous prototyping methods. Often, a plenary prototyping method can be used because the full application has, presumably, been divided into components that are sufficiently small. Unless user-interface requirements are closely specified beforehand, each module is likely to support its own. Walters and Nielsen (1988) suggest that in applications where the interface is a critical aspect of the system one of the previous prototyping methods should be used. Both of those methods bring the man-machine interface to testing and evaluation sooner and provide for a single, uniform interface.

Uniformity can be much more difficult to accommodate when each subcomponent is separately developed as in a vertical prototype. However, it can be possible to incorporate a critical man-machine interface into the vertical prototype also. To do so, specifications must be carefully defined a priori or, alternatively, the interface may be imbedded in the supervisor program.

6.3.2 Prototype Development Considerations

There are also hardware and additional software considerations that must be addressed during prototyping. What specific hardware and software will be used for initial coding? This decision must consider: (1) human and financial resources, (2) efficiency needs for the hardware and software (i.e. how fast and flexible must the computing environment be), and (3) extensibility of the initial prototype to advanced prototype or delivery system. Let's look at each of these issues in a little more detail. Their order below does not imply any ranking of importance or any required sequence.

Not all of the problem-solving methods described in Chapter 2 are easily implemented in all prototyping environments. Some shells and some languages may make implementation of particular methods difficult or awkward. Also, some methods require more time, programming expertise, and money to code in a particular programming language or shell. If any restrictions in these areas are expected, it may be necessary to limit one's choice of implementation techniques on the basis of implementation software. If not, the prototyping environment (shell or language) should be considered *after* representation, reasoning, and interface methods have been specified.

6.3.2.1 Human Resources. Different software environments require different levels of programmer expertise. Many of the shells which are available are advertised as very easy for computer programmers to use, and they may be easy for non-programmers, also. However, as shells become more powerful they also become more complex. The most complex shells often provide access to the underlying language in which they are written. This puts them in the same class as general purpose programming languages. *It is not advisable to begin learning a programming language at the same time as developing a large and complex* *application.* Rather, programming expertise should be sought if it becomes apparent that a substantial coding effort will be needed. When the developer and programmer are the same individual, then software that is already familiar to the developer should be used if possible.

When an application relies on the expertise of human specialists, then these individuals are part of the human resource, also. Experts were mentioned in Chapter 4 during the system planning phase and were discussed at length in Chapter 5 with regard to knowledge acquisition. We have noted in those chapters that experts should be knowledgeable, available, interested, and committed. They are critical both as knowledge sources and as valuable test pilots for successive prototypes.

6.3.2.2 Software and Hardware Resources. Initial purchase prices of development environment software vary by several orders of magnitude (Harmon and others 1988). In general, the flexibility and complexity of this software increases with increasing price. At the same time, a flat learning curve also accompanies highly complex software, thereby increasing the length of time required to reach an acceptable level of proficiency. However, the relative price/performance ratio is not the same for all products (Harmon and King 1985) [Note: while this reference and the software it examines may be dated, the basic conclusion here remains valid for current software tools, as well.]. Consequently, some software tools provide greater efficacy for the dollar. General purpose programming languages have the greatest flexibility and are generally the least expensive. But they often force the developer/programmer to create all necessary code for knowledge representation, inference, and user interfaces. PROLOG is an exception in that it provides a built-in inference mechanism. However, unless native PROLOG logic programming methods are used (see Chapter 8), it may be necessary to create or borrow a toolkit of KBS development components (see Chapter 9).

One of the critical considerations for development software should be: does the development environment support the features that the application requirements, knowledge model, and human factors imply? If the answer is no, then some of these specifications must be changed or more powerful software should be sought. It should be emphasized that system developers should avoid extensive sacrifices of system requirements to accommodate inexpensive software. The result of such compromises may be a final product, i.e. system, of limited usefulness.

If prototyping software is expected to be used for subsequent delivery, then those costs must also be figured into the final cost of the software. Most shell vendors have separate software environments for development and for delivery. Delivery software allows the developer to distribute his or her knowledge base with a version of the development environment that executes a knowledge base but does not allow modification of it. These run-time, delivery systems are copyrighted and must often be purchased under either a per unit or an unlimited copy agreement. On the other hand, languages are either interpreted (similar to the development environment of shells) or compiled (i.e. capable of creating executable, stand-alone modules). Languages that can be compiled allow a developer to deliver run-time versions of a system without incurring any additional costs associated with delivery software.

An assessment of financial resources needed for development should also include hardware, of course. Specific hardware required by development software products has become much less of a concern in the last few years. Most software runs on a variety of hardware platforms with fairly good compatibility between different machine versions. Even the large shells have scaled-down versions that run on small workstations or personal computers. If the chosen development machine is already in place and does not represent a special purchase, hardware becomes less of an important cost consideration.

In some instances, more important concerns might be the speed and utility of a development machine. As the power of computers continues to increase, while prices fall, concerns about access to a sufficiently powerful development machine become a fading issue. With more powerful hardware comes more sophisticated software (although, there is generally a time lag involved). Most computer operating systems now typically provide a sophisticated windowing and graphics environment. These are much easier and more pleasant for system development; a large amount of information can be displayed on the screen at one time and the programmer can move easily between different displays. Also, relatively spontaneous response to program changes afforded by such environments can make development more efficient. This is particularly noticeable in interpreted environments where a program may run somewhat slowly. On the other hand, program changes can be incorporated instantly without recompilation and re-linking (a time consuming process).

6.3.2.3 Extensibility of the Prototype. When developing the initial prototype, some thought should be given to final system delivery and how a prototype might be upgraded to a full system. If the initial prototype environment is drastically different from that which is planned for an advanced prototype or delivery system, there may be a substantial re-tooling cost that needs to be accepted and planned for. On the other hand, if few non-standard knowledge representation and inferencing features are used in the initial prototype, it is likely that extensions and re-coding will not be particularly expensive. Oftentimes a small shell on a PC may be useful and sufficient for an initial prototype, whereas a language and a workstation may be better able to accommodate the needs of an advanced prototype. The rapidity with which an initial prototype can be developed using a small shell may more than compensate for any time and expense required to re-code it for an advanced version. Serious KBS developers should not be reluctant to utilize a variety of tools throughout a system development life cycle.

6.4 SUMMARY

We have presented the concept of a knowledge model and illustrated it using RP-FMAS. We have discussed the importance of the user interface and human factors engineering. Guidance was offered in the selection of knowledge representations, control methods, and knowledge-based system architectures. The available AI programming options were categorized and generalized and the results of recent surveys of actual AI applications were reviewed. Choices in these areas were justified for the implementation of the extended example of RP-FMAS.

Having settled on PROLOG as the major AI implementation language for RP-FMAS, the next chapter serves as a brief introduction to programming in PROLOG. Chapter 8 and 9 present different implementations of the knowledge model for RP-FMAS described in this chapter.

CHAPTER 7

Programming Knowledge Systems in PROLOG

Make rules, not programs!

Tore Amble

After planning an application and developing the *problem definition* in chapter 4, acquiring the domain knowledge from sources in chapter 5, designing the *knowledge model* and *human factors model* in chapter 6, and selecting PROLOG as our major AI development tool, we need to pause before continuing to develop prototype systems in chapter 8 and 9 (Figure 8–1). In the present chapter, we discuss the strengths and weaknesses of PROLOG and provide an introduction of the main elements of the PROLOG language. If the reader has never programmed in any language before, we suggest this chapter be skipped because it will convey little useful information. On the other hand, readers who are programmers in other languages, should be able to understand the essence of PROLOG from this introduction. Those already familiar with PROLOG programming may regard this chapter as a useful review.

Before we introduce the PROLOG language elements and their use for KBS development, let us briefly review the major advantages of this language.

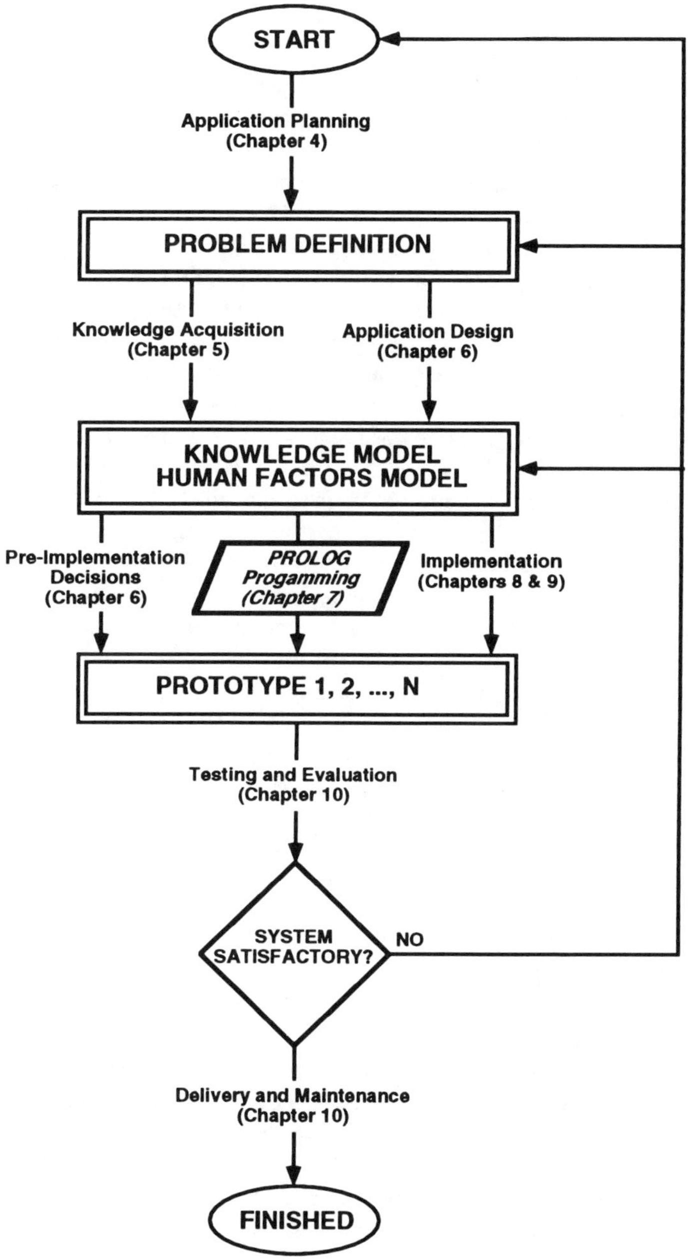

Figure 7–1. Before examining knowledge system implementation using PROLOG in chapters 8-9, we provide a brief overview of PROLOG in this chapter.

7.1 WHY PROLOG ?

PROLOG, invented in 1972 by Alain Colmerauer and his colleagues at the University of Aix-Marseille, in Marseille, France, is an implementation of logic programming (Hogger 1984). A syntactic variant of Marseille PROLOG called Edinburgh PROLOG (Clocksin and Mellish 1981) has become a de facto standard to which many commercial PROLOG implementations adhere. The name, PROLOG, stands for *pro*gramming in *log*ic. PROLOG programming has a significantly different "feel" to it than programming in such languages as FORTRAN or PASCAL because PROLOG's syntax and semantics are based on logic rather than on algebra and mathematics.

The evolution of computer languages has moved away from low-level languages, in which a programmer specifies *how* a problem is to be solved (procedural programming), and toward high-level languages that focus attention on representing *what* is to be solved (declarative programming). PROLOG is a language that encourages the programmer to describe situations and problems rather than worry about how to control the solution process step by step. PROLOG is well suited as a tool to write knowledge-based systems because it has a built-in, goal-driven inference engine as part of the language. Furthermore, it is powerful and flexible, and has implementations that can produce fast, compiled, executable programs that can be distributed royalty-free (Arity Corporation 1988, Logic Programming Associates 1994).

PROLOG offers several major advantages to KBS developers:

1. Any knowledge representation method can be implemented in PROLOG and used to describe the knowledge required to solve a problem. Because the major bottleneck in KBS development is the building of the knowledge base, it stands to reason that the gap between an expert's representation of the domain knowledge and the representation in the knowledge base should be minimized. PROLOG offers maximum flexibility for knowledge representation.
2. Clauses are independent, with separate logical meaning. This makes programs modular and understandable and they are more easily maintained.
3. Inference and control mechanisms are flexible, whether you use the default mechanisms, which are a part of the language, or build new ones yourself.
4. Relatively inexpensive and fast interpreters and true compilers are available for all the popular microcomputers, minicomputers, and mainframe computers. This makes PROLOG programs portable when delivering your own applications.
5. Excellent facilities exist for interfacing to the host's operating system and to other computer languages, databases, spreadsheets, or data files.

There is no doubt that learning a new language requires a large commitment of time and energy. However, if a general purpose programming language is important for the application development environment, we believe that this time and energy is well spent learning PROLOG for the development of KBS.

7.2 INTRODUCTION TO PROLOG

This section delves into writing actual program code for implementing knowledge systems in PROLOG. It is intended as a review of the language and not a primer for readers unfamiliar with PROLOG. The basics of the language are covered and some natural resource examples are included. Those readers with no prior experience with PROLOG are referred to an excellent tutorial by Young (1988) and books by Townsend (1987a), and Smith (1988). Books at the intermediate skill level are by Yin (1987), Rowe (1988), Weiskamp and Hengl (1988), Townsend (1987b), Covington et al. (1995), Marcus (1988), and Shafer (1987). Advanced books are available from Malpas (1987), Sterling and Shapiro (1986), Schnupp and Bernhard (1987), Amble (1987), and Bratko (1986). For more details on building KBS using PROLOG, we highly recommend the book by Merritt (1989).

7.2.1 Basic language elements

The syntax of PROLOG is relatively simple and based on the concept of defining a problem as "objects" and their relationships. (Note: Our use of the term "object" here does not refer to the objects of object-oriented programming.) A PROLOG program consists of a set of assertions together with a set of goals. The assertions may be thought of as the axioms of a theory stated in first order predicate logic and the goals as theorems to be proved true or false (Lazarev 1989). In another sense, PROLOG programs may be viewed as consisting of assertions as facts and rules (clauses) and questions to be answered, i.e. goals to be solved. PROLOG "rules" are generally distinct from rules in the rule base sense, although they could be the same, as we'll see later. From a programming standpoint, it may be most useful to consider a PROLOG program in this light, such as:

```
<fact>.
<fact>.
<fact>.
<fact>.
<rule>.
<rule>.
<rule>.
<goal>.
<goal>.
```

Part of the PROLOG programming environment is an inference engine that reads each fact, clause, and goal in a PROLOG program. The inference engine then attempts to prove each goal by syntactically matching it with a specified fact or with new facts deduced from the rules. In the remainder of this section we will often refer to PROLOG by "PROLOG does this" or "PROLOG does that". When we use this terminology we are referring to the actions of the inference engine as a sort of pattern-matching "being" that performs a few simple operation over and over.

Let us first examine the syntactical elements used to make the assertions. *Terms* are the lowest level building blocks in PROLOG. There are three types of terms—constants, variables, and structures—out of which all program data, and programs themselves, are constructed.

7.2.1.1 Constants. Constants are defined objects. The three main types of constants are atoms, strings, and numbers. Atoms are symbols which may contain letters, digits, or other alphanumeric characters. If an atom begins with an uppercase letter or a digit then it must be enclosed in single quotes or double quotes depending upon the particular PROLOG implementation used. Single quotes distinguish some atoms from variables or numbers. Examples of atoms are:

```
forest        '2.3'        'Red pine'        →        cubic_foot
```

To improve the legibility of PROLOG programs, it is useful to give atoms meaningful names like, *cubic_foot*. Because atoms must be contiguous characters (no spaces), underscores are used for readability to separate the English words composing an atom.

Clocksin and Mellish PROLOG treats string constants as lists of ASCII characters; this convention is awkward for a programmer. String constants are included in a number of PROLOG implementations as an extension of Clocksin and Mellish PROLOG to enhance the ease of working with text-based constants. For example, in Arity PROLOG, strings are textual constants enclosed in dollar signs, as for example:

```
$The basal area is very high.$
```

Numbers are the third type of constant and include both integers and real numbers, for example:

```
0        -77        1.234
```

7.2.1.2 Variables. Variables are terms that have no pre-defined value. They are represented in PROLOG by symbols that start with capital letters or the underscore character.

```
Basal_area        _Mean_dbh        X        _
```

The underscore by itself is a special variable called the "anonymous" variable. The anonymous variable is useful in situations where a variable must be specified as part of the syntax, but as a programmer you don't care what the value of that variable really is.

One of the powerful mechanisms in PROLOG for deducing new facts is variable instantiation. Goals are questions that PROLOG tries to answer based on assertions that have been previously specified. As solutions are sought for these goals, *variable* terms within a goal are instantiated with values from the list of known assertions that match the goal. These variables and their values can then be used in subsequent goals. This allows information to be passed between goals much like parameters are passed between procedures in traditional programming languages. At times it may not matter what value a variable has been instantiated to, we may not intend to use that variable in other goals. In those situations the anonymous variable could be used as a place holder.

As an example of variable instantiation, consider the following facts which are present in the system. Although we have not talked about PROLOG *clauses* yet, each of these facts is a clause. Each fact stores the age for a particular forest stand. For example, the first fact says that "the age of stand 564 is 25 years."

```
stand_age(stand564, 25).
stand_age(stand660, 30).
stand_age(stand329, 10).
```

Now, we may wish to ask the system whether there exists a stand whose age is 30 years. To do so we would pose the following goal to the system.

```
stand_age(X, 30).
```

PROLOG's pattern matcher looks for facts that match our goal statement. In the process, it instantiates the variable *X* with the value *stand660*. We can then use the instantiated variable *X* in subsequent goals to do certain things. This pattern matching process underlies much of what PROLOG does.

When knowledge base rules contain variables rather than constants, the same KB rule can be used to process any number of individual cases. Although we haven't talked about rules in PROLOG yet, here's an idea of how variables can be useful in rules. For example:

RULE 1

IF	*STAND* is a stand of type *TYPE*
AND	(*AGE* of *STAND*) > (*ROTATION_AGE of TYPE*)
THEN	Harvest *STAND*

where the capitalized words are variables (except IF, AND, and THEN). This way of writing a rule is much more generally useful than:

RULE 2
IF The stand is a red pine stand
AND age > 120
THEN Harvest this stand

We can use RULE 1 for any type of stand and any current age and rotation age if we can look up the specific values for those variables in a database or spreadsheet file. RULE 1 uses unification and *modus ponens* as inference methods, whereas RULE 2 uses only the latter method. The combined effect of the former results in a more generally applicable rule, one that can obtain its facts external to the knowledge base. By generalizing a KBS in this way, we can use the same knowledge base to give us advice for many different types of forest stands.

7.2.1.3 Structures. Structures are general purpose data types used to express a relationship between other terms (objects). The general form of a structure is:

$$f(a_1, a_2, \ldots, a_n),$$

where f is called the *functor* or *predicate* and the a_i represent its arguments. The name f of a functor must be an atom. The number of arguments that a functor possesses is called its *arity*. Together, the functor and its arity uniquely define the structure, this is denoted as *f/n*, e.g. **grows_on/2** in the example below. The arguments can be any type of term, either constants, variables, or other structures. This feature allows for nested structures and an endless variety, for example:

```
tree(Species, Age, Height, Volume, biomass(Foliage, Stem, Roots)).
grows_on(Species, Site).
```

The above examples are merely declarative statements in a particular subject area. PROLOG attaches no meaning to these terms, when presented with a goal to prove, it methodically searches for terms that match patterns in the goal. PROLOG is merely a *syntactic pattern matcher*, but it does that task very well.

In most instances, goals to be solved will contain variables as in the examples above. PROLOG can only prove goals that eventually reduce to facts. Facts can only be constants or structures containing constants. Therefore, in order to prove a goal containing variables by using facts containing no variables, PROLOG must associate, or *unify* variables and with facts. Suppose we have the following assertions in our program:

```
grows_on(red_pine, sandy_soil),
grows_on(red_pine, mesic_soil),
grows_on(white_pine, mesic_soil),
grows_on(jack_pine, sandy_soil).
```

These facts represent statements in our domain, that is, "red pine grows on sandy soil", "red pine grows on mesic soil", etc. Then, we try to solve the following goal:

```
grows_on(X, mesic_soil).
```

PROLOG sets up a temporary association between a variable and possible instances of that variable, instances that are stored as part of facts in the knowledge base or that can be inferred from clauses. In this example, PROLOG tries to match the goal with the first fact. Both have the same number of arguments (arity), the functors match, and *X* can be unified with *red_pine*, but the second argument of the fact, *mesic_soil*, does not match with the second argument of the goal. PROLOG then tries the next fact. Now, all conditions are met for a proper match, so the goal has been solved, with *X* instantiated to *red_pine*.

This unification mechanism eventually results in variables becoming instantiated with all possible constants in eligible facts until one of the instantiations matches the goal. This process should become clearer as we introduce rules, below, and describe how they work.

7.2.1.4 Lists. Lists are a very general and powerful data structure. A list is an ordered sequence of elements. Lists can be of any length and contain any combination of atoms, strings, numbers, variables, or structures. Syntactically, lists are written as a sequence of terms separated by commas and enclosed in square brackets, for example:

```
[deer, grouse, moose]        []
```

The symbol '[]' is known as the empty list. A list may also be viewed as a structure consisting of its first element, the *head*, and the rest of the list, known as the *tail*. The head of a list is designated as such by separating it from the tail by a vertical bar '|', for example:

```
[oak|[red_pine, maple, beech]]
```

Viewing a list as a head and a tail (the rest of the list) has some advantages that will hopefully become apparent when we introduce recursion, below. A list is a sequential data structure that does not allow random access in the sense that an array would allow in a procedural language like FORTRAN. However, because lists are recursively defined they provide a powerful and compact means for performing many different operations and one does not need to know in advance the number of elements that a list will eventually contain.

Thinking recursively, the *tail* of the tree list, above, could also be written as a *head* and *tail*:

```
[red_pine|[maple, beech]]
```

Similarly, the *tail* of this reduced list could be re-written:

```
[maple|[beech]]
```

and again, where the *tail* of this final list is the empty list.

```
[beech|[]]
```

Because the empty list is the last element of every list, when the *tail* of our list is an empty list, we are at the end. We can put these *heads* and *tails* back together to form the original list in *head* and *tail* syntax:

```
[oak|[red_pine|[maple|[beech|[]]]]]
```

which represents the same list as if we had written it:

```
[oak, red_pine, maple, beech]
```

7.2.1.5 Assertions As Facts. Facts and rules are known as *clauses*. A fact is an unconditional truth; it only contains constants, no variables. A fact states the existence of a particular relationship between any number of "objects":

```
shot(john, deer).
is_a(buck, male_deer).
answer(yes).
present.
```

These facts might represent the following statements in our domain: "John shot a deer", "A buck is a male deer", "The answer is yes", and *present* is true. All clauses in PROLOG must be terminated with a period, '.'; we will end all our clauses with a period from now on.

A fact does not have to be a true statement about the real world. If you asserted in the knowledge base that "rabbits have antlers",

```
have(rabbits, antlers).,
```

then PROLOG would not know that this is a false fact. As in most knowledge base languages, only the explicitly represented facts are known and these are interpreted as true regardless of the reality in the real world. This is referred to as the **closed world assumption**. That is, whatever has not been explicitly specified in the knowledge base may be assumed to be false.

7.2.1.6 Assertions As Rules. A rule is a statement of conditional truth. It may be seen as equivalent to an if-then clause:

```
IF site index is greater than 60
AND species is red pine
THEN site is suitable.
```

This conditional assertion could appear in a PROLOG rule as:

```
site(suitable):-
      greater_than(site_index, 60),
      species(red_pine).
```

We've added indentations and line breaks for readability; there is no reason we could not have written it all on one line. The comma, ',', symbol is used to represent AND and the ':-' symbol represents IF. If we desired to represent OR (disjunction), we would use the ';' symbol. The clause, *site(suitable)*, is the then-part of the rule and is referred to as the "head" of the clause. PROLOG only permits the head of a clause to be a single term, no conjunctions or disjunctions of terms may be used. Beyond this restriction, the term in the head of the clause may be arbitrarily complex.

7.2.2 Goals and Basic Operations

We've mentioned goals in several of the preceding pages, we'll now be a little more specific and use an example for illustration. A goal is just another clause that functions like a question. PROLOG attempts to satisfy, that is, prove a goal to be "true". A goal is true when it is matched with facts in the knowledge base, either explicitly or as the result of applying a rule. When presented with a goal to prove, PROLOG's pattern matcher searches through the facts and rules in the knowledge base in an attempt to find that goal. If a goal contains variables, then those variables will become instantiated with the values found in the knowledge base. For example, suppose our KB has the following clauses:

```
stand(site_index, red_pine, 65).
site(suitable, X) :-
      stand(site_index, X, N),
      N >= 60.
```

An interpretation of these clauses might be "A stand has site index 65 for red pine" and "IF a stand has site index N for species X and N is greater than 60, THEN the site is suitable for growing species X." The first clause is a fact and the second, a rule. If our goal is to find all tree species for which a site is suitable for growing those trees, we would use the goal *site(suitable, Y)*. It does not matter that we have used *Y* in our goal question while the clause in the KB has an *X*; these variables will be unified (i.e. made equivalent) by PROLOG. In the process of establishing *site(suitable, Y)*, PROLOG would try to prove the goal *stand(site_index, X, N)*. Because there exists a fact in the KB which can be associated with *stand(site_index, X, N)*, *X* would be unified with the value *red pine* and *N* would be unified with the value *65*. This value for *N* is then used in the next clause, *N >= 60*. Because this comparison is true, the clause *site(suitable, X)* would be proven true with the unification of *X* with *red_pine*. If we wished to store this new information into the KB, we could do so by adding the predicate "asserta", such as *asserta(site(suitable, X))*, to the clause **site/2**.

```
stand(site_index, red_pine, 65).
site(suitable, X) :-
      stand(site_index, X, N),
      N >= 60,
      asserta(site(suitable, X)).
```

If the proven clause was inserted in the KB, then the next time we needed to prove *site(suitable, X)*, it would match a fact, namely *site(suitable, red_pine)*, already present in the knowledge base. The rule **site/2** would not have to be used for this clause unless we needed other instantiations for that clause, e.g., *site(suitable, white_pine)*.

7.2.2.1 Backtracking. PROLOG does not know in advance which clause from a set of clauses will satisfy the current goal, but it does know how to back out of blind alleys. This process is called *backtracking*. In its single-minded attempt to satisfy goals, PROLOG tries each clause with the same predicate name in the order in which the clauses appear in the knowledge base. If a clause does not lead to success, it backs up, undoes its previous work (i.e. instantiations), and tries the next clause on the list. Let's use the previous example with the addition of one other clause.

```
stand(site_index, jack_pine, 55).
stand(site_index, red_pine, 65).
site(suitable, X) :-
    stand(site_index, X, N),
    N >= 60.
```

Now, when PROLOG tries to solve the goal *site(suitable, Y)*, it unifies *stand(site_index, X, N)* with the first clause. *N* then becomes unified with *55*, which fails the comparison *N >= 60*. PROLOG anthropomorphically senses a dead end, so it backs up (backtracks) to the previous structure **stand/3** in the third clause and tries to find another unification for *stand(site_index, X, N)*. When it uses the second clause for the new instantiation, things turn out as before and the goal is solved with the identical instantiation of *X* to *red_pine*.

This search strategy is called *depth first search* because it involves going as far along each path as possible before backing up and trying another path. It has the advantage of searching through all possible states of the current goal before considering others, but it can also spend a large amount of time investigating blind alleys. If all attempts to satisfy a goal fail, then the goal fails.

7.2.2.2 Controlling Backtracking: Cut & Fail. Backtracking is built into PROLOG as a method for searching for all possible solutions, if necessary. It is part of the inner workings of the pattern matcher that executes PROLOG programs. This section covers some of the ways that the programmer may control the actions of PROLOG's implicit search mechanism. Special clauses (*cut, fail, repeat*) are available to give the programmer this capability. When the programmer has some control over how goals are solved, programs can be much more efficient. Although programs become harder to debug because they depart from the strictly declarative paradigm and move toward procedural processing. In practical programs, this compromise is often unavoidable.

The *cut* operator, '!', limits the search process to taking only certain paths to a solution. Backtracking is not permitted past the point where the cut occurs in a clause body. Let us use the clauses from before with the addition of the cut operator to the above rule and the addition of one more rule for **site/2**:

```
stand(site_index, jack_pine, 55).
stand(site_index, red_pine, 65).
site(suitable, X) :-
    stand(site_index, X, N),
    !,
    N >= 60.
```

```
site(suitable, Z) :-
    stand(site_index, Z, N),
    Z = jack_pine,
    N >= 50.
```

As in the backtracking example previously, *stand(site_index, X, N)* in the first rule will unify first with the clause *stand(site_index, jack_pine, 55)*. Then, the cut succeeds (it always succeeds) and the *N >= 60* comparison fails. Backtracking will attempt to back up and re-instantiate new values for variables, but the cut prevents backtracking back to the **stand/3** predicate. PROLOG re-does the *N >= 60* comparison as part of backtracking, but, of course, it fails again. When a cut operator appears in a clause that fails, then no other clauses will be attempted for solving the goal *site(suitable, Y)*, hence, PROLOG fails to solve the goal. Normally, without the cut in the first rule, the second rule would be tried upon failure of the first. Because PROLOG blindly tries rules as it encounters them, the order of rules for any *particular* predicate *does* make a difference. Order *does not*, however, make any difference in the order of *different* predicates. That is, in the above example, the two clauses for **stand/3** could appear after **site/2** and the effect would be the same.

The *fail* predicate always causes a goal to fail. There are situations where it is desirable to have more than one clause tried for the same goal even when one of them has already succeeded. Using the fail predicate ensures that all clauses for that goal will be tried. Again, we will use the immediately preceding example with the following change to the first **site/2** rule:

```
site(suitable, X) :-
    stand(site_index, X, N),
    N >= 60,
    fail.
```

As in the backtracking example above, PROLOG will reach the solution *site(suitable, red_pine)* via backtracking. Then, however, PROLOG gets to the *fail* predicate, which never succeeds. Using backtracking PROLOG tries to re-instantiate **stand/3** again, but there are no more clauses for **stand/3** so this **site/2** clause fails. The second **site/2** rule is then tried, with success, and the goal succeeds with *site(suitable, jack_pine)*. Without the *fail* predicate inserted into the first rule, the second rule would not have been tried, unless **site/2** was a clause in another rule in which backtracking occurred.

To see this, let us add the following rule for the predicate **regenerate/1** and remove the fail predicate from the rule above; altogether, we then have:

```
regenerate(X) :-
      site(suitable, X),
      seedlings_available(X).
stand(site_index, jack_pine, 55).
stand(site_index, red_pine, 65).
site(suitable, X) :-
      stand(site_index, X, N),
      N >= 60.
site(suitable, Z) :-
      stand(site_index, Z, N),
      Z = jack_pine,
      N >= 50.
```

Here, we are saying that we will regenerate a site with a particular species if the site is suitable for that species and if seedlings are available. It may happen that the site is suitable for red pine, but no red pine seedlings are available, then we would like to deduce what other species are suited for this site. In this example, PROLOG will first find that red pine is suitable. For the sake of example, assume that there are no seedlings available for red pine planting, i.e. *seedlings_available(red_pine)* fails. Then, PROLOG will backtrack in the regeneration rule to re-instantiate **site/2**. Because it cannot re-satisfy **site/2** for any instantiations by using the first rule, it will then try the second rule for **site/2** and instantiate its second argument with the value *jack_pine*. This instantiation is carried back to the regeneration rule where it can be used in the **seedlings_available/1** predicate. Because backtracking may occur within any rule (unless specifically limited) it is apparent that the actual search for solutions to a goal can be a long and confusing search process.

7.2.2.3 Recursion. PROLOG offers two ways to perform repetitive computations, backtracking and recursion. A recursive clause is one which uses itself in its own definition. Recursion consists of three parts: (1) one or more termination conditions (usually success and failure are included), (2) a recursive call to itself with a reduced version of the problem, and (3) a way to accumulate or collect the components of the solution (this part is optional because it does not apply to all problems). Any iterative algorithm can be expressed recursively. A simple iterative loop in BASIC can be defined as:

```
FOR I= 1 TO 5
PRINT I, SQR(I)
NEXT I
```

It can also be defined recursively in PROLOG as:

```
print_square_roots :- print_square_roots_helper(1).
print_square_roots_helper(I) :- I > 5, !.
print_square_roots_helper(I) :-
      R is sqrt(I),          % Anything after a '%' sign
                             % is a programmer's comment.
                             % Here the 'is' operator unifies
                             % the integer variable R with the
                             % square root of I.
   write(I),                 % write is a special predicate
   write('  '),              % that outputs to a standard I/O
                             % device, i.e. the monitor.
   write(R), nl,             % nl is a special predicate to
                             % write a new line to the standard
                             % output device.
   NewI is I + 1,
   print_square_roots_helper(NewI).
```

Here, the **print_square_roots_helper/1** clause does the work and is the recursive clause. There is only one termination condition that must be checked. Because the desired result of this clause is the side effect of printing, there is no need to accumulate and return a solution. The second **print_square_roots_helper/1** clause is used each time until $I > 5$, then the first **print_square_roots_helper/1** clause succeeds and the cut prevents further solutions to the goal *print_square_roots_helper(1)*. As you can see, in this example, we were not interested in receiving any return values in the goal *print_square_roots*, there were no arguments in the goal. Rather, we were interested in the side effect of the goal, that printed the square roots of the numbers 1–5.

Recursion is also frequently used to effectively operate on lists. For example, suppose we have a list containing the names of tree species that grow on a particular site. Given a particular species name, we may want to know if it is present in the list and, hence, if it grows on the site in question. The following recursive definition should do it:

```
on_site(X, []) :- fail.
on_site(X, [X|T]).
on_site(X, [H|T]) :- on_site(X, T).
```

We might try to solve the goal *on_site(sugar_maple, Species_list)*, where *Species_list* has previously been instantiated to a list of species. The first clause contains the terminating condition for failure, i.e. an empty list indicates that we have not found the species *X* on the list *Species_list* and therefore the search fails. When the list is not empty we check to see if the first element of the list is the species we are looking for. If it is, then we are done, the second clause has succeeded. This is the terminating condition for success. If it is not, then we try to prove a version of the clause containing a list that is one element shorter, i.e. with the tail of the current list, *T*; this is the recursive step. Again, we do not need to accumulate a solution in this example because our goal is a predicate that tests whether some condition holds or not. A BASIC program performing this same function would have to be significantly longer.

Beside being theoretically important in the fields of mathematics and logic, recursion also allows programmers to create code that is compact, elegant, and (given the proper software and hardware) efficient. It is important for PROLOG programmers to utilize tail recursion where possible because it is very efficient, but we will not explain this topic further here and direct readers to Covington (1995).

7.2.3 Global Control Procedures

The meaning of a PROLOG program can be understood either declaratively or procedurally. The declarative meaning emphasizes the static nature of relationships between "objects." The order of subgoals in a rule does not affect the declarative meaning of the rule. Despite the fact that one of the strengths of PROLOG is its ability to work out many of the procedural details on its own, the declarative approach is rarely sufficient to produce practical knowledge-based systems (Bratko 1986). There is a need to control program execution at both the global and local levels. The procedural meaning emphasizes the sequence of steps taken by the interpreter to evaluate a goal. In this view, the order of subgoals is significant.

Any automated reasoning system consists of a system of logic, which is usually viewed as declarative in nature, plus a control strategy that governs the search for solutions, the procedural portion of the PROLOG program.

7.2.3.1 Forward chaining. When an application problem has just a few facts available but the number of goal states is very large and difficult to enumerate, then forward chaining may be the most efficient reasoning control mechanism. Facts are taken in order of their appearance as clauses. A forward chaining process then finds all rules whose antecedents contain a predicate expression that can be matched to it. When all antecedents of a rule have been found "true", then the conclusion of the rule is asserted as a true fact and the process starts all over again. In this manner, the program moves closer and closer to a conclusion. Although PROLOG interpreters don't automatically forward chain, it is not hard to implement this procedure. Further discussion can be found in Chapter 2 (section 2.4.2.1) and an example implementation of a forward chaining meta-interpreter in PROLOG can be found in Chapter 9.

*7.2.3.2 **Backward chaining.*** Backward chaining or goal-directed reasoning, the normal functioning of the PROLOG interpreter, is often a good control structure when there are many more facts than final conclusions (goals). In backward chaining, the process initiates a root goal to prove. PROLOG then uses the clause order in the knowledge base as a way of selecting individual rules and facts. If a rule or fact fails along any given line of reasoning, then backtracking is used to attempt others until either the goal is proven true or all paths are exhausted and the goal fails. Further discussion can be found in Chapter 2 (section 2.4.2.2) and implementation details in Chapter 8 and 9.

7.2.4 Local Control Procedures

Local control structures are used either to change the order in which rules are considered for evaluation or for imposing a sequential ordering to make a practical program work properly.

*7.2.4.1 **Procedural rules.*** As has been mentioned, PROLOG derives much of its strength from its declarative features, but in order to write practical programs procedural control rules are needed. Often, a basic need of a program is to specify a sequential series of operations which are very order dependent. For example:

Step 1: clear the screen
Step 2: write "Please enter the age of this deer."
Step 3: read the age from the keyboard
Step 4: evaluate a clause that helps determine the size based on age
Step 5: report the size to the user
Step 6: wait until the user is ready to continue

This sequence of steps can be formulated in a procedural control rule as follows:

```
run :-
        cls,
        write('Please enter the age of this deer →'),
        read(Age),
        get_size(Age, Size),
        nl, nl,
        write('The approximate size of the deer is '),
        write(Size), nl,
        write('Please press any key to continue.'), keyb(_,_).
```

7.2.4.2 Meta-logical rules. Meta-level rules are used to augment the built-in PROLOG inference engine. One common need is to change the sequence in which rules are evaluated from the order in which they appear in the knowledge base to some order that is more a function of domain related logic. Suppose we wish to give preference to the rule that has been used successfully most recently. The meta-rule to carry out this logic might be formulated as (Rowe 1988):

```
site(good1) :-
    species(red_pine, Site_index),
    Site_index > 60.
site(good2) :-
    soil_type(sandy).
solve_site_rule :-
    clause(site(good1), R1),
    clause(site(good2), R2),
    used(site(good1), R1, T1),
    used(site(good2), R2, T2),
    T1 > T2,
    site(good1).
solve_site_rule :-
    site(good2).
```

Here the predicate **used/3** is a fact asserted after each successful rule application stating the time the rule was used. In this case, if rule *site(good1)* was used more recently than rule *site(good2)*, then *T1* > *T2* and *site(good1)* is preferred to *site(good2)*. The **clause/2** predicate can be used to find the left-hand side and the right-hand side of any rule in the knowledge base. If *site(good2)* had been used more recently, then the first *solve_site_rule* would fail and the second one would be fired.

Other applications of meta-rules to program or strategy control include explanation facilities, and debugging aids. See Chapter 9 for further illustrations on using meta-programming to implement interpreters that perform various functions.

7.2.5 Knowledge Representation in PROLOG

The previous sections include some of the basic elements of PROLOG and how they work in a knowledge system. Again, our presentation was not intended to be a comprehensive account of the language or its use. The remainder of this chapter implements the knowledge representations introduced in Chapter 2. There exists no one way to implement these knowledge structures; we present some of the common methods as examples only. These topics may be fairly advanced for readers new to PROLOG. Readers are encouraged to acquire some minimal proficiency in PROLOG before examining the examples in this section.

Regardless of how knowledge is represented, the declarative logic in the knowledge base must be kept separate from the control structures so that new knowledge can be added to the system without modifying the system as a whole. Beyond this necessity, any of the five major knowledge representation methods previewed in Chapter 2 may be used for storing domain specific facts and logic. PROLOG implementations of these methods are illustrated below.

7.2.5.1 Predicate Logic. The use of predicate logic as a knowledge representation scheme is called logic programming and is the basis for PROLOG. Thus, declarations of facts and rules in this form can be executed by the PROLOG inference engine without further work. Let us represent the knowledge encoded in the rules of Figure 2-10 into predicate logic formalism:

```
causative_agent(pales_weevil, 0.35) :-
        entire_seedling(discolored),
        stand_located(pales_weevil, high_frequency_county),
        not branch_flagging(present).
entire_seedling(discolored) :-
        stand_size_class(seedling),
        needle_injury_type(discoloration),
        affected_foliage(old_and_new),
        needle_damage_distribution(throughout_crown).
stand_size_class(seedling) :-
        stand(height, Height),
        Height <= 4.5.
```

We assume here that many of the clauses, e.g., **stand/2** can be determined by other assertions, either based on user input or on deductions. We might explicitly included a clause to ask the user for average stand height when it is not known. Then this fact is added to the knowledge base so when it is needed again in some other clause, the user does not have to be queried again and again. Alternatively, we could have a control rule somewhere that would force a query to the user when information about any fact cannot be deduced otherwise.

7.2.5.2 Production Rules. If we build a goal-driven, rule-based KBS shell using PROLOG, then the rules of Figure 2-10 take on a slightly different form. The form of the rules in the knowledge base is dependent upon how we program the shell. Using the "exshell" program from Luger and Stubblefield (1993), the rules would be:

```
rule((causative_agent(pales_weevil) :-
     entire_seedling(discolored),
     stand_located(pales_weevil, high_frequency_county),
     not branch_flagging(present)), 0.35).
rule((entire_seedling(discolored) :-
     stand_size_class(seedling),
     needle_injury_type(discoloration),
     affected_foliage(old_and_new),
     needle_damage_distribution(throughout_crown)), 1.0).
rule((stand_size_class(seedling) :-
     stand(height, Height),
     Height < = 4.5), 1.0).
askable(['What is the average height of trees in the stand', H],
     (height, H)).
```

The additional structure of the form *rule((P :- G), CNF)* is necessary for a set of meta-logical rules to automatically build a 'how' and 'why' explanation system. As noted above, this shell also contains meta-logical rules to automatically ask the user for information that is not known and has been defined as askable by the clause **askable/2**.

7.2.5.3 Frames. There are two main approaches to representing frames in PROLOG (Lazarev 1989). Some systems rely on lists as the basic structure (Luger and Stubblefield 1993) and others rely on simple predicates (Cuadrado and Cuadrado 1988). We will use the second approach to illustrate how to define frames in PROLOG.

A frame is composed of slots; slots contain one or more facets with one value per facet. Each slot commonly has at least a "value" facet or a "default" facet. Additionally, any number of other facets, named *demons* could be defined, some examples might be: "if_needed", "if_added", "if_removed", etc. These demons are typically rules which are evaluated when a particular condition exists. Through this mechanism, rules and frames are integrated. Inheritance is tracked through the use of "is_a" and "ako" (a kind of) slots. "Is_a" represents a member-class relationship while "ako" represents a subclass-class relationship (Lazarev 1989).

Let us represent a frame-slot-facet combination with the predicate:

```
frame(Name, Slot, Facet, Value)
```

The "Continuous_Forest_Inventory_Tree" frame and its instance, the "Tree_24" frame in Figure 2-4 will be used to illustrate how to represent a frame in PROLOG.

```
frame(cfi_plot_tree, ako, value, tree).
frame(cfi_plot_tree, contained_in, value, cfi_plot).
frame(cfi_plot_tree, measurement_years, value, [1974, 1979, 1984, 1989]).
frame(cfi_plot_tree, measurement_variables, value, [dbh, height, crown_ratio,
   species, live_dead_status]).
frame(cfi_plot_tree, dbh, default, calc_median_plot_dbh).
frame(cfi_plot_tree, height, default, calc_median_plot_height).
frame(cfi_plot_tree, height, if_needed, height_measurement_rule).
frame(cfi_plot_tree, crown_ratio, default, 0.5).
frame(cfi_plot_tree, species, default, stand(Cover_type)).
frame(cfi_plot_tree, live_dead_status, default, live).
frame(cfi_plot_tree, event_sequence, value, measure_cfi_plot_tree_script).
frame(tree_24, is_a, value, cfi_plot_tree).
frame(tree_24, contained_in, value, cfi_plot_#215).
frame(tree_24, measurement_years, value, [1984, 1989]).
frame(tree_24, dbh, value, 4.5).
frame(tree_24, height, value, 29).
frame(tree_24, crown_ratio, value, 0.6).
frame(tree_24, species, value, 'Pinus banksiana').
frame(tree_24, live_dead_status, value, live).
frame(tree_24, X, Y, Z) :-
      frame(tree_24, is_a, value, Parent),
      frame(Parent, X, Y, Z).
```

The last clause for the "tree_24" frame facilitates inheritance from the parent frame *cfi_plot_tree*. Any slot-facet-value combination that has not been previously specified for "tree_24" can be obtained from one of its parents if that combination exists. So, in this case, the slot-facet-value combination from *frame(cfi_plot_tree, event_sequence, value, measure_cfi_plot_tree_script)* can be inherited to establish the clause, *frame(tree_24, event_sequence, value, measure_cfi_plot_tree_script)*. While the representation of frames using multiple clauses means that the knowledge base will have a large number of facts, it has the advantage of shortening any references to a particular frame. Only the pertinent slots and facets of the frame need to be referenced in a clause and not the entire frame structure.

7.2.5.4 Objects with Messages. To implement object-oriented programming in PROLOG we can define predicates called:

```
object(Name, Method_list).
```

The *Name* of the object may be a compound structure that contains all the local data items that are defined for the object and might be needed by its various methods. Alternatively, as we have done below, the *Name* contains no variables. Local data variables are stored as methods; values of those variables may be obtained by other methods by sending the appropriate message to the object, even to the same object itself. The second argument to the functor *object* is a list of methods that the object knows. Let us take an abbreviated version of the two frames above and translate them into two objects and their methods. This implementation is adapted from Covington et. al. (1995) with some modification.

```
object(cfi_plot_tree,
    [
    is_a(tree),
    measurement_yrs([1974, 1979, 1984, 1989]),
    measurement_vbls([dbh, height, crown_ratio, species, live_dead_status]),
    (dbh(D) :- calc_median_plot_dbh(D)),
    (basal_area(Tree_number, Area) :- send(Tree_number, dbh(Dbh)), Area is
         0.005454 * Dbh^2),
    crown_ration(0.5),
    (species(Cover_type) :- stand(species, Cover_type)),
    live_dead_status(live)
    ]).
object(tree_24,
    [
    is_a(cfi_plot_tree),
    measurement_yrs([1984, 1989]),
    dbh(4.5),
    height(29),
    crown_ratio(0.8),
    species('P. banksiana')
    ]).
is_a(Object, Object)          % TRY IMMEDIATE OBJECT
is_a(Object, Object1) :-      % TRY ALL PARENTS
```

```
        object(Object, X),
        get_parent(X, P),
    Object1 = P.
    is_a(Object, Object1) :-      % TRY ALL ANCESTORS OF PARENTS
        object(Object, X),
        get_parent(X, P),
        is_a(P, Object1).
    get_parent([X Z], Y) :-       % PARENT IS FIRST METHOD IN LIST
        arg(1, X, Y),             % GET ARGUMENT OF FIRST METHOD
        is_a(Y).
    get_parent([X Z], Y) :-       % RECURSE ON REMAINING
        get_parent(Z, Y).         % ELEMENTS
    send(Object, Message) :-
        is_a(Object, Object1),
        object(Object1, Method_list),
        get_method(Message, Method_list, Method),
        call(Method).
    get_method(Message, [First_method Rest], Method) :-
        fact_or_rule(Message, First_method, Method), !.
    get_method(Message, [_ Rest], Method) :- get_method(Message, Rest, Method).
    fact_or_rule(Message, Message, true).
    fact_or_rule(Message, (Message :- Body), Body).
```

The predicates *is_a* and *get_parent* handle the search for objects that are able to respond to a particular message. Methods may be inherited from ancestor objects, so all parent and other ancestor objects must be searched until the method is found. The search proceeds from closest ancestor (parent) up to the root object (the root object is the ancestor of all objects in an object-oriented implementation. The predicates *send*, *get_method*, and *fact_or_rule* send messages and evaluate methods associated with those messages. It is interesting to note that the method *basal_area* contains a message to another object, which is, in fact, the object which received the original message for the basal area calculation. So, the message *send(tree_24, basal_area(tree_24, Y))* causes the message *send(tree_24, dbh(Dbh))* to be broadcast also.

7.2.5.5 Semantic Networks. Some of the earliest attempts to represent knowledge tried to build structures that represented the meaning of words; this resulted in a semantic network (Parsaye and Chignell 1988). Many important concepts, such as inheritance, were first identified through these efforts. The semantic network represented in Figure 2-6 can be formulated in PROLOG as:

```
has_attribute(stand, location).
has_attribute(cfi_plot, location).
has_attribute(plant, leaves).
is_a(red_pine_plantation, stand).
is_a(tree_23_on_plot_16, tree).
is_a(tree, plant).
element_of(tree, cfi_plot).
element_of(tree, stand).
```

These clauses describe the structure of the network and its declarative knowledge. To be useful, we should include some clauses that permit inferencing over this representation. Two such rules were presented in Section 2.1.4 and can be translated into PROLOG as:

```
has_attribute(X, Z) :-
    is_a(X, Y),
    has_attribute(Y, Z).
element_of(Y, Z) :-
    is_a(X, Y),
    element_of(X, Z).
```

At the present time, semantic networks on their own are used in relatively few systems (Parsaye and Chignell 1988). However, in combination with other representations they possess a great deal of flexibility and power. We will not be discussing them further in the remainder of this text.

Readers interested in becoming more familiar with PROLOG concepts and programming should consult one of the excellent texts devoted to the language. Many of those sources contain KBS shells written in the PROLOG language, e.g. Covington et al. (1995). These simple shells can be useful tools for quickly coding a proof-of-concept system. Later, additional application enhancements can be included into prototype systems, such as in Chapter 9, or another development environment can be adopted.

7.3 SUMMARY AND ADDITIONAL READINGS

It can be argued that PROLOG is conceptually similar to an expert system shell because it has a built-in backward chaining inference engine, manages its own working memory, and uses predicate logic as its native knowledge representational form (Fig. 2–1). However, it lacks a user-interface, explanation facilities, and facilities for self-modification of the knowledge base. These missing elements must be developed by the programmer in order to arrive at a complete knowledge-based system.

In Chapter 8, we will develop our initial prototype using predicate logic and the necessary extensions to PROLOG to produce a simple, first-level working knowledge-based system. In Chapter 9, we will show how to attach compiled, procedural programs to a KBS and how to implement the blackboard framework architecture. For further practice, the book by Covington et al. (1995) contains a large number of example problems for which PROLOG programs have been developed.

CHAPTER

8 An Initial Prototype Using Native PROLOG

The good news is that there is considerable opportunity for computer-based expert systems to improve on human decision-making in those domains where the knowledge and decision-making processes can be captured and applied.

Raymond Kurzweil

After defining the problem in Chapter 4, designing the knowledge model and user interface model in Chapter 6, we can turn our attention to implementing the knowledge model (Figure 8–1). This model is detailed in section 6.1.1 (Figure 8–2). Having decided to implement the knowledge model in PROLOG, we must still choose between two major paradigms. On the one hand, we can implement the knowledge model using *native PROLOG* as the inference mechanism and predicate logic as the knowledge representation method. This approach is conceptually the easiest and has resulted in many useful expert systems (Roach et al. 1985, Rauscher et al. 1990, Thomson 1989). The other approach, is to use *meta-programming in PROLOG*. Meta-programming in PROLOG can be used to create a toolkit of PROLOG modules; each module implements some KBS feature. Unfortunately, the toolkit approach is more complex than native PROLOG programming and therefore requires a higher level of expertise from the developers.

In this chapter, we will present and discuss the concepts and methods that will allow an entry level PROLOG programmer to develop an expert system in

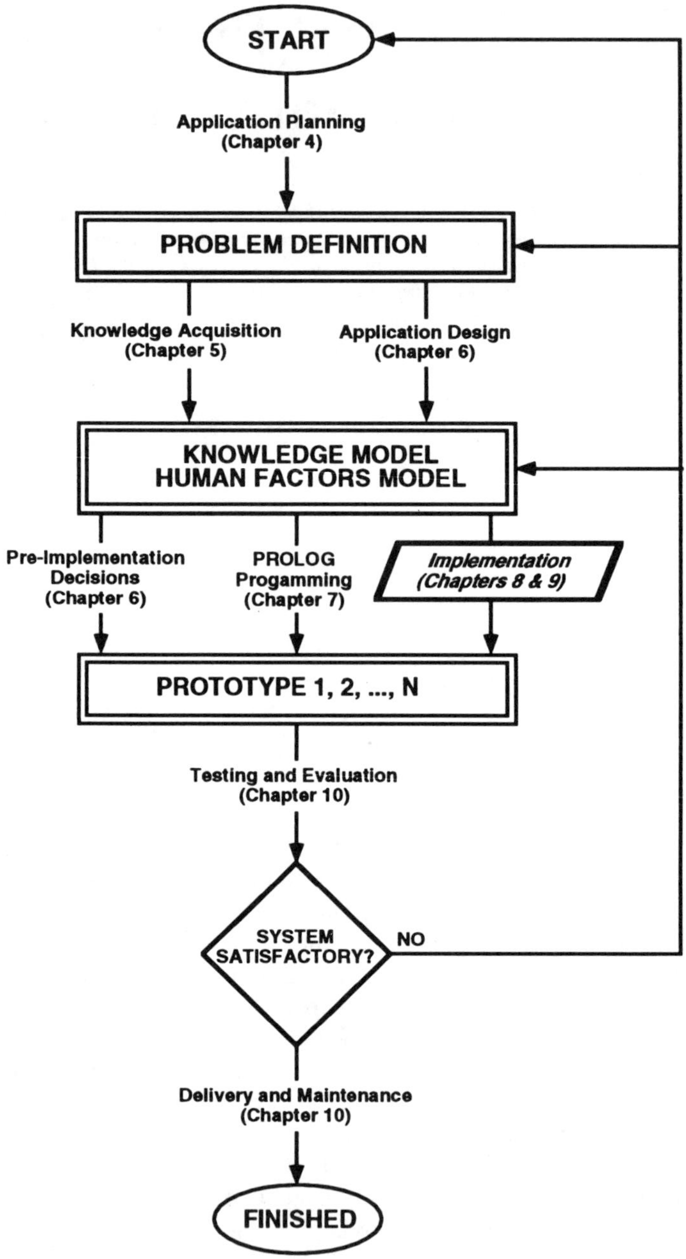

Figure 8–1. This chapter describes the coding of an initial prototype using native PROLOG.

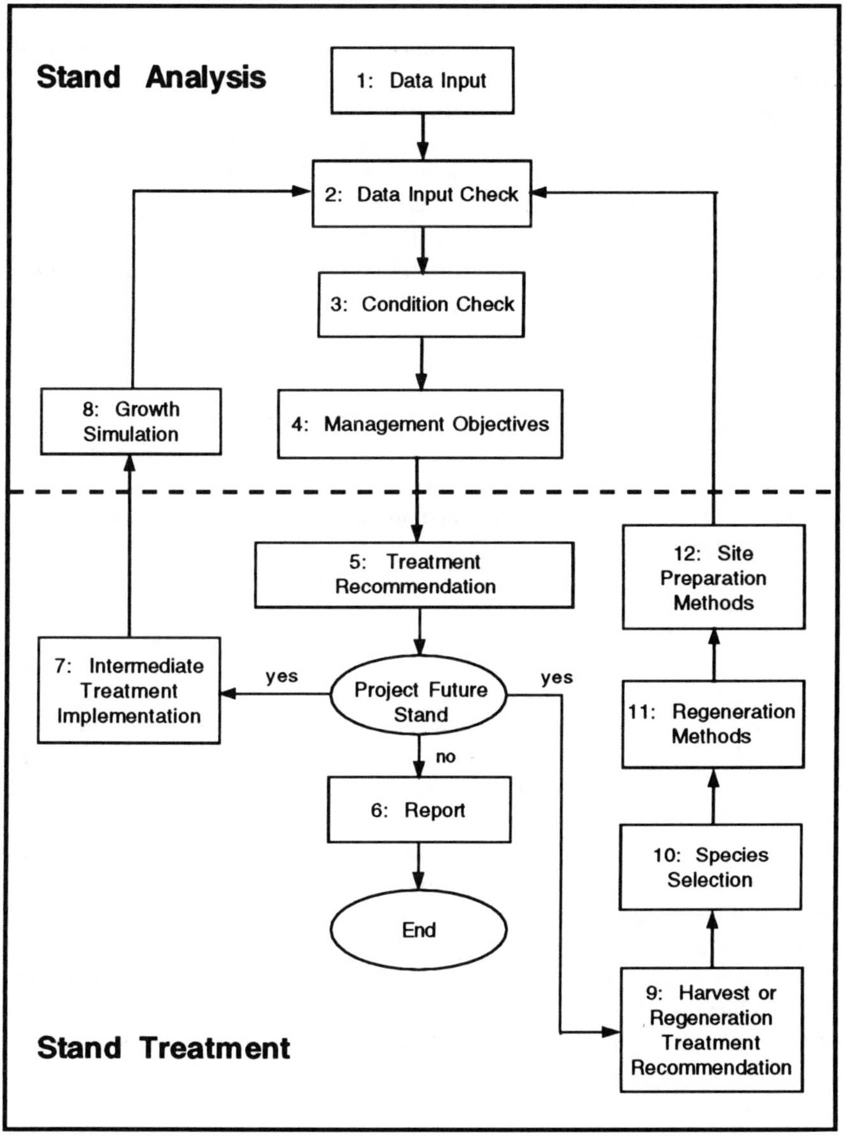

Figure 8–2. The sketch of the RP-FMAS knowledge model illustrated here provides an operational model of even-aged red pine forest management.

native PROLOG. We will continue to use the Red Pine Forest Management Advisory System (RP-FMAS) introduced earlier as our implementation example. Readers unfamiliar with forest management theory and practice should review the background for RP-FMAS in section 4.4 and the knowledge model presented in section 6.1.1. Rauscher et al. (1990) may be consulted for additional detail (in addition, see the Frontmatter for access to a hypertext version of the Red Pine knowledge model). The focus of this chapter is how to build a prototype knowledge-based system using native PROLOG. The forest management example is important only to provide substance to this presentation. In Chapter 9, we will cover developing a meta-PROLOG toolkit using the same RP-FMAS example.

8.1 KNOWLEDGE-BASE SYSTEM ARCHITECTURE

In section 6.2, we established that all but the simplest knowledge-based systems need three logical levels to provide adequate control and explanation flexibility. We labeled these levels as:

Level 1: the domain level;
Level 2: the tactical control level; and
Level 3: the system control level

The domain level contains the domain knowledge represented declaratively without embedded control logic. The tactical control level provides a flexible, logic-based method for controlling the inference process. Explicitly separating domain and control knowledge enables their independent modification, it provides the opportunity to explain control as well as domain inferences, and it allows domain knowledge to be developed for representational adequacy without having to worry about execution efficiency (Jackson et al. 1989). It is furthermore becoming clear that the problem of context evaluation, i.e. recognizing the conditions for which a particular KBS module should be activated, demands control at the system architecture level. System control guides which particular modules operate (see section 6.2). We will use these levels as a conceptual framework to guide our presentation in this and the next chapter.

8.2 THE DOMAIN LEVEL

The most important function of the domain logic level is to implement the knowledge representational scheme(s) of the KBS. First-order predicate logic, used as the knowledge representation scheme in this chapter (see sections 2.1.5 & 7.2.5.1), is the most widely used knowledge representation scheme in logic programming and PROLOG is an ideal language for implementing it (Luger and Stubblefield 1993). PROLOG has a built-in backward chaining inference engine capable of executing predicate logic statements directly consequently requiring no further work on the part of the developer. Knowledge representation in predicate logic is simple and directly understandable.

A natural language rule might be formatted as:

```
IF     first premise        IF     site index is high
AND    second premise       AND    soil texture is sandy
AND    ...                  AND    ...
THEN   conclusion           THEN   red pine is suitable
```

Using a first-order predicate logic clause in PROLOG, this rule would read:

```
conclusion :-               suitable(red_pine) :-
    first_premise,              site_index(high),
    second_premise,             soil_texture(sandy),
    ...                         ...
```

If these constructs look strange to you, please refer to Luger and Stubblefield (1993, Chapter 6) or any of the PROLOG programming books recommended in section 7.2. Because the terms "rule" and "clause" are used ambiguously throughout the PROLOG literature, we need to clarify our usage. By rule we intend to mean a particular type of knowledge representation as illustrated above by the IF-THEN natural language structure. Rule structures can be implemented in many different shells, computer languages, and mathematical systems. In the present case, we have illustrated above how to implement a natural language rule using a first-order predicate logic clause in PROLOG. Having made this clear, it becomes legitimate to refer to the natural language as well as the predicate-logic implementation as a "rule" when we wish to focus on the knowledge representational meaning. If we wish to focus on the computer language PROLOG, then we will use the term "clause".

Facts, in predicate logic, are merely clauses with a head but no body (see section 7.2.1.5). In RP-FMAS (Rauscher *et al.* 1990), we have defined the major fact structure as follows:

```
stand(attribute,value,certainty_factor)
```

Some examples of facts are:

```
stand(treatment,'Harvest',100).
stand(age,16,100).
stand(size,'Sapling',100).
```

The natural language interpretation of the first fact above is: the object *stand* has a *treatment* attribute with a value of *'Harvest'* and a certainty factor of 100 (completely true). This particular choice of a fact structure is essentially arbitrary with the only constraint being that once selected, the fact structure must be the same within a particular knowledge-base system.

Rules are used to define relationships among facts (see section 7.2.1.6). Let's examine some of the rules found in module #5 of RP-FMAS and described in detail in the *Knowledge Model* of Chapter 6. Module #5 contains the predicate logic rules that determine the appropriate treatment recommendations for a subject stand:

```
Case 1: recommendation for weeding
test_rec :-
    stand(size,Size,_),
    Size == 'Seedling',
    stand(stocking,Level,_),
    Level == 'Above maximum',
    asserta((stand(recommendation,'Weed',100))).
This rule can be interpreted as follows:
    IF stand size is seedling
    AND stand stocking is above maximum
    THEN recommendation is weed
    AND save this recommendation to the working memory
```

The clause *stand(size,Size,_)* is processed by the PROLOG interpreter so that the knowledge base is searched for a matching predicate. Suppose the first fact in the knowledge base is *stand(size,'Seedling',100)*, then the variable Size will equal the value *'Seedling'* which will make the second clause of the rule be true. The underscore character "_" is the anonymous variable described in section 7.2.1.2.

```
Case 2: recommendation is thinning
test_rec :-
    stand(dbh,Dbh,_),
    Dbh > 5.0,
    stand(ba,Ba,_),
    stand(thin_ba_x,X,_),
    stand(max_ba,Max_ba,_),
    Temp is X * Max_ba,
    Ba > Temp,
    asserta((stand(recommendation,'Thin',100))).
This rule can be interpreted as follows:
    IF the average dbh of the stand is greater than 5.0
    AND assign Temp the value of the thinning basal area multiplier multiplied by
        the maximum basal area
    AND the basal area of the stand is greater than Temp
    THEN recommendation is 'Thin'
    AND save this recommendation to working memory
```

One major principle in developing a rule-based system is to make the rules as declarative and informative as possible. To do this, we eliminate from the rules, as much as possible, all procedural interference. In other words, ideally rules should describe a state of knowledge and not contain procedural statements (how the knowledge is to be applied) within them. For example, *stand(size,Size,_)* is a procedural predicate because its only function in the above rule is to find a value for the variable *Size*. This type of action is call *extra-logical* or procedural because it is not defined in the theory of first-order predicate logic, the basis for PROLOG (Lloyd 1984). In contrast, the predicate *Size == 'Seedling'* performs a logical comparison operation to see if the variable Size is or is not logically equal to *'Seedling'* and by its truth or falsity it determines a state of knowledge. Ideally, rules should declare what facts and relationships are true. They should not provide instructions about how to prove things true.

Rules should also be order independent, that is to say, it should be possible to scramble the rules in the knowledge base in any order without affecting their logic. In practice, order independence of rules is not always easy to achieve because it causes a loss in execution efficiency. An example of order dependent clauses is provided in section 8.3.2 – User Interface Management. The developer must decide on the relative value of logical clarity versus execution speed. Many of these problems of logical clarity are forced upon us by the selection of native PROLOG as our reasoning mechanism (see section 8.6 for a further discussion).

For those familiar with logic programming, it is immediately clear that asserting a fact into the working memory as done in the rule examples above is also an extra-logical, procedural activity and thus violates the declarative nature of the domain rules. We could, of course, develop a new predicate, call it **save_fact/3** and define it as follows:

```
save_fact(Attribute, Value, Certainty_factor) :-
    asserta((stand(Attribute, Value, Certainty_factor))).
```

But that really does not help much conceptually and is less efficient from an implementation point of view. The real problem lies with the fact that using native PROLOG makes it practically impossible to clearly separate declarative from procedural activities in the knowledge base. This is one reason why meta-PROLOG programming is an attractive alternative as we will discuss in section 8.6 of this chapter.

8.3 THE TACTICAL CONTROL LEVEL

The domain level discussed in the previous section focuses on knowledge representation issues. Once we have defined our rule structure, we need to provide for a reasoning mechanism to manipulate those rules in order to satisfy goals (see section 2.4 for more detail). When using native PROLOG as the inferencing mechanism, we can simply rely on the backward chaining (see section 2.4.2.2) inferencing procedures built into PROLOG. Other portions of a knowledge-based system, such as explanation management (see section 3.1), user interface management (see section 3.2), and uncertainty management (see section 3.5), must be implemented by the developer in the native PROLOG language (Merritt 1989). We will now present one way to implement each of these features in order to flesh out a reasonably complete initial prototype.

8.3.1 Inference Control Management

As mentioned already, in RP-FMAS we used the built-in backward chaining inference procedure of native PROLOG thus saving ourselves much work (Rauscher *et al.* 1990). The use of backward chaining is discussed in section 2.4.2.2. In addition to the built-in PROLOG inference engine, we need two other types of clauses which help us control the inferencing process, namely *meta-rules* and *procedural-rules* (see section 6.2.2).

The function of a meta-rule is to direct the sequence of reasoning rather than to perform it. For example, each of the major modules contains a goal which calls a meta-rule such as the following one from module #2:

```
main :-
    checking_complete.
checking_complete :-
    check_age,
    check_si,
    check_height,
    check_dbh,
    ...
    calc_size,
    calc_stocking.
```

PROLOG's backward chaining system gets the goal **main/0** to satisfy. So the first, and in this case only, step is to see if the predicate **checking_complete/0** is true or false. The first step in that is to see whether the predicate *check_age* is true or false. In module #2, there happen to be 8 *check_age* declarative domain rules (see the Red Pine hypertext system referenced in the Frontmatter) and the backward chaining system is allowed to find the true one without further control by meta-rules. Meta-rules allow us to perform the inevitable control functions of the system explicitly thus keeping domain rules as declarative as possible. The impact of meta-rules is to make the inferencing process more efficient and flexible than would be possible relying only on the backward chaining process (Sterling and Shapiro 1986).

The purpose of procedural clauses is to execute some bookkeeping, file transfer, or screen management function required by the program. Procedural clauses are not directly involved with reasoning about domain knowledge nor do they function as control clauses. For example:

Step 1: clear the screen
Step 2: write "Please enter the age of this tree."
Step 3: read the age from the keyboard
Step 4: evaluate a clause that helps determine the size
Step 5: report the size to the user
Step 6: wait until the user is ready to continue

This sequence of steps can be formulated in a procedural clause as follows:

```
run :-
    cls,
    write('Please enter the age of this tree →'),
    read(Age),
    get_size(Age,Size),
    nl,nl,
    write('The approximate size of the tree is '),
    write(Size),nl,
    write('Please press any key to continue.'),keyb(_,_).
```

In each PROLOG module, we should clearly separate the meta-rules, procedural-rules, and declarative rules into different sections of code as is illustrated in section 8.5 below. It is common practice to place the declarative rules, which make up the domain knowledge base, into a separate file which is accessed by the rest of the PROLOG code at run-time. This explicit separation enhances program clarity and makes program maintenance considerably easier.

8.3.2 User Interface Management

Given that a system performs its task well, the user interface is likely to determine its success or failure in the application arena. It is therefore wise to study the techniques for effective user-interface (Simpson 1986) and screen format (Galitz 1985) design. It must be kept in mind that developing a good user interface system in PROLOG is a problem that has an infinity of solutions. We will present one method in this chapter and a radically different one in the next. It is expected that each developer will modify our solutions to meet their particular needs.

When the same or similar processes need to be accomplished numerous times in a knowledge-based system, it is efficient to develop a generic toolkit module.

We will present one such toolkit system in this section. The first major task of a user interface is to "ask" the user for input. In RP-FMAS, we identified 8 types of questions that the advisory system typically needs to manage. The predicates **question/2**, **question/4**, or **question/5** perform this function. Let's see how to employ these predicates. Using the example predicate from the previous section, we can start out the program with the **main/0** predicate which calls **checking_complete/0** and then **check_age/0** is activated.

```
main :-
    checking_complete.
checking_complete :-
    check_age,
    check_si,
    check_height,
    check_dbh,
    ...
    calc_size,
    calc_stocking.
check_age :-
    question(age,_,_,0,Age).
```

To simplify the discussion, let's assume we just want to ask the user the age of the stand. Therefore, **check_age/0** simply calls the **question/5** predicate. First, we know that the variable *Age* is a numeric variable which places it into cases 5–8, discussed below. The user's answer would be returned in the variable *Age*. If we were looking for a numeric comparison, cases 5 & 6, meaning that we want to see if the age is between 10 and 20 years, we would have used the **question/4** predicate as follows:

```
question(age,10,20,0)
```

If we wanted a text comparison, we would formulate the query as:

```
question(age,'Young')
```

If we wanted a text retrieval from the user, we would query:

```
question(age,Age_class)
```

and the variable *Age_class* might return the value *'Young'*. Each of the 4 types of questions occurs in two modes. If the value already is known in the knowledge base then the first mode in each of the 4 cases will execute without ever asking the user. This is to ensure that we never ask the user the same question twice. The user is only queried if we need to know the value for a variable and it cannot be found in the knowledge base or in the working memory.

We suggest that you study the code excerpt below and try running it in your own PROLOG interpreter until you understand its operation. Definitions for the **places/3** and **check_number/5** predicates can be found in section 8.5 below. Notice also that each set of 2 clauses is order dependent (see section 8.2 above). If we reverse the order, then the user is always asked a question regardless of whether the answer is already in the knowledge base or not. One could, of course, combine these two clauses into a single, more complex clause such as:

```
question(Name,Answer) :-
    nonvar(Answer),
    ( stand(Name,Input,_)
    ;                              % ← logical OR operator
    ask(Name,Input),
    asserta((stand(Name,Input,100)))
    ),!,                           % ← the cut operator
    Answer == Input.
```

Keeping these two functions separated is easier for novice PROLOG programmers to understand and it does not use up much extra processing time.

case 1: A comparison of text is required and the data is already present in the knowledge base (KB).

```
question(Name,Answer) :-
    nonvar(Answer),
    stand(Name,Input,_), !,
    Answer == Input.
```

case 2: A comparison of text is required but the data is not in the KB, hence, the user must be asked, and the result stored in the KB for future use.

```
question(Name,Answer) :-
    nonvar(Answer),
    ask(Name,Input),
    asserta((stand(Name,Input,100))), !,
    Answer == Input.
```

case 3: A text answer must be retrieved and the data is already in the KB. No comparison is required.

```
question(Name,Answer) :-
    var(Answer),
    stand(Name,Answer,_).
```

case 4: A text answer must be retrieved but the data is not in the KB, hence, the user must be asked, and the result stored in the KB for future use.

```
question(Name,Answer) :-
    var(Answer),
    ask(Name,Answer),
    asserta((stand(Name,Answer,100))).
```

case 5: A numeric comparison is required and the data is already present in the KB. Here a minimum and maximum value must be specified to provide the allowable range. If equality is to be tested, then the minimum and maximum values should be the same number.

```
question(Name,Min,Max,_) :-
      stand(Name,Input,_), !,
      Input >= Min,
      Input =< Max.
```

case 6: Numeric comparison is required and the data is not in the KBS, hence, the user must be asked and the result stored in the KB. The number of digits in the final result must also be specified. If the number of digits is "0", then the number will be returned as an integer.

```
question(Name,Min,Max,N_digits) :-
      repeat, cls,
      ask(Name,Instring),
      check_number(Name,Min,Max,Instring,Inreal), !,
      places(Inreal,N_digits,Answer),
      asserta((stand(Name,Answer,100))).
```

case 7: A number must be retrieved and the data is already in the KB. It will be formatted with the specified number of digits.

```
question(Name,_,_,N_digits,Answer) :-
      stand(Name,Inreal,_),
      places(Inreal,N_digits,Answer).
```

case 8: A number must be retrieved but the data is not in the KB, hence, the user must be asked and the formatted result saved in the KB.

```
question(Name,_,_,N_digits,Answer) :-
      repeat, cls,
      ask(Name,Instring),
      check_number(Name,_,_,Instring,Inreal), !,
      places(Inreal,N_digits,Answer),
      asserta((stand(Name,Answer,100))).
```

Notice that the various **question** predicates create a unified question management subsystem for RP-FMAS. We are not claiming that this system is in some sense an "optimal" solution to implement the typical *ask-the-user* function; but it gets the job done and is simple to program and explain.

With this generic predicate structure in place, we can now show how the user interface module interacts with the user. As an example illustrating how the system works, we can use a sequence from the intermediate treatment implementation module (Figure 8–2, #7) of RP-FMAS. When the treatment is *Thin*, we suggest a residual basal area for the stand after thinning. The user then is asked to concur with this suggestion or change the value to a more suitable one. This process starts with a rule that estimates the residual basal area and asserts it into the working memory:

```
implement_rec('Thin') :-
      stand(product,Product,_),Product = 'Large sawtimber',
      stand(min_ba,Minba,_),
      stand(max_ba,Maxba,_),
      stand(ba,Ba,_),
      (Minba+Maxba)/2 < Ba/2,N1 is Ba/2, places(N1,0,N),
      asserta((stand(residual_ba,N,100))),
      check_resid_ba.
```

The last clause of this predicate calls the next clause **check_resid_ba/0** which checks with the user to see if the residual basal area is agreeable:

```
check_resid_ba :- cls,
      stand(ba,Ba,_),
      stand(residual_ba,Residba,_),
      display_window(rpex28),
      tmove(8,49),write(Ba),
      tmove(10,49),write(Residba),
      display_yes_no_dialog(rpq6),
      get0_noecho(Result),
      case([ Result=89 → Ans='yes', Result=78 → Ans='no']),
      Ans == 'no',
      asserta((temp(residual_ba,Residba))),
      retract(stand(residual_ba,_,_)),
      question(residual_ba,_,_,0,New_Residba),
      cls.
check_resid_ba.
```

The basal area *Ba* and the current residual basal area *Residba* are fetched from the working memory. A window is displayed (Figure 8–3) and these values are written to the screen in their correct locations (the **tmove/2** predicate moves the cursor to a particular row-column position). A yes-no dialogue window is displayed which waits for the user to make a selection to agree or disagree with the suggested residual basal area. If the user agrees (*Ans = yes*), then the predicate fails and backtracking causes the second clause of the predicate to succeed and the program continues.

If the user wishes to change the residual basal area (*Ans = no*), then the current residual basal area is saved in working memory under a temporary predicate and its **stand/3** value is removed. The user is then asked to supply a new value for the residual basal area using the **question/5** predicate defined above. The **ask/2** predicate is called to retrieve input from the user:

Module Name: Non-Harvest Treatment Implementation

Residual Basal Area Acceptance Check

The current basal area is . . . . 120 sq. ft.

The recommended basal area is . . . . 109 sq. ft.

Do you accept the recommended basal area that this stand will be thinned back to?

No
Yes

Figure 8–3. An example window explains run-time dependent information and requests user input.

```
ask(residual_ba,Response) :-
     cls,
     temp(residual_ba,Residba),
     display_window(rpex55),
     tmove(9,12),write('The suggested residual basal area after'),
     tmove(10,12),write('thinning is '),write(Residba),write('sq.ft/ac. Enter your'),
     tmove(11,12),write('preferred value. (Enter ? for explanation)'),
     tmove(13,37),read_string(255,Response).
```

The screen in Figure 8–4, is displayed, the text and variable values written to the screen, and the user is prompted for an input value using the **read_string/2** predicate. This value is processed by the **check_number/5** and **places/3** predicates. If a number is returned then it will be saved to working memory by the **question/5** predicate. If a "?" was input then **check_number/5** executes the accompanying explanation screen **explain/1**.

```
explain(residual_ba) :- display_window(rpc07).
```

The various display predicates, **display_window/1** and **display_yes_no_dialog/1** are predicates that put information on the screen and handle single-choice-

Please enter the requested numeric information.

The suggested residual basal area after thinning is 109 sq. ft./ac. Enter your preferred value. (Enter ? for explanation)

-->

Figure 8–4. An I/O window prompts the user for numeric input. Access to the explanation subsystem is also available to the user.

selections by the user. The definition of these predicates are so system dependent that a particular implementation would be meaningless to most readers. Please refer to the source code of RP-FMAS (see the Frontmatter for the location of source code) to study our particular way of defining these predicates.

8.3.3 Explanation Management

The reader may remember that in Chapter 3 we introduced the role of explanation management in knowledge-based systems. This section identifies how explanations are implemented in native PROLOG in RP-FMAS. Two types of explanations will be covered in this section: (1) 'why' is this question being asked, and (2) 'how' did the system arrive at this conclusion (see section 3.1.2.1).

The 'why' type of explanation is managed by the **explain/1** predicate attached to each question being asked. The purpose here is to provide an interpretation of the question and some reasoning about why the question needs to be answered. The current implementation provides a single static text screen supplemented as needed by run-time determined values retrieved from the working memory. The screens are called in the **explain/1** predicate.

```
explain(harvest2) :-
    repeat,
    stand(age,Age,_), stand(rotage,Rotage,_),
    display_window(rpex46),
    tmove(7,41),write(Age),
    tmove(8,51), write(Rotage),
    tmove(24,20),write('Press [RETURN] to continue'),
    keyb(_,N1), N1 == 28, !.
```

The stand age and rotation age are fetched from working memory, a file with a static, pre-designed explanation is displayed, the current age and rotation age are written to the correct locations on the screen (tmove(7,41) moves the cursor to row 7, column 41 before writing the age), and the user is prompted to press the carriage RETURN key. Pressing any other key will cause backtracking to occur until the **repeat/0** predicate is encountered causing another run through each of the following predicates. That is, the user must eventually press the RETURN key to get past this screen display. This is in effect a DO-UNTIL loop in programming terminology. The **explain/1** predicate above shows how to combine a static, explanation screen 'rpex46' (Figure 8–5) with dynamic, run-time dependent information into a user support system.

An effective explanation of the inference process, that is an explanation of the 'how' type, can be developed using text statements, which if combined in the proper order make an understandable justification of the reasoning process

```
You have selected to:  end this rotation and harvest this stand.

We recommend this selection with a confidence level of:  100

BECAUSE:  The age of this stand [ 104 years ] reached or passed
          the planned rotation age [ 104 years ].

Do you wish to exit the advisory system at this point?

                                                  No
                                                  Yes
```

Figure 8–5. An explanation screen may be coupled with a user input request.

(Marcus 1986, Lucas 1988, Merritt 1989). We feel this type of explanation can be more meaningful for the user than a simple regurgitation of the sequence of rules which have fired so far. Riordan and Carden (1990) illustrate another method for associating explanatory text with general principles rather than particular rules that have fired. For a further discussion of this issue, please return to Chapter 3. We use the **text/1** predicate to perform this function. For example, using one of the treatment recommendation clauses given above:

```
test_rec :-
      stand(dbh,Dbh,_),Dbh > 5.0,
      stand(ba,Ba,_),
      stand(thin_ba_x,X,_),
      stand(max_ba,Max_ba,_),
      Ba > X * Max_ba,
      asserta((stand(recommendation,'Thin',100))),
text(recommendation('Thin')).
text(recommendation('Thin')) :-
      M1 = 'Because this stand is greater than 5" dbh, and',
      M2 = 'the basal area is greater than the recommended maximum,',
      M3 = 'thinning is recommended.',
      report_text([M1,M2,M3]).
```

The process, then, invokes the predicate **text/1** from the rule which is fired. Next, **text/1** creates a list of text lines which explains the reasoning of the fired rule in plain English and calls the predicates **report_text/1** and **write_list/2** which append the explanatory text to the trace file 'rptrace.doc'. The trace file is available to the user for editing and saving for future reference at the end of every consultation session.

8.3.4 Uncertainty Management

RP-FMAS-1989 (the 1989 version of RP-FMAS) is a deterministic, not a probabilistic, system and does not use inexact reasoning methods in the commonly accepted ways (Giarratano and Riley 1994, Graham and Jones 1988). We recognize that reaching decisions with absolutely certain knowledge and data is an ideal that is impossible to achieve in reality. And yet when experts teach their expertise or condense it into "manager's handbooks" (Benzie 1977) there is rarely any explicit mention of uncertainty. Psychological testing indicates that people seldom employ explicit, quantitative, and rigorous methods for making likelihood or certainty estimates when making decisions (Penzias 1989). Transforming the Red Pine Manager's Handbook into RP-FMAS did not require the extensive use of uncertainty. When uncertainty was used, it was used merely as a way to index the degree of belief in a rule if it successfully fired. There are others who look upon uncertainty manipulations in expert systems with skepticism (Covington et al. 1988; pp. 328-330). We will, however, present implementation techniques for using the standard inexact reasoning methods in the next chapter.

To illustrate how we treated uncertainty in RP-FMAS, let's look at the treatment recommendation module (Figure 8–2, #5). The degree of belief in *Harvest* depends upon how close the age of the stand is to the rotation age as follows:

Case 1: Stand age equal to or greater than rotation age; CF = 100

```
test_rec :-
    stand(age,Age,_),
    stand(rotage,Rotage,_),
    Age >= Rotage,
    asserta((stand(recommendation,'Harvest',100))).
```

Case 2: Stand age within 95% of rotation age; CF = 90

```
test_rec :-
    stand(age,Age,_),
    stand(rotage,Rotage,_),
    Age >= 0.95*Rotage,Age < Rotage,
    asserta((stand(recommendation,'Harvest',90))).
```

Case 3: Stand age within 90% of rotation age; CF = 80

```
test_rec :-
    stand(age,Age,_),
    stand(rotage,Rotage,_),
    Age >= 0.9*Rotage,Age < 0.95*Rotage,
    asserta((stand(recommendation,'Harvest',80))).
```

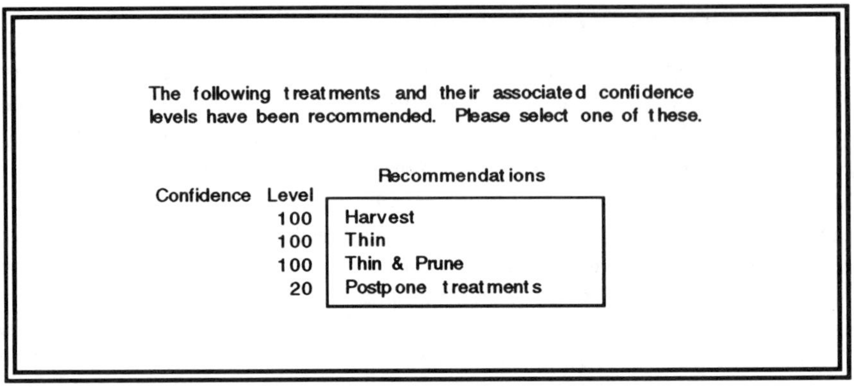

Figure 8–6. Confidence values can be used to express degrees of belief to the user.

Case 4: Stand age within 85% of rotation age; CF = 70

```
test_rec :-
    stand(age,Age,_),
    stand(rotage,Rotage,_),
    Age >= 0.85*Rotage,Age < 0.9*Rotage,
    asserta((stand(recommendation,'Harvest',70))).
```

Case 5: As long as a stand is merchantable it is eligible for harvest but only let it show up when no other harvest is recommended therefore this predicate is position dependent; CF = 20

```
test_rec :-
    not stand(recommendation,'Harvest',_),
    check_merchantability(Time),Time == 'Now',
    asserta((stand(recommendation,'Harvest',20))).
```

This way of indexing uncertainty allows us to show the user the varying degree of belief we have in the various recommendations from which the user must choose (Figure 8–6). Admittedly, these values are simply set and not calculated from previous uncertain values and may be improved as the need arises in later versions of RP-FMAS.

8.4 SYSTEM CONTROL LEVEL

Implementing the blackboard architecture (discussed in Chapter 2) requires (1) a supervisory module, (2) a collection of autonomous, specialty modules and (3) a blackboard data structure with an agreed upon data format (implemented as a file in RP-FMAS) (Figure 6–14).

The function of the supervisory module is to examine the blackboard file, 'b-board.kbs', and decipher which module should be executed next. The implementation of the supervisory module begins with the **main/0** predicate:

```
main :-
    initialization,
    introduction,
    repeat,
    reconsult('b-board'),
    route,
    known(end), !,
    cls.
```

The first predicate initializes the necessary data files and the second presents introductory screens to the user. We then set up a repeat-fail loop that can only succeed if the consultation is to be ended. It works in the following way, **route/0** is the main predicate in the supervisory module; it executes specialty modules. The **known/1** predicate is used by the **route/0** predicate to determine which specialty module to call next. If *known(end)* is true, then the user wants to end the consultation and the **main/0** predicate becomes true. If *known(end)* is false, then backtracking occurs until the **repeat/0** predicate is encountered and forward execution starts again with the predicate **reconsult/1**. The changed blackboard is accessed and **route/0** is once again called to decide which specialty module to call next.

Let us look at a typical set of **route/0** predicates:

```
route :-
    not known(input),
    shell(fmsinput),
    asserta((known(input))).
route :-
    not known(input_analysis),
    shell(fmsdesc),
    asserta((known(input_analysis))).
route :-
    not known(m_objective),
    shell(fmsprod),
    asserta((known(m_objective))).
route :-
    not known(treat_rec),
    shell(fmsrec),
    asserta((known(treat_rec))).
```

The sequence is always the same. Each time **route/0** is invoked, the knowledge base is searched from beginning to end until the first **route/0** predicate evaluates to true. A consequence of this simplicity is that the **route/0** predicates are order dependent. A **route/0** predicate can succeed only if the working memory of the supervisory module does not contain a **known/1** predicate corresponding with that rule. For example, if *known(input)* is false then *not known(input)* is true and the input processing module "fmsinpuut" is executed by the **shell/1** predicate. The input processing module is a completely independent executable program that can be written in any convenient computer language and can use any knowledge representation scheme (see section 6.2.3). The only requirement is that it be able to read-from and write-to the blackboard file. Independent modules may also be written as separate modules within the main program.

The data structure of the blackboard must be uniform so that all modules can access it easily. RP-FMAS is concerned with describing and manipulating single stands. Using the fact structure defined above, we can create the blackboard as a set of order-independent native PROLOG facts of the form **stand/3**:

```
stand(treatment,'Postpone treatments',100).
stand(rotage,108,100).
stand(thin_ba_x,0.8,100).
stand(dbh_goal,16,100).
stand(product,'Multiple products',100).
stand(stocking,'Within recommended levels',100).
stand(tpa,2037.0,100).
stand(dbh,3,100).
stand(height,21.0,100).
stand(covertype,'Red pine',100).
stand(age,16,100).
```

The blackboard contains all of the important attributes and values that the system knows about the stand at any given time. This is also the only information available to and needed by the 12 autonomous modules that make up the RP-FMAS system.

It will be necessary for each module to write the updated version of the blackboard predicates out to the 'b-board.kbs' file for later use by other modules. The **output/0** predicate, defined in the next section, was written to perform this standard function.

8.5 PUTTING IT ALL TOGETHER

To illustrate how the predicates discussed above together constitute a complete knowledge base system, we provide the full source code for module #7 of the knowledge model. The purpose of this module is to change the stand description thus simulating the implementation of the selected intermediate treatment. The knowledge base of this module contains the logic to simulate implementation of treatments. For example, if the recommendation is to "thin" the stand, then *basal area* would be reduced by 1/2 and the other affected variables, such as *trees per acre*, are also recalculated. If thinning is selected, then the only action taken is to suggest a *residual basal area* value to the user. The user can either accept this value or choose to enter a new *residual basal area* value. The blackboard will not actually be changed, i.e. thinning implemented, until the growth and yield model is called (see the operational flowchart depicted in Figure 8–2).

```
/*
'* * * * * * * * * * * * * * * * * * * * * * * * * * * * * * * * * * *
'* FOREST MANAGEMENT ADVISORY SYSTEM
'* * * * * * * * * * * * * * * * * * * * * * * * * * * * * * * * * * *
'* Program......:  rpimp.ari
'* Function......:  implement the treatments selected ie. make
'*                  any changes necessary to the stand records
'* Author.........:  H.M. RAUSCHER
'* Date............:  12/9/88
'* Last Update.:  10/26/94
'* Comments...:  treatments covered are: release, weed, clean,
'*                  prune, thin, do nothing, thin & prune
'
* * * * * * * * * * * * * * * * * * * * * * * * * * * * * * * * * * */

% CONTROL SECTION -----------------------------------------------

main :-cls,
     reconsult('b-board.kbs'),
     stand(treatment,Name,_),
     implement_rec(Name),
     output.
% PROCEDURAL SECTION --------------------------------------------

% these predicates create a unified question system for
% the Forest Management Advisory Systems
% the place for certainty factors is reserved but not used

% case 1: text comparison if in knowledge base
question(Name,Answer) :-
     nonvar(Answer),
     stand(Name,Input,_), !,
     Answer == Input.
```

```
% case 2: text comparison if not in knowledge base
question(Name,Answer) :-
    nonvar(Answer),
    ask(Name,Input),
    asserta((stand(Name,Input,100))), !,
    Answer == Input.

% case 3: text retrieval if in knowledge base
question(Name,Answer) :-
    var(Answer),
    stand(Name,Answer,_).

% case 4: text retrieval if not in knowledge base
question(Name,Answer) :-
    var(Answer),
    ask(Name,Answer),
    asserta((stand(Name,Answer,100))).

% case 5: numeric comparison if in knowledge base
question(Name,Min,Max,_) :-
    stand(Name,Input,_), !,
    Input >= Min,
    Input =< Max.

% case 6: numeric comparison if not in knowledge base
question(Name,Min,Max,N_digits) :-
    repeat, cls,
    ask(Name,Instring),
    check_number(Name,Min,Max,Instring,Inreal), !,
    places(Inreal,N_digits,Answer),
    asserta((stand(Name,Answer,100))).

% case 7: numeric retrieval if in knowledge base
question(Name,_,_,N_digits,Answer) :-
    stand(Name,Inreal,_),
    places(Inreal,N_digits,Answer).

% case 8: numeric retrieval if not in knowledge base
question(Name,_,_,N_digits,Answer) :-
    repeat, cls,
    ask(Name,Instring),
    check_number(Name,_,_,Instring,Inreal), !,
    places(Inreal,N_digits,Answer),
    asserta((stand(Name,Answer,100))).

% predicates to handle checking a number out
% case 1: Instring is "?"
check_number(Name,Min,Max,Instring,Number) :-
    $?$ == Instring,
    explain(Name), !,
    fail.
```

```
% case 2: comparison needed
check_number(Name,Min,Max,Instring,Number) :-
    nonvar(Min),
    float_text(Number,Instring,_),
    Number >= Min,
    Number =< Max.

% case 3: number retrieval needed
check_number(Name,Min,Max,Instring,Number) :-
    var(Min),
    float_text(Number,Instring,_).

% case 4: error checking and try again
check_number(_,Min,Max,_,_) :-
    cls,
    write('A number between '),write(Min),write(' and '),
    write(Max),write(' was expected.'),nl,nl,
    write('Press [RETURN] and please try again'),nl,
    keyb(_,_),
    fail.

% routines to provide the correct type of number
places(Inreal,N_digits,Number) :-
    N_digits = 0,
    Number is integer(Inreal+0.5).
places(Inreal,N_digits,Number) :-
    Number is round(Inreal,N_digits).

% assume dbh remains the same and tpa has changed, so need to redo ba
recalc_ba :-
    retract(stand(ba,_,_)),
    stand(dbh,Dbh,_), Dbh > 0,
    stand(tpa,Tpa,_), Tpa > 0,
    Iba is Tpa*Dbh*Dbh*0.005454,
    Ba is round(Iba,0),
    asserta((stand(ba,Ba,100))).

% assume dbh remains the same and ba has changed, so need to redo tpa
recalc_tpa :-
    retract(stand(tpa,_,_)),
    stand(ba,Ba,_),Ba > 0,
    stand(dbh,Dbh,_),Dbh > 0,
    Itpa is Ba/(Dbh*Dbh*0.005454),
    Tpa is round(Itpa,0),
    asserta((stand(tpa,Tpa,100))).
/* ------------------------------------------------- */

% check with user to see if the residual basal area is agreeable
check_resid_ba :- cls,
    stand(ba,Ba,_),
```

```
    stand(residual_ba,Residba,_),
    display_window(rpex28),
    tmove(8,49),write(Ba),
    tmove(10,49),write(Residba),
    display_yes_no_dialog(rpq6),
    get0_noecho(Result),
    case([ Result=89 → Ans='yes',
        Result=78 → Ans='no']),
    Ans == 'no',
    asserta((temp(residual_ba,Residba))),
    retract(stand(residual_ba,_,_)),
    question(residual_ba,_,_,0,New_Residba),
    cls.
check_resid_ba.
ask(residual_ba,Response) :-
    cls,
    temp(residual_ba,Residba),
    display_window(rpex55),
    tmove(9,12),write('The suggested residual basal area after'),
    tmove(10,12),write('thining is '),write(Residba), write('sq.ft/ac. Enter your'),
    tmove(11,12),write('preferred value. (Enter ? for explanation)'),
    tmove(13,37),read_string(255,Response).
explain(residual_ba) :- display_window(rpc07).
% save the stand/3 and known/1 predicates to disk file 'b-board.kbs'
output :-
    create(H,'b-board.kbs'),
    findall(stand(P,Q,R),stand(P,Q,R),L),
    out1(H,L),
    findall(known(P),known(P),L1),
    out1(H,L1),
    close(H).
out1(H,[]) :- !.
out1(H,[Head Tail]) :-
    writeq(H,Head),write(H,.),nl(H),
    out1(H,Tail).
% KNOWLEDGE BASE SECTION -------------------------------------------
% every implementation predicate must fail except one which must succeed.
% there must be an overall trap to see that there is some predicate
% for all treatments.
% implementation means changing something in the stand description
% or adding something to it
% implementation for release; assume 100% release effectiveness
implement_rec('Release') :-
    stand(free_to_grow,_,_),
```

```
    retract(stand(free_to_grow,_,_)),
    stand(tpa,Tpa,_),
    asserta((stand(free_to_grow,Tpa,100))).
% assume 50% removal
implement_rec('Weed') :-
    stand(tpa,Tpa,Cnf),Tpa > 0,
    Itpa is Tpa/2, places(Itpa,0,Xtpa),
    retract(stand(tpa,_,_)),
    asserta((stand(tpa,Xtpa,Cnf))),
    recalc_ba.
% assume 50% removal
implement_rec('Clean') :-
    stand(ba,Ba,Cnf),Ba > 0,
    New_ba is Ba/2, places(New_ba,0,X_ba),
    retract(stand(ba,_,_)),
    asserta((stand(ba,X_ba,Cnf))),
    recalc_tpa.
implement_rec('Prune') :-
    stand(tpa,Tpa,Cnf),
    Tpa >= 50,
    asserta((stand(trees_pruned,50,Cnf))).
% if <50 tpa, then 30 of the best are pruned
implement_rec('Prune') :-
    stand(tpa,Tpa,Cnf),
    Tpa < 50,
    asserta((stand(trees_pruned,30,Cnf))).
% there is no consideration yet for thinning from above or below, ie. the thinning
   method, row, above, below
implement_rec('Thin') :-
    stand(product,Product,_),Product = 'Large sawtimber',
    stand(min_ba,Minba,_),
    stand(max_ba,Maxba,_),
    stand(ba,Ba,_),
    (Minba+Maxba)/2 < Ba/2,N1 is Ba/2, places(N1,0,N),
    asserta((stand(residual_ba,N,100))),
    check_resid_ba.
implement_rec('Thin') :-
    stand(product,Product,_),Product = 'Large sawtimber',
    stand(min_ba,Minba,_),
    stand(max_ba,Maxba,_),
    stand(ba,Ba,_),
    (Minba+Maxba)/2 >= Ba/2,N1 is (Minba+Maxba)/2,places(N1,0,N),
    asserta((stand(residual_ba,N,100))),
    check_resid_ba.
```

```
implement_rec('Thin') :-
    stand(product,Product,_),Product == 'Large sawtimber',
    stand(min_ba,Minba,_),
    stand(ba,Ba,_),
    Minba > Ba/2,
    asserta((stand(residual_ba,Minba,100))),
    check_resid_ba.

implement_rec('Thin') :-
    stand(product,Product,_),Product = = 'Large sawtimber',
    stand(min_ba,Minba,_),
    stand(ba,Ba,_),
    Minba =< Ba/2,N1 is Ba/2,places(N1,0,N),
    asserta((stand(residual_ba,N,100))),
    check_resid_ba.
% to implement thin and prune, simply implement each independently
% note that pruning has no consequences so far in the system
implement_rec('Thin & Prune') :-
    implement_rec('Prune'),
    implement_rec('Thin').

implement_rec('Postpone treatments').

implement_rec(What) :- cls,
    write('There is no implementation for the treatment '),nl,nl,
    write(' '),write(What),nl,
    keyb(_,_),
    asserta((known(error))).
```

This completes our presentation of RP-FMAS using native PROLOG. Interested readers are referred to the Frontmatter for the locations of the complete source code files and stand-alone-executable modules. The next section examines some of the PROLOG alternatives to using native PROLOG for KBS implementation. Those readers that have no interest in going beyond native PROLOG may skip the remainder of this section and also Chapter 9.

8.6 FROM NATIVE PROLOG TO META-PROLOG PROGRAMMING

The field of logic programming has expanded tremendously in the last few years and its application to knowledge-based system development has become much easier and conceptually more clear (Jackson et al. 1986). Logic programming is ideal for problem solving because (1) it allows all reasoning to be performed explicitly and not hidden in data or control structures and (2) there is a practical language, PROLOG, for implementing problem solvers, i.e. knowledge-based systems (Sterling 1984). The simplicity, elegance, high expressiveness, and unifying effect of logic programming has been instrumental in its successful application as a powerful problem solving paradigm (Lloyd 1984). In addition, native PROLOG provides much of the inferencing machinery needed for knowledge-based system development. PROLOG has proven suitable for the representation of knowledge in most of the formalisms used in AI, i.e. production rules, semantic nets, frames, objects (Lee 1986). Even neural nets can be implemented and executed using PROLOG once these nets have been trained.

Having justifiably praised PROLOG, it is time to qualify that praise somewhat. Using only native PROLOG to develop knowledge-based systems may be unsatisfactory except in simple cases. The three essential features of most knowledge-based systems not found in native PROLOG are (Sterling and Beer 1989):

1. little flexibility and accessibility in controlling the inferencing method;
2. no explicit mechanism for reasoning with uncertainty;
3. no explicit explanation facilities;
4. non-standard user interface management support tools.

To overcome these shortcomings, developers of KBSs have used PROLOG to extend and modify itself. It is precisely this ability to *enhance* the PROLOG language that has contributed in no small degree to its successful use in solving many diverse problems. Features of PROLOG that allow this modification are (Rossi 1986):

1. no distinction between "data" and "program"
2. meta-level predicates such as *clause, call, listall*
3. flexible operator definitions

The question that has not yet been cleanly resolved is how to enhance PROLOG without severing its ties to the clear and powerful foundation provided by logic (Rossi 1986).

Four major approaches for extending PROLOG for use as an KBS development language have been proposed so far (Torasso and Console 1989):

1. direct implementation
2. meta interpretation
3. extending the native PROLOG language
4. front end preprocessing systems

Each of these approaches has its problems. The method of direct implementation (Clark and McCabe 1982) augments each predicate in the knowledge base with additional arguments to implement such features as explanation, uncertainty, and additional control mechanisms. The objection to this method rests upon its violating the generally agreed upon dictum that control and domain knowledge be kept strictly separate (Kowalski 1979, Shapiro 1983, Sterling and Beer 1989, Torasso and Console 1989). Direct implementation maintains the efficiency and structure of the knowledge base but sacrifices modularity and clarity (Sterling and Beer 1986). The entire knowledge base must be altered to incorporate a new feature because the knowledge representation structure has to be redesigned to implement these extensions. This is a clearly undesirable situation (Rossi 1986).

A second approach is based on the use of meta interpreters (Bowen and Weinberg 1985, Sterling and Shapiro 1986). The importance of this approach is that the control component is explicitly separated from the knowledge base (Rossi 1986). The meta interpreter is used to implement alternative control strategies and to add features like explanation and reasoning with uncertainty. Separating control and domain knowledge makes it possible to modify the meta interpreter without changing the knowledge base and vice versa (Jackson et al. 1989). Meta-interpreter enhancements have been studied and categorized as structural, contextual, or behavioral (Sterling and Beer 1989). On the negative side, the extra level of interpretation may impose an unacceptable slowdown in execution speed. A technique called partial evaluation of meta interpreters (Takeuchi and Furukawa 1986, Safra and Shapiro 1986) has recently been developed to cope with the inefficiency problem of meta-interpreters at the cost of a rather daunting increase of complexity. Using the partial evaluation technique effectively relegates this method to expert-level PROLOG programmers.

The third approach is to extend the native PROLOG language to make it more directly useful for representing and manipulating knowledge. The resulting extended PROLOG should have increased flexibility and efficiency while still maintaining its logical roots and declarative style. Ordinary PROLOG would be a subset of the extended version (Rossi 1986). There is no shortage of proposals for such extensions in the logic programming literature (see Torasso and Console 1989, p. 174) some of which have already been implemented as enhanced PROLOG languages: IC_PROLOG (Pereira 1982), Epilog (Porto 1982), PROLOG/KR (Nakashima 1982), ESP (Chikayama 1983), CPU (Leonardi et al. 1989). No extension has as yet gained wide-spread acceptance as a better KBS implementation language than PROLOG.

The final approach to enhancing PROLOG is called front-end preprocessing (Torasso and Console 1989). Given a knowledge representation formalism, i.e. production rules, frames, etc., a high level language for representing both domain knowledge and the desired inference strategy to be used in a specific case can be defined. Such a high-level description can then be automatically translated by a preprocessing program into native PROLOG code. Proponents claim that this method is flexible, the resulting systems are efficient, and the preprocessor can be implemented in PROLOG. On the negative side, the preprocessor is very complex and the resulting code may be difficult to interpret.

It is not yet clear, which of the four approaches will eventually dominate. Direct implementation can be eliminated from consideration on the basis of its serious practical as well as theoretical disadvantages. Extending the native PROLOG interpreter/compiler is not an option for the typical expert system developer and can therefore also be eliminated. Front-end preprocessing is the newest of the methods and not enough is known about it to consider it a contender for implementation at the present time. The elimination of each of the other three, leaves meta-interpretation as the method of choice.

This approach is fairly straight forward. Implementation is feasible for a "typical" PROLOG programmer. Unfortunately, many argue that it is usually necessary to couple a meta interpreter with partial evaluation to avoid a substantial loss in operating speed. Because a great jump in complexity occurs when a partial evaluation method is used, developers must be "expert" PROLOG programmers to successfully apply this method. Both Takeuchi and Furukawa (1985) and Sterling and Beer (1986) report reductions of speed by a factor of 40 without partial evaluation methods. On the other hand Black and Manly (1987) report that their meta-interpretation system without partial evaluation ran 2/3 as fast as native PROLOG. It is probably safe to say that unless applications are very large and deal with real-time operations, then meta interpretation without partial evaluation will execute fast enough.

One should certainly avoid a single, large PROLOG KBS with hundreds and maybe even thousands of rules. Such a system is bound to be slow, even without the additional burden of a meta interpreter. On the other hand, if the system is highly modular and each module is relatively small and independently executable, then execution speed tends to be adequate (Rauscher et al. 1990). System control architectures such as blackboards, intelligent agents, and actors have been developed to coordinate the performance of such a set of modules. Furthermore, a meta interpreter control mechanism can be written to make each particular system module find its answers by considering a minimum number of domain rules, thus increasing run-time efficiency. Finally, micro-computer generations seem to have about a 4-5 year cycle. Each new generation increases processing speed by a factor of 10 to 20. In our experience, computing speed is much less of a concern than software development complexity. For these reasons, the meta-interpreter approach without the partial evaluation technique seems the most reasonable for "enhancing" native PROLOG for KBS problem solving applications at this time.

8.7 SUMMARY

The three logical levels that make up the architectural structure of most KBSs are the domain level, the tactical control level and the system control level (or architecture). Using these levels as an organizational framework, a fully functional knowledge-based system using native PROLOG has been presented. Native PROLOG provides the inferencing mechanism and predicate logic serves as the knowledge representation scheme. To illustrate how these pieces fit together, the source code of one complete module from RP-FMAS was included. Although native PROLOG has many attractive features for implementation of KBSs, not the least of which is its simplicity, there are several major deficiencies. Native PROLOG does not provide any explicit structures for knowledge-based system explanations, uncertainty, and multiple inference control methods. There are four possible alternatives for "enhancing" PROLOG to allow a more elegant inclusion of these features. Writing meta-interpreters in PROLOG itself seems the most attractive option for increasing the power of PROLOG for the development of knowledge-based systems. A variant of this approach uses meta PROLOG to develop a toolkit of interchangeable KBS components. These components, then, make up modular pieces of a knowledge-based system "shell". This approach will be explained in the next chapter.

CHAPTER 9

A PROLOG Toolkit Approach to Developing Forest Management Knowledge-Based Systems[1]

If you want to change the world, go develop new tools. Brilliant people will always find innovative approaches to interesting problems. But its the tools that set the spectrum of opportunities.

Professors William Dreyer and Leroy Hood

In Chapter 8, native PROLOG was used to develop a complete knowledge-based system. Some of the implementation problems generated by the decision to limit ourselves to native PROLOG were discussed. In this chapter, we will look at how to use meta-PROLOG programming methods to build forest management KBS using a toolkit called DSSTOOLS. We will continue to use the Red Pine Forest Management Advisory System (RP-FMAS) introduced earlier as our implementation example, however, the focus is really on the toolkit. We have rewritten the original RP-FMAS using the PROLOG toolkit presented in this chapter. Readers unfamiliar with forest management theory and practice should review the background for RP-FMAS in section 4.3 and the knowledge model presented in section 6.1.1. Rauscher et al. (1990) and the

[1] We want to acknowledge Donald E. Nute, Guojun Zhu, and Yousong Chang, members of the Artificial Intelligence Laboratory at the University of Georgia, for their contributions to the DSSTOOLS development project described in this chapter.

RP-FMAS hypertext document (see the Frontmatter for file location) may be consulted for additional detail.

The PROLOG toolkit approach, described in this chapter, makes the most sense for developers who expect to be involved in the development of a large and evolving KBS or who expect to be involved in the development of a series of systems in a variety of domains. On the other hand, for more casual experimentation in support of learning activities or for rapid prototyping exercises, the native Prolog programming approach described in Chapter 8 is probably more practical.

9.1 THE KBS TOOLKIT

Many authors have called for toolkit packages that can be tailored to specific applications with a minimum of new programming (Chandrasekaran 1986, Cherubini et al. 1989, Geneserth and Ginsberg 1985, Suzuki 1988). A few PROLOG toolkits have been reported in the literature (Oliveira 1984, Lee 1986). Unfortunately, none of these previously developed PROLOG KBS toolkits are readily available or in wide-spread use at the present time.

The Red Pine FMAS described in this chapter is a decision support system written in PROLOG using the DSSTOOLS toolkit. DSSTOOLS lets developers concentrate on solving the domain problem rather than on writing computer code by making available a library of reusable software code. The toolkit modules can be invoked with small amounts of additional code. They also encapsulate some of the more difficult components of an KBS.

Toolkits are typically associated with a programming language and require at least a moderate level of familiarity with that computer language. For example, the C Language Integrated Production System (CLIPS) toolkit, developed by NASA, is written in the C language and meant to be used within a C language program (Giarratano and Riley 1994). CLIPS is a forward-chaining, rule-based KBS development toolkit. It does not directly support backward chaining, frames, or object-oriented programming (Giarratano and Riley 1994). An important difference between CLIPS and DSSTOOLS is that it is easier to develop alternative inference engines for KBS applications in PROLOG than in C.

The toolkit approach is actually a variation of the native Prolog programming approach, used in Chapter 8, together with a particular design philosophy based on taking the long view about system development. The basic idea of the toolkit is that developers collect pieces of generic code that can be used repeatedly in the same or several systems. When two very similar tasks have to be performed, the toolkit design philosophy dictates that a generic routine should be developed that can do both tasks. The resulting routine is likely to be reusable. As more tasks of a similar nature are discovered, the routine is modified to incorporate them. In this way, the routine eventually evolves into a generic "tool", a module that can be used to perform a class of similar tasks. Toolkit developers attempt to encapsulate the most technical and difficult application functions into easily usable software modules so that the greatest number of application developers can successfully use them.

The toolkit approach has other similarities to the native Prolog programming approach of Chapter 8. The developers write many of the routines themselves and they usually have access to source code for any tools they borrow from other developers. It is also like the native code approach because the developers have the opportunity to modify any part of the code in whatever way is necessary to get a good fit with the domain. The toolkit approach has the added advantage that once the technically challenging toolkit modules have been developed by PROLOG programming experts, it is easier for the intermediate skill level PROLOG programmer to use them in new applications.

With the toolkit approach, system design becomes a matter of determining the tools that will be needed for the project, selecting the system architecture that will allow the different tools to be integrated, and either constructing or acquiring the tools needed. There is a growing repository of code for inference engines and other components of KBSs, making it increasingly easy to acquire items for the system developer's toolkit by modifying existing routines rather than by constructing these routines from "scratch" (Covington *et. al.* 1988, 1995; Shoham 1994).

9.2 DSSTOOLS SYSTEM ARCHITECTURE

In Chapter 7 we made the point that the function of controlling the flow of the KBS should be clearly separated from the function of dealing with the domain knowledge. Failure to adhere to this guideline results in systems that are difficult to understand and maintain (Davis 1980, Sterling 1984). To enforce this design principle, each module of RP-FMAS belongs to one of three logical levels: domain level, tactical control level, or system control level.

The general structure of a forest management advisory system constructed using DSSTOOLS is shown in Figure 9–1. The components of the system that constitute DSSTOOLS itself appear shaded in the diagram. The PROLOG components shown in Figure 9–1 belong to the following logical levels:

Domain Level:	Knowledge Bases
Tactical Control Level:	Reasoning Engines Interface Tools Domain Control Modules
System Control Level:	Blackboard

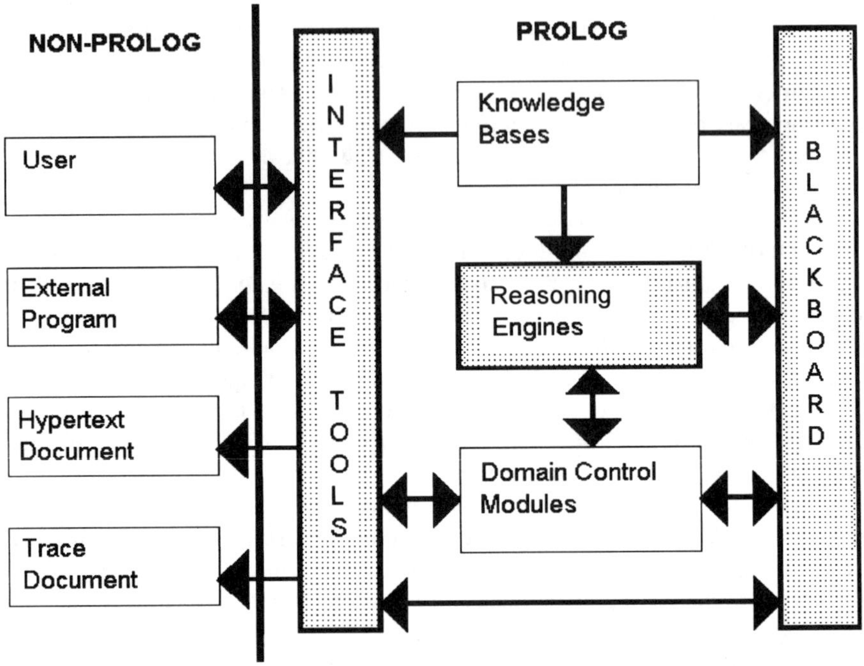

Figure 9–1. The general structure of a KBS constructed using DSSTOOLS includes both PROLOG and nonPROLOG components Shaded components contain the toolkit modules.

In RP-FMAS, the domain level contains the knowledge of how to manage red pine forests in the form of rules. These rules are found in text files so that users can view them easily and make changes as they gain more experience with knowledge-based systems. These text files can be found in the location spedifical in the Frontmatter of this book. Facts describing a particular red pine stand of interest are also part of the domain level. See section 9.3 for a more detailed discussion of the domain level.

The tactical control level contains knowledge of (1) how to reason with the rules in the knowledgebases; (2) how to obtain input and guidance from the user; (3) how to explain to the user what the system is doing and why; and (4) how to deal with uncertainty in our knowledge. Section 9.4 provides a detailed discussion of the tactical control level in RP-FMAS.

Finally, the system control modules determine how the various tactical modules need to interact to perform system-wide tasks. For example, it is the job of system control to implement the user's decision to start and stop the execution of the program. Section 9.5 discusses the implementation of the system control module in more depth.

9.3 DOMAIN LEVEL

The domain level of the knowledge-based system is concerned with the rules which contain the knowledge about red pine forest management and the facts which identify a specific red pine stand for which decision support is desired.

9.3.1 Fact Structure

Facts are used to store current, instance dependent knowledge in the working memory of the KBS. A uniform structure for all facts must be chosen and then consistently followed because the suite of toolkit modules are all designed to work with a particular fact structure. The following fact structure was used in RP-FMAS:

```
fact(Statement, Certainty, Origin, Time_stamp)
```

where:

Statement = any legal PROLOG assertion

Certainty = any defined symbols for establishing the degree of belief: 95, –.25, true, false, unknown.

Origin = describes the origin of the fact: from the user, from a database file, calculated by an external procedure, inferred from a rule, etc.

Time_stamp = the time when this fact was asserted to the working memory.

An example of a fact in this format is:

```
fact(tpa(800),100,user,102416)
```

which means that: (1) it is known that the variable trees per acre (tpa) has the value of 800; (2) we know this with a 100% certainty; (3) the origin of this fact was the user; and (4) the user entered this fact into the system at 10:24:16 AM.

To reiterate: the particular definition of fact structure a developer chooses is really a matter of personal taste, but once chosen, it must be strictly adhered to. Changing this fact structure will mean changing elements of most of the other toolkit modules in RP-FMAS.

9.3.2 Rule Structure

Rules are used to define relationships between facts and are the heart of a rule-based system. As with facts, the original structural definition of rules is rather arbitrary. But once defined, a single rule structure must be faithfully followed because all the modules of the toolkit will expect to encounter the same structure. Any variance will lead to a system error. Rules are defined in DSSTOOLS in the following way:

```
rule(Type,Environment, Rulename, CF, Body, Short_explanation, Long_explana-
    tion).
```

where:

Type = *fc* for forward chaining; *bc* for backward chaining (see section 9.4.1)

Environment = the context name for a group of rules

Rulename = the unique name of the rule

CF = any defined symbols for establishing the degree of belief: 95, –.25, true, false, unknown.

Body = the 'If..Then..' statement that makes up the rule

Short_explanation = a short explanatory statement to help the user understand the meaning of the rule

Long_explanation = a pointer to a hypertext chunk that provides the user with an in-depth explanation of the rule and its scientific origin.

Both forward chaining and backward chaining rules have the same structure except for the Body, i.e. the form of the 'IF..THEN' statement. The Body of a forward chaining rule looks like:

```
rule(fc,stand_evaluation,age_1,true,
    (   goal(age),
        ht(H),
        si(S),
        A is (H/S)*50,
        then,
        age(A),
        remove(age)
    ),
    'If height and site index are known, age can be estimated by: H/S * 50.', 53).
```

In contrast a backward chaining rule looks like:

```
rule(bc,recommendations,rec_1, 100,
    (   recommendation(_Regenerate_) :-
        size(_Seedling_),
        stocking('Below minimum_)
    ),
    'If the stand is a seedling stand that is poorly stocked then it should be
        regenerated.', 53).
```

Rules should be complete, self-explanatory, logical chunks of knowledge. That is, rules should be statements about the application domain, in this case, red pine forest management. As in the examples above, they should define how to find the value for "age" or how to determine when "Regenerate" is an appropriate recommendation. Each rule then becomes a small piece of logic about the domain that can be understood in isolation and changed in isolation regardless of anything else in the system. To increase clarity, names, i.e. symbols such as "site index", used in the rules, should be chosen to communicate their logical meaning. The meaning of the rules should be easy to understand for those familiar with the domain of discourse.

Readers are referred to the Frontmatter for the location of all the knowledge base files in RP-FMAS.

9.4 THE TACTICAL CONTROL LEVEL

The tactical control level provides a flexible, logic-based method for controlling the operations of each individual module. Each module must be able to perform the following functions:

Inference control management,

User interface management,

Explanation management, and

Uncertainty management

Inference control deals with directing the sequence of reasoning of the module. User interface control concerns the management of the flow of information from the user to the system. Explanation management addresses the flow of information from the system to the user. Uncertainty management allows the use of inexact facts and rules in the reasoning process.

9.4.1 Inference Control Management

The DSSTOOLS modules incorporate as many of the generic tasks associated with inference control as possible. For example, the two inferencing tools, the forward and backward chaining tools, are both in a toolkit module. The application developer is isolated from the technical details of these two meta-interpreters. All that is necessary is for the developer to understand how to use them to support the reasoning process. For example, the domain control module for finding the treatment recommendation (Figure 9–2, Module 3) is written as a single PROLOG clause:

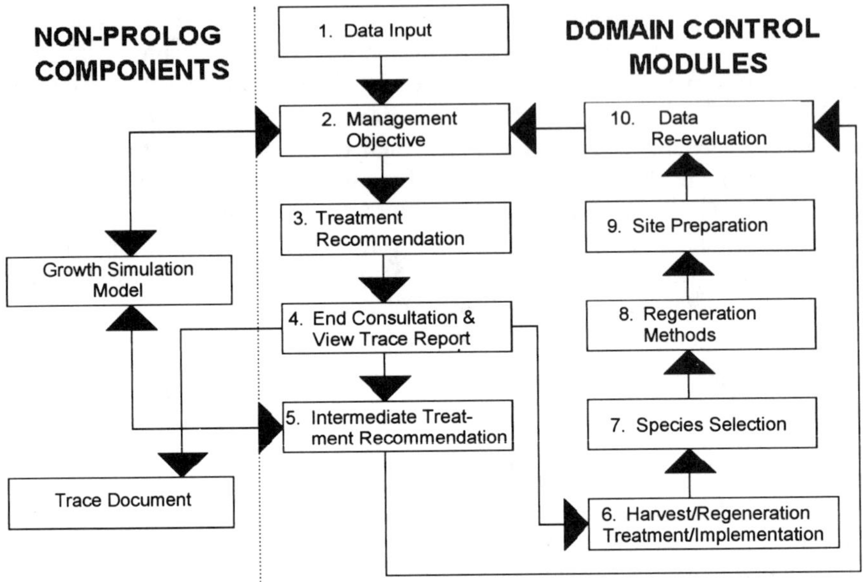

Figure 9–2. A revised version of the operational model flow diagram of Figure 6 depicts the functional nature of the domain control modules.

```
bb_module(recfind) :-
    solve_backward_chainer(merchantability,Value,red_pine_merch,[one,save,trace]),
    solve_backward_chainer(recommendation,Value,red_pine_recs,[all,save,trace]),
    query(user_decision,'Treatment Recommendation',Decision).
```

bb_module/1 is a control predicate which defines the recommendation module as one of the modules recognized by the system control level (see Section 9.5). **solve_backward_chainer/4** is a call to the backward chaining meta-interpreter to perform an analysis on the declarative knowledge base of the domain level. In this case, we need to determine what value the red pine stand in question has for the merchantability variable. So we call the backward chainer and let it look at the merchantability rules to find this value and store it in the working memory as well as return it in this clause. Next, we can call the backward chainer to find all the possible recommendations for this stand. Once all the recommendations have been found, we can use the **query/3** predicate to ask the user to decide which one to implement. See section 9.4.2 for a discussion of the **query/3** predicate.

It is obvious that there is a lot of work going on here, but the application developer is not the one doing it! Generic tools are provided in DSSTOOL in the form of PROLOG predicates which do most of the work.

9.4.1.1 Composing PROLOG Meta-Interpreters. A PROLOG meta-interpreter is a PROLOG program for interpreting and computing statements in PROLOG (Sterling and Beer 1986). No single meta-interpreter for PROLOG will be ideal for all purposes, thus it is desirable that a toolkit provide a suite of different meta-interpreters (Sterling and Lakhotia 1988). For example, the native PROLOG interpreter provides backward chaining inferencing using a depth-first search methodology. We might want to perform inferencing using forward chaining logic or use fuzzy logic for inexact reasoning. We might want to perform breadth-first searches of the domain space rather than depth-first. A different meta-interpreter will be needed for each of the previous examples.

In order to make it easier for the intermediate-level PROLOG programmer to effectively use this advanced technology, a toolkit needs to supply a library of meta-interpreter modules from which to pick appropriate flavors (Sterling and Lakhotia 1988). In the following sections, we will present several different versions, called "flavors", of meta-interpreters for performing the many functions needed by a complete KBS.

The common starting point is the "Vanilla Meta-Interpreter" (Sterling and Shapiro 1986):

```
solve(true).
solve((A,B)) :-
    solve(A),
    solve(B).
solve(A) :-
    sys(A),
    call(A).
solve(A) :-
    clause(A,B),
    solve(B).
```

The reader unfamiliar with the functions of a meta-interpreter, should try the "Vanilla Meta- Interpreter" and trace out its behavior until a complete understanding of its performance has been gained. [The source code of the vanilla meta-interpreter can be located by referring to the Frontmatter.] It is almost impossible to understand the execution details and power of this simple meta-interpreter without running it and studying the execution details.

9.4.1.2 A Backward Chaining Meta-Interpreter. By using the knowledge representation structure discussed in section 9.3, the domain section, we can modify and enhance the vanilla meta-interpreter introduced above. The backward chaining meta-interpreter (BCHAIN) defined below, starts with a query and works backward through the rules to find a way to answer the query.

```
solve_backward_chainer(Goal,Value,Environment,[Howmany,Save,Trace]) :-
    solve_b(Goal,CF,Source).
solve_b(A,CF,Source) :-
    fact(A,CF,Source,_),
    member(CF,[true,100]).
solve_b(A,_,_) :-
    A,
    !.
solve_b((A,B),_,_) :-
    !,
    solve_b(A,_,_),
    solve_b(B,_,_).
solve_b((A;B),_,_) :-
    !,
    (solve_b(A,_,_)
    ;
    solve_b(B,_,_)).
solve_b(A,CF,RuleID) :-
    rule(bc,Environment,RuleID,CF,(A :- Premises),_),
    one(solve_b(Premises,_,_)).
```

This interpreter is capable of solving for *Goal* using backward chaining on the rule structure discussed in section 9.3 where declarative rules may contain:

not A	negation
A , B	conjunction
A ; B	disjunction (if A is true, B is not tried)
A	system predicate or a user defined native PROLOG clause
A	is the Head of a declarative rule

and A and B are legal PROLOG predicates. The "!" is the PROLOG cut (see section 7.2.2.2). The cut prevents backtracking and thus improves efficiency in the operation of the above backward chaining meta-interpreter.

9.4.1.3 A Forward Chaining Meta-Interpreter. The backward chaining meta-interpreter (BCHAIN) described above, starts with a query and works backward through the rules to find a way to answer the query. The forward chaining meta-interpreter (FCHAIN) described in this section starts with the data in the working memory and finds rules that are satisfied by this data. Typically, we want to employ different sets of forward chaining rules at different times. We do this by organizing the rules into "environments" within the knowledge base. When FCHAIN is called to execute within a module, the environment is specified together with a set of goals. FCHAIN fires and performs the action or "then" part of the rule of the first rule it finds in the environment whose conditions are satisfied by facts in the working memory. It then removes any goals from its agenda that the actions it has just taken satisfy and will not fire any other rules for that goal. Then it looks for another rule to fire. It continues in this manner until it can find no more rules in the environment that can be fired.

The PROLOG code for FCHAIN is:

```
call_forward_chainer(GOALS, UNMET_GOALS, ENVIRONMENT) :-
     set_goals(GOALS),
     set_unmet_goals(UNMET_GOALS),
     solve_f(ENVIRONMENT).

solve_f(Environment) :-
     rule(fc,Environment,RuleID,CF,Body,Explanation),
     process(RuleID,CF,Body),
     !,
     solve_f(Environment).
solve_f(Environment) :-
     abolish(goal/1).
process(RuleID,CF,(then,Actions)) :-
     !,
     perform(RuleID,CF,Actions).
process(RuleID,CF,(Condition,Conditions)) :-
     !,
     satisfy(Condition),
     process(RuleID,CF,Conditions).

satisfy(goal(Goal)) :-
     !,
     goal(Goal).
satisfy(test(Pred,Values)) :-
     list(Values),
     !,
     fact(Pred(Value),_,Source,_),
     Source = null,
     member(Value,Values).
satisfy(test(Pred,Value)) :-
     fact(Pred(Value),_,Source,_),
     Source = null.
satisfy(test(X)) :-
     !,
     X.
satisfy(Pred(Value)) :-
     fact(Pred(Value),_,Source,_),
     Source = null,
```

```
        !.
satisfy(X) :-
        X.
perform(RuleID,CF,(Action,Actions)) :-
        !,
        perform(RuleID,CF,Action),
        perform(RuleID,CF,Actions).
perform(_,_,remove(Goal)) :-
        !,
        retract(goal(Goal)).
perform(_,_,goal(Goal)) :-
        !,
        assert(goal(Goal)).
perform(RuleID,CF,Fact) :-
        update_fact(Fact,CF,Rule).
set_goals([]).
set_goals([First Rest]) :-
        assert(goal(First)),
        set_goals(Rest).
set_unmet_goals([]).
set_unmet_goals([First Rest]) :-
        fact(First(_),_,Source,_),
        Source = null,
        !,
        set_unmet_goals(Rest).
set_unmet_goals([First Rest]) :-
        assert(goal(First)),
        set_unmet_goals(Rest).
```

FCHAIN is a short but intricate piece of PROLOG code that can best be understood by actively experimenting with its functions. The code is available by referring to the Frontmatter of this book and we suggest most readers would profit from examining it with a test knowledge base under run-time conditions. GOALS is a list of predicates for which facts are to be inferred. If there are already facts in the working memory for some of these predicates, they will be replaced. UNMET_GOALS is a list of predicates for which facts are to be inferred ONLY IF there are not already facts for these predicates in the working memory. ENVIRONMENT identifies the set of forward chaining rules to be used by the forward chaining inference engine. No values are returned. Instead, appropriate facts will be asserted to the working memory.

9.4.2 User Interface Management

The user interface management module should provide a unified, yet flexible, way to query the user for input and to report the output to the user.

Interface development is time consuming, requires a significant amount of code (up to 60% of total system code), and requires skill in graphic arts and educational psychology that most knowledge engineers do not possess (Rewerts et al. 1990). PROLOG-based query-the-user facilities have been developed by Hammond (1984), Lee (1986), Hein et al. (1987), Covington et al. (1988), and Mao and Pao (1988). One of the difficulties with PROLOG is the lack of standards for user interface features which have prevented the development of generally available standard input-output modules (Meritt 1989). Each PROLOG implementation has more or less unique clauses that control input and output.

In DSSTOOLS, we have defined a "generic" query predicate to handle all input requests to agents outside of the knowledge-based system, which may include users, databases, and other programs. We have designed **query/3** to obtain input from a quantitative simulation model as well as from the user. Several standard types of input are commonly required:

1. query(user_form,Title,Environment).

The PROLOG toolkit contains a generic screen input routine that will build a fill-in-the-blanks form for the user to complete (Figure 9–3). The developer can include information in the knowledge base that will determine:

any introductory information for the form,
the label for each field,
the required data type for each field,
minimum and maximum values for numeric fields,
rules for computing values for some fields when the user provides partial information,
explanatory information for each field

The generic form tool manages this information, informing the user when the information provided is insufficient or violates some parameter. It also provides the user an explanation for any field upon request. The entire form is automatically sized and formatted to fit the specifications identified in the knowledge base.

2. query(user_buttons,Title,Value).

This variant builds an input screen with labeled push buttons. The customized information comes from the knowledge base. The user selects one of the push buttons and its value is returned to the tactical control module for processing.

3. query(user_diagram,Title,Value).

This variant builds an input screen using a user specified graphic diagram in the PCX graphic format. Push buttons are also generated using information supplied in the knowledge base (Figure 9–4).

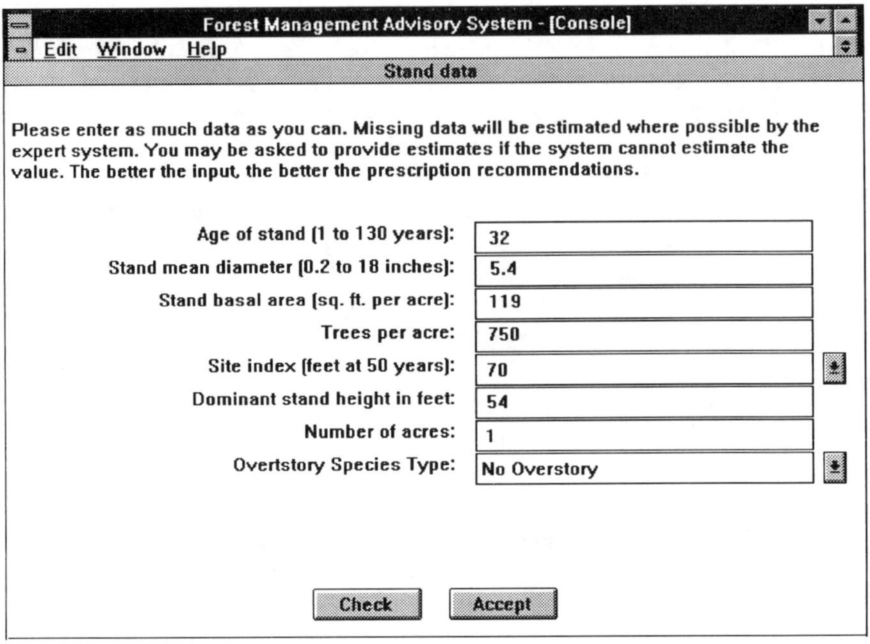

Figure 9–3. The stand data entry screen allows users to enter data in a template fashion. Because this image is a screen capture, its quality is not ideal.

4. query(user_decision,Title,_).

This variant builds an input screen with a list of treatment recommendations together with their degrees of belief (Figure 9–5). This list is dynamically generated by the RP-FMAS and as such is a specific input form for this program. Inevitably, its not possible to generalize all functions that a specific application program will need to perform. In such cases, the application developers must fall back on the underlying PROLOG in order to modify the toolkit modules.

5. query(model,Name,OutputList).

Finally, we can obtain values from a simulation model using the same method. RP-FMAS uses a quantitative simulation model to estimate future growth and yield of red pine stands. The simulation model is written in Microsoft Visual Basic and communication between the PROLOG knowledge-based system and the Visual Basic simulator is by means of ASCII data files. This method is easy to program and understand. Specifically, the application

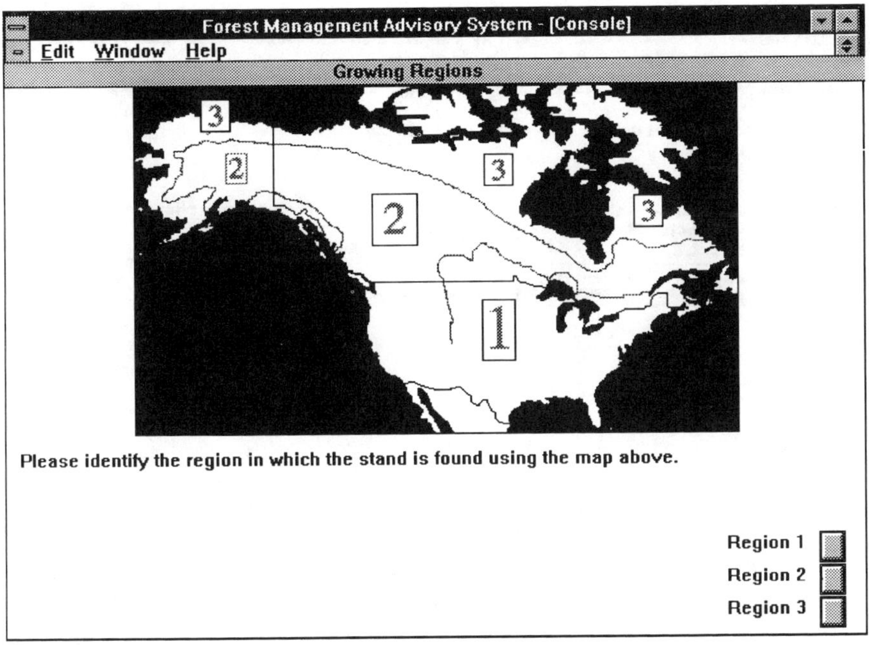

Figure 9–4. A button-screen using a bitmap graphic for user data entry of tree growth zones can be a very useful and informative way to obtain input from the user. Because this image is a screen capture, its quality is not ideal.

developer can put a query to the model into any of the tactical control modules, as for example:

```
query(model,'Red Pine Growth', OutputList, InputList).
```

This statement calls up the following statement in the toolkit:

```
query(model,Modelname,OutputList,InputList):-
    run_model(Modelname,OutputList,InputList).
```

The definition for clause **run_model/2** executes the external model and manages the passing of data back and forth between PROLOG and the external model using data files as explained above. This is a very generic definition and by defining **run_model/2** appropriately, can be used to get values to and from any external model.

Because each variant of the **query/3** tool behaves in a very similar fashion, we will illustrate how they all work by looking in more detail at the structure

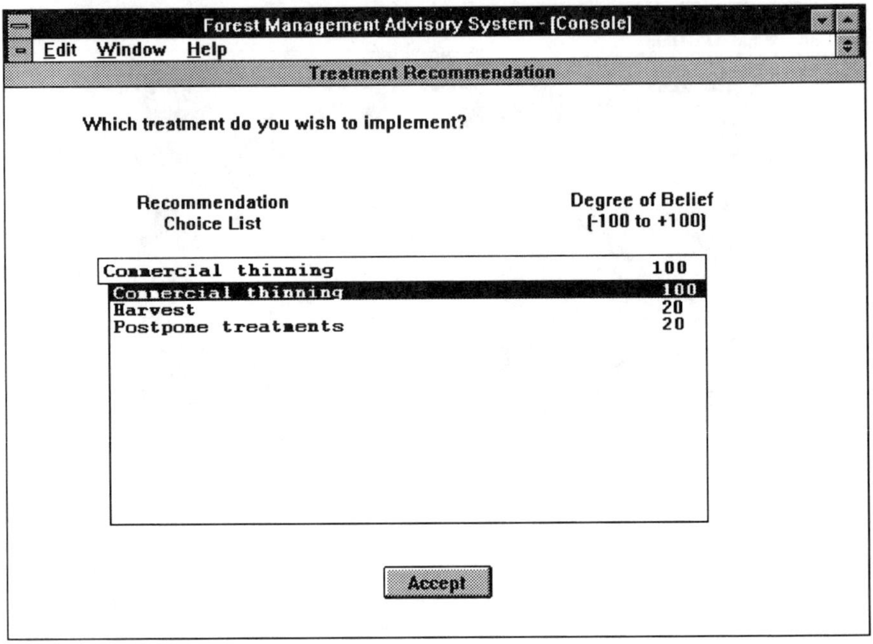

Figure 9–5. A ranked recommendation selection screen prompts for user data entry based on feedback to the user. Because this image is a screen capture, its quality is not ideal.

of only one of them. The initial call that the developer places into the domain control module might be as follows:

```
query(user_form,'Residual Basal Area',Value).
```

This query routine builds an input screen with push buttons labeled using information from a developer-supplied input list defined in the knowledge base in the predicate **input_data/4**. It returns the number of the button that the user selected. An example of what an application developer would place into the knowledge base is:

```
input_data(_Residual Basal Area_,
     ['The current basal area is', ba,
     'sq. ft. The recommended residual basal area is',
     residual_ba, 'sq. ft. Do you wish to change the estimated residual', 'basal
        area?'],
     ['Yes',
     'No'],
     Value).
```

Notice that the current values of the variables "ba" and "residual_ba" are automatically found and inserted into the text that the user will see on the screen. A button screen would be generated and the user would be asked to select one of the available buttons. The number of the button would be returned in *Value* (1 if Yes; 2 if No). By learning a few rules concerning the format of **input_data/4**, an application developer can quickly generate almost all the input screens any program will need.

9.4.3 Explanation Management

One of the defining features of knowledge-based systems is their extensive capability to explain system reasoning. The objective of explanation management is to allow the user to build an understanding of what is happening during the execution of the program and why things are happening the way they are. The DSSTOOLS currently provides four explanatory features:

Explain Further
Facts
Hypertext
Trace

9.4.3.1 Explain Further. Whenever a user is confronted with an input request or is viewing a known fact or rule, it is possible for the user to request further explanations. These explanations are available in short form or in depth. The short form is first offered to the user. This takes the form of a simple natural language text sentence which attempts to explain why the input is needed or the origin of a fact or rule. This sentence is provided by the application developers and is located in the knowledge base files.

The extended explanation is found in a hypertext system which accompanies the knowledge-based system. The Hypertext system can be accessed from the beginning as a stand-alone repository of organized domain-level knowledge, as explained in section 9.4.3.3 below, or a very specific section of the total Hypertext system can be accessed in a context-sensitive fashion. Let's review an example rule:

```
rule(bc,recommendations,rec_1, 100,
    (recommendation('Regenerate') :-
        size('Seedling'),
        stocking('Below minimum')
    ),
    'If the stand is a seedling stand that is poorly stocked then it should be
        regenerated.', 53).
```

The sentence in bold is the short explanation that would be presented to the user to explain this rule further.

The number "53" following this explanatory sentence is a pointer to a particular location in a Hypertext system. We use the Dynamic Data Exchange (DDE) Microsoft Windows standard to communicate between the two Windows-based applications, the PROLOG knowledge-based system and the HyperWriter[tm] hypertext system. A typical scenario begins with the user requesting further explanation about a question being asked or a rule that has been used. The PROLOG knowledge base module involved contains a pointer to a particular section of the hypertext system that is relevant to that particular question or rule. The DDE communication module of the toolkit finds that pointer and sends a message to the hypertext system to go to that particular section and display it for the user. The user is free to use the entire services of the hypertext document and browse at will to any or all parts of it. Upon quitting the hypertext document, the user is returned to the exact location in the knowledge-based system execution point from which the request to the hypertext was generated. In this fashion, the background context for rules can be presented and the rule can be grounded in the scientific literature through standard reference citations.

9.4.3.2 Facts. All the facts currently known by the system can be viewed by the user at any time (Figure 9–6). Furthermore, the user may highlight any particular fact to find an explanation of how that fact was found to be true. If the fact came from a rule, an in-depth explanation in hypertext may also be selected as explained above (Figure 9–7).

9.4.3.3 Hypertext. As has been discussed in Chapter 2, human knowledge is encoded in many ways. One of the most recent and useful representations is in the form of hypertext. In RP-FMAS we use a hypertext authoring system called HyperWriter[tm] (Ntergaid 1994) to organize and present the theory and practice of red pine forest management. It runs on IBM-PC compatible microcomputers under the Windows operating system.

Selecting this option from the explanation system will cause the hypertext system to execute from the beginning. The user will be able to browse through the available information and return to the PROLOG knowledge-based system upon quitting the hypertext document. Having a hyperdocument accompanying a knowledge-based system is a very powerful combination of tools. The knowledge-based system encodes that portion of the scientific knowledge pertinent to the decision process at hand while the hypertext document contains the larger background knowledge that is needed for a deeper understanding of the domain.

9.4.3.4 Trace. The trace explanation feature is implemented as a text file which contains run-time specific information. As each fact is established or each rule is successfully used, this information is placed in the trace file in an easily readable style. All intermediate reports are also copied into this trace file for

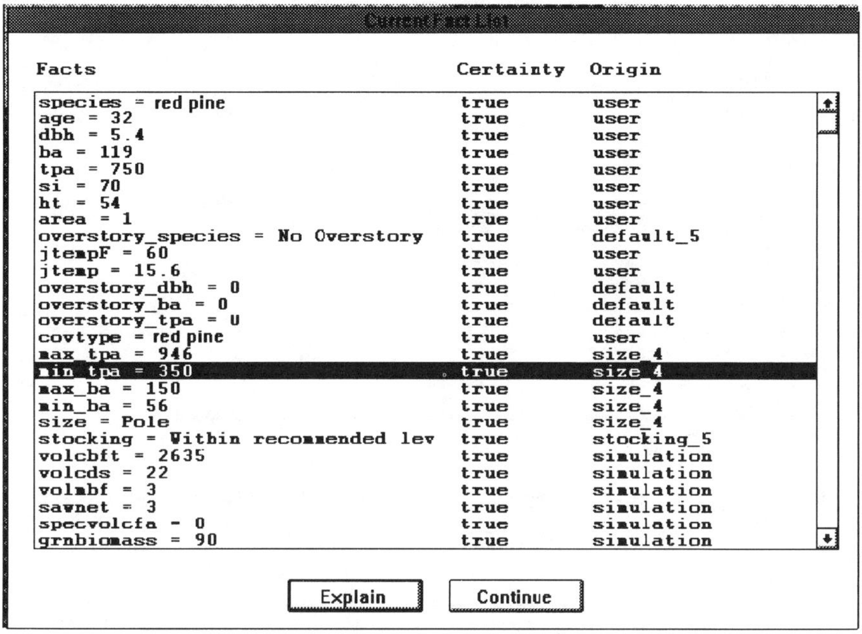

Current Fact List

Facts	Certainty	Origin
species = red pine	true	user
age = 32	true	user
dbh = 5.4	true	user
ba = 119	true	user
tpa = 750	true	user
si = 70	true	user
ht = 54	true	user
area = 1	true	user
overstory_species = No Overstory	true	default_5
jtempF = 60	true	user
jtemp = 15.6	true	user
overstory_dbh = 0	true	default
overstory_ba = 0	true	default
overstory_tpa = 0	true	default
covtype = red pine	true	user
max_tpa = 946	true	size_4
min_tpa = 350	true	size_4
max_ba = 150	true	size_4
min_ba = 56	true	size_4
size = Pole	true	size_4
stocking = Within recommended lev	true	stocking_5
volcbft = 2635	true	simulation
volcds = 22	true	simulation
volmbf = 3	true	simulation
sawnet = 3	true	simulation
specvolcfa = 0	true	simulation
grnbiomass = 90	true	simulation

Explain   Continue

Figure 9–6. One of the explanatory features of DSSTOOLS is a fact list which the user can access at any time. Because this image is a screen capture, its quality is not ideal.

future reference. The user can select this option to review the contents of the trace file at any time during program execution.

At the end of the program, the user is given the opportunity to edit the trace file by adding additional explanatory material as desired. By electing to rename the trace file to some other filename, the user can save a record of any particular run for future reference.

9.4.4 Uncertainty Management

Developing a meta-interpreter for uncertainty management is complex but good guidelines have been developed (Shapiro 1983, Sterling and Shapiro 1986, Safra and Shapiro 1986, Covington et al. 1988, Yalcinalp and Sterling 1990). An excellent treatment of uncertainty and its computation in KBSs can be found in Giarratano and Riley (1994). An uncertainty management module should offer the standard options: classical probability, Bayesian probability, certainty factors, belief theory, and fuzzy logic. In order to accommodate projects where uncertainty may not be needed, a certainty of 1 in any of the above schemes should degenerate to the standard PROLOG program without certainty (Shapiro

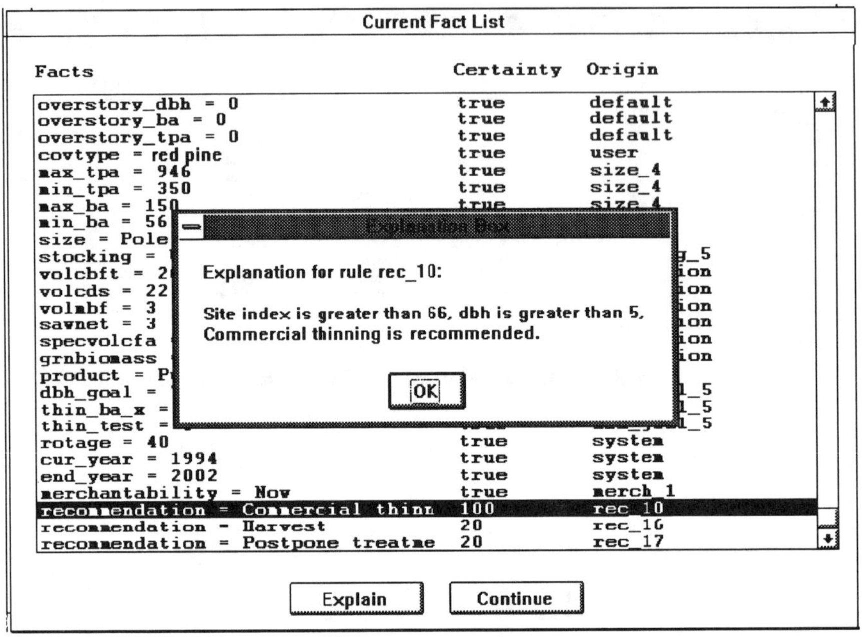

Figure 9–7. In addition to displaying a fact list, the toolkit also allows the user to request an explanation for a particular fact. Because this image is a screen capture, its quality is not ideal.

1983). It should be noted that uncertainty is not necessarily appropriate for every application (Helm et al. 1985).

We reason and make decisions in the face of uncertainty every day. We would like our AI systems to be able to do the same. Coping with uncertainty in KBSs is complex and by no means a solved problem (Parsaye and Chignell 1988). There are three basic aspects of inexact information:

1. it may be incomplete, but not inherently uncertain
2. it may be inherently uncertain
3. it may be vague, with no consistent interpretation.

The second case has received by far the most attention. Many methods for representing uncertainty have been developed and are currently in use. These methods, though useful, fail to address the thorny issue of how does one determine just how uncertain a piece of knowledge really is. Once this has been accomplished, we need to develop reasonable approaches to manipulating uncertainties. Thus coming to grips with inexact knowledge involves two major steps:

1. quantifying uncertain knowledge
2. combining the results of multiple pieces of uncertain knowledge to arrive at the uncertainty of complex events.

Uncertainty methods are extremely sensitive to the particular problem being solved. Most commercial packages for KBS development offer at best a limited variety of methods for handling uncertainty, which may or may not fit the needs of a particular problem (Lecot and Parker 1986). An open toolkit with general purpose predicates that guide the developer at the meta-level provides a much more practical option.

This section presents the design and use of a customizable inexact reasoning toolkit module patterned after the Inexact Reasoning Module (IRM) developed by Lecot and Parker (1986). Its implementation will then be presented using the backward meta-interpreter developed earlier.

9.4.4.1 Implementing Uncertainty. There are various ways in which uncertainty can be implemented but they all have to deal with the same basic situations (Merritt 1989):

1. rules whose conclusions are uncertain;
2. rules whose premises are uncertain;
3. user entered data which is uncertain;
4. combining uncertain premises with uncertain conclusions;
5. updating uncertain working storage data with new, also uncertain information; and
6. establishing a threshold of uncertainty for when a premise is considered known or true.

The uncertainty about a fact, A, may be expressed as U(A). If A is acquired from the user, then U(A) must also be provided by the user. If A is acquired through the inferencing procedure, then U(A) must be obtained as described below. Every A must be accompanied by its uncertainty estimate U(A).

The first task is to determine the uncertainty of the antecedents for a rule (Marcus 1986). Given a single inference rule of the form:

if E then H

the uncertainty associated with the rule may be expressed as U(H|E). U(H|E) is estimated by the domain expert for each rule and must be present in the knowledge base. U(E) is the uncertainty associated with the premise, E, of the rule. In general, it is likely that there is more than one premise clause, E, so we must be able to combine the certainty of all the premise clauses, which are connected by one of the three Boolean operators: *and*, *or*, *not* (Lecot and Parker 1986). Thus regardless of the uncertainty paradigm we choose to use in our system, we will need to define the predicates:

```
cf_and(CfA, CfB, CfC) :- definition for conjunction
                                        CfA and CfB = CfC
cf_or(CfA, CfB, CfC) :- definition for disjunction
                                        CfA or CfB = CfC
cf_not(CfA,CfB) :- definition for negation
                                        not CfA = CfB
```

The next task is to determine the uncertainty of the rule itself and combine it with the combined uncertainty of the antecedents (Marcus 1986). The general definition we need is one that can be used to estimate U(H), the uncertainty of the conclusion, given U(E) the combined estimate of uncertainty of all the premises and $U(H|E)$, the uncertainty associated with the rule itself. Let us call this problem the single-rule combination problem.

The last step is to determine whether the conclusion occurs more than once in the knowledge base and, if so, combine the uncertainties of the multiple conclusions into a global uncertainty (Marcus 1986). For example, given several rules in the knowledge base with the same conclusion, i.e. :

if $U(E_1)$ then H with $U_1(H|E_1)$
if $U(E_2)$ then H with $U_2(H|E_2)$

we want to know U(H) following the application of both rules. Let us call this problem the multiple-rule combination problem.

In order to reduce computer code, it is common that both the single-rule and multiple-rule combination problem are implemented in the same combination rule. Thus in general, we not only compute $U_1(H|E_1)$ independently of $U_2(H|E_2)$ but also compute a combined, global U(H) in the same predicate. Thus we need to define a predicate similar to the following pseudo-predicate:

```
cf_combine(Prior_U(H),U(E),U(H|E),U(H)) :-
```

where the first three arguments are inputs and the last one is output. If we have only a single rule for which we need a combined U(H) then we set Prior_U(H) = 0.

To deal with multiple conclusions, the KBS should record in the knowledge base the CF of each conclusion reached for which the possibility of multiple conclusions exists. Then, whenever the system draws that conclusion again, the new CF can be combined with the old one (Marcus 1986). For example, the **update_cf/4** predicate can be defined to look in the knowledge base for the presence of some *Goal* and update its CF value based on its current value and the *CF* of the new evidence and the rule that just concluded *Goal.*

```
update_cf(Goal,CF_newE,CF_newrule,CF_combined) :-
    fact(Goal,CF_old,X,Y),
    cf_combine(CF_old,CF_newE,CF_newrule,CF_combined),
    retract(fact(Goal,_,_,_)),
    asserta(fact(Goal,CF_combined,X,Y)).
update_cf(Goal,CF_newE,CF_newrule,CF_combined) :-
    get_time_stamp(Time),
    ct_combine(0,CF_newE,CF_newrule,CF_combined),
    asserta(fact(Goal,CF_combined,deduced,Time)).
```

We also need to define a threshold value and a test predicate which can determine whether a given certainty factor exceeds a threshold value. The threshold value determines whether a fact or rule is to be considered true. There are two cases: where the CF is positive and where the CF is negative. The negative case can be explained as: "it is true or known that E is false".

```
above_threshold(CfA) :-
    threshold(T),
    (CfA > T
    ;
    CfA < -T).
threshold(X).                  % where X = 0.2, 20, etc. whatever is appropriate
```

The above predicates, provide the necessary and sufficient definitions for any uncertainty paradigm. Several supporting predicates may also be useful from time to time.

Another useful predicate defines a table which links the numeric CF values to the symbols true, false, unknown, and unknown_true, unknown_false. The last two are equivalent to the value in **threshold/1**. Using this scheme and a minor change in the **above_threshold/1** predicate above, it is possible to provide different values for the positive and negative thresholds (Lecot and Parker 1986).

```
cf_value(true, 1).
cf_value(unknown_true,0.2).
cf_value(unknown,0).
cf_value(unknown_false, -0.3).
cf_value(false, -1).
```

Finally, it may be useful to provide a certainty phrase table to allow expressing CF in simple words. Such translation may make the explanations more digestible to people. The predicate pattern is:

```
cf_phrase(Word, Typical_CF, LowCF, HighCF).
cf_phrase(definite, 1.0, 1.0, 1.0).
cf_phrase(definitely, 1.0, 1.0, 1.0).
cf_phrase(positive, 0.95, 0.9, 1.0).
cf_phrase(undoubtedly, 0.85, 0.8, 0.9).
cf_phrase(unknown, 0, -0.2, 0.2).
cf_phrase(dubious, -0.3, -0.4, -0.2).
cf_phrase(unlikely, -0.6, -0.7, -0.3).
cf_phrase(improbable, -0.8, -0.95, -0.7).
cf_phrase('definitely not', -1, -1, -0.95).
```

9.4.4.2 Paradigms for Manipulating Uncertainty. After values have been assigned to the facts and rules of an KBS, the task is to use some paradigm for inexact reasoning (a.k.a.: likelihood calculation, probability combination, or truth value propagation) (Parsaye and Chignell 1988). It must be clearly realized at the outset that there is no fully general mathematical approach to the problem of evidence combining—even worse, the problem can be proven mathematically intractable (Ross 1988). There are, however, some methods that have been shown to work in practice. It is these that KBS applications use.

There are several competing paradigms for managing uncertainty in KBSs:

1. Classical and Bayesian Probability
2. Certainty Theory—MYCIN Approach
3. Fuzzy Logic

We will briefly introduce each of these and present the implementation details using the general purpose predicates developed above. The reader should be aware that there are other competing schemes besides the ones covered in this report.

9.4.4.2.1 CLASSICAL AND BAYESIAN PROBABILITY. The Bayesian approach to probability relies on the concept that one should incorporate the prior probability of an event into the interpretation of a situation (Parsaye and Chignell 1988). Bayes' theorem provides a mathematically correct method for combining prior probabilities with the probability of current events to update the prior probability in light of the new events. The use of Bayesian probability for inexact reasoning in KBSs is controversial and filled with practical difficulties as well. One may, however, adopt a pragmatic attitude and consider the Bayesian probability method as just another method of combining inexact values, thus cutting its ties to its mathematical legitimacy. Used in this manner, it is a practical alternative and frequently used.

```
cf_and(CfA,CfB,CfC) :- CfC is CfA * CfB
cf_or(CfA,CfB,CfC) :-
    Case 1: CfC is CfA + CfB – CfA * CfB
        (independence assumption 'OR' Ross (1988))
    Case 2: CfC is maximum(CfA, CfB)
        (borrowed from Fuzzy Logic because of problems with
        Case 1 'OR' (Giarratano and Riley 1994))
cf_not(CfA,CfB) :- CfB is 1 – CfA
Single Rule Combination:
    cf_combine(0,CfA,CfB,CfC) :- CfC is CfA * CfB
Multiple Rule Combination:
    cf_combine(CfA,CfB,CfC,CfD) :- Temp is CfB * CfC,
    CfD is CfA + Temp – CfA * Temp
Combining Positive and Negative CF:
    Multiple Rule Combination is the sum of positive and negative CF
```

9.4.4.2.2 MYCIN-LIKE CERTAINTY FACTORS. Certainty Theory represents uncertainty as a degree of belief in a fact or rule rather than the probability that a fact or rule is true as is the viewpoint in Probability Theory. Certainty factors, used in the MYCIN KBS (Shortliffe and Buchanan 1975), are an ad-hoc approach to inexact reasoning. The system is not based in any mathematical tradition but it has been shown to be flexible, powerful, and useful.

Certainty Theory calculates and tracks the belief in a fact separate from the disbelief in a fact. Then, after all the evidence for both belief and disbelief have been accumulated separately, their difference is the final CF (see Section 3.5). The following treatment propagates only a single belief value, not two. A good understanding of this PROLOG code, however, should enable anyone to easily implement a more strict interpretation of Certainty Theory.

```
cf_and(CfA,CfB,CfC) :- CfC is minimum(CfA, CfB)
cf_or(CfA,CfB,CfC) :- CfC is maximum(CfA, CfB)
cf_not(CfA,CfB) :- CfB is 1 – CfA
Single Rule Combination:
    cf_combine(0,CfA,CfB,CfC) :- CfC is CfA * CfB
Multiple Rule Combination:
    cf_combine(CfA,CfB,CfC,CfD) :- Temp is CfB * CfC,
    CfD is (1 – CfA) * Temp + CfA
Combining Positive and Negative CF:
    Multiple Rule Combination is sum of positive and negative CF
```

9.4.4.2.3 FUZZY LOGIC. Fuzzy Logic was invented in the mid-1960's as an alternative to two-valued logic and probability theory (Bellman, et al. 1966 Zadeh 1965). Two-valued logic and set theory are viewed as, special cases of a more general multi-valued theory of logic. Mathematical probability theory is claimed to be inappropriate and is replaced by a system called possibility theory (Zadeh 1978). A good discussion may be found in Giarratano and Riley (1994).

```
cf_and(CfA,CfB,CfC) :- CfC is minimum(CfA, CfB)
cf_or(CfA,CfB,CfC) :- CfC is maximum(CfA, CfB)
cf_not(CfA,CfB) :- CfB is 1 – CfA
Single Rule Combination:
    cf_combine(0,CfA,CfB,CfC) :- CfC is CfA * CfB
Multiple Rule Combination:
    cf_combine(CfA,CfB,CfC,CfD) :- Temp is CfB * CfC,
CfD is maximum(CfA,Temp)
Combining Positive and Negative CF:
    Multiple Rule Combination is the CF with largest absolute value
```

A very attractive fuzzy logic implementation for KBSs has been developed by Leung et al. (1989). A description is omitted here in the interest of brevity.

9.4.4.3 Application Considerations. Some discussion of the above definitions is in order. Starting with the independence assumption for *and* and *or* combinations in Bayesian probability:

```
cf_and(CfA,CfB,CfC) :- CfC is CfA * CfB
cf_or(CfA,CfB,CfC) :- CfC is CfA + CfB – CfA*CfB
```

The independence assumption is very difficult to justify and in many cases it does not hold (Ross 1988). The Bayesian *or*, Case 1 earlier, yields results which are, in general, much too high. A conservative approach uses instead, Case 2 *maximum(CfA,CfB)*, which says: the combined CF must be no worse than that of the strongest evidence. Conservative definitions for *or* and *and* would be:

```
cf_or(CfA,CfB,CfC) :- CfC is maximum(CfA, CfB)
cf_and(CfA,CfB,CfC) :- CfC is CfA + CfB –1          %if CfC > 0, else 0
```

Use the conservative *or* when there is strong positive correlation between evidence. Use the conservative *and* when there is strong negative correlation between the evidence. The conservative *or* is also useful when many a priori facts exist in the knowledge base whose CFs are close to 0 (Ross 1988). For example, unless you have other reasons for supporting it, an expensive piece of evidence to obtain might be assigned a probability of .1 initially until other, cheaper to determine, events have been disproved.

There is, of course, a liberal approach also. The liberal approach uses the maximum CF possible with the evidence at hand (Ross 1988). Use the liberal *and* when one piece of evidence implies all the others. Use the liberal *or* when different pieces of evidence prove the same conclusion in ways that cannot hold simultaneously. Use the liberal *and* when there are strong positive correlations between evidence and the liberal *or* when there are strong negative correlations—the opposite advice as for the conservative formulas.

```
cf_or(CfA,CfB,CfC) :- CfC if CfA + CfB          % if CfC < 1, else 1
cf_and(CfA,CfB,CfC) :- CfC is minimum(CfA,CfB).
```

It is perfectly acceptable to use any and all of these versions uniformly or in combination as long as they meet the needs of your problem. It is even to be expected that certain parts of a problem will require one method while others require another. If one is not sure which method to use, there is nothing wrong with taking a weighted average of all three major approaches. Note that the relationship among the 6 formulas is: (Ross 1988):

conservative *and* < independence *and* < liberal *and*
< conservative *or* < independence *or* < liberal *or*

The best approach of all, of course, is to ensure that the domain rules are strictly independent. Rule ordering, domain control rules, and the use of control facts can make this possible. There is nothing wrong with having a unified chunk of rules in which their ordering is part of the domain knowledge. To ensure that this ordering is obvious to someone examining the knowledge base, it should be clearly specified by an offset comment. Also, any dynamic explanation of those rules should include this ordering as part of the explanatory information. The second approach places specifications in the knowledge base that tell particular control rules the ordering of certain rule sets, just as we placed user-interface management clauses in the knowledge base to describe how input screens should be constructed or just like we used *environments* for the forward chaining interpreter. Placing control facts right inside of rules can also be used, but it is the least elegant approach and contradicts our aim to keep the domain rules uncontaminated with nondomain features.

It is also perfectly acceptable to create customized formulas of your own. Customizing requires care, however. Formulas should have the following properties (Ross 1988):

1. smoothness: the combined CF should never make abrupt jumps in value as input CF vary smoothly;

2. consistency: the combined CF should never give values outside the range of the conservative and liberal method;
3. commutative: the order of the inputs should not matter;
4. associative: grouping of inputs should not affect the output CF.

Finally, negation causes problems. Sometimes we want negative evidence to count positively toward the evidence of a result. For example, we might have a rule:

```
organism(tree,CfC) :-
    stem(woody,CfA),
    not herbaceous(true, CfB).
```

It would be best to avoid use of the negation and just use the positive *herbaceous(true,CfB)* with a negative *CfB*. Some systems split evidence for and against a hypothesis into two parts and combine them finally in the end. As noted previously, this is done in a strict application of Certainty Theory. It also seems to reflect the way in which many people weigh evidence.

9.4.4.4 Meta-Interpreter Modifications for Uncertainty. The modifications of the backward chaining meta-interpreter are relatively straightforward and have been worked out by Sterling and Shapiro (1986). The code below incorporates Bayesian probability inexact reasoning.

```
solve_b(A,Cf) :-
    fact(A,Cf,_,_), above_threshold(Cf).
solve_b(not A,Cf) :-
    not(solve_b(A,Cf1)),
    cf_not(Cf1,Cf).
solve_b((A,B),Cf) :-
    solve_b(A,Cf1),
    solve_b(B,Cf2),
cf_and(Cf1,Cf2,Cf), above_threshold(Cf).
solve_b((A;B),Cf) :-
    (solve_b(A,Cf1)
    ;
    solve_b(B,Cf2) ),
    cf_or(Cf1,Cf2,Cf), above_threshold(Cf).
solve_b(A,Cf) :-
    rule(_,(A :- B),Cf1), above_threshold(Cf1),
    solve_b(B,Cf2),
    update_cf(A,Cf2,Cf1,Cf), above_threshold(Cf).
% Inexact Reasoning Management Predicates
%    --------------------------------------
```

```
cf_and(CfA,CfB,CfC) :-
    minValue(CfA,CfB,CfC).
cf_or(CfA,CfB,CfC) :-
    maxValue(CfA,CfB,CfC).
cf_not(CfA,CfB) :-
    CfB is -CfA.

ct_combine(0,CfA,CfB,CfC) :-
    CfC is Cfa * CfB.
cf_combine(CfA,CfB,CfC,CfD) :-
    Temp is CfB*CfC,
    cf_calc(CfA,Temp,CfD).

cf_calc(CfA,CfB,CfC) :-
    CfA >= 0,CfB >= 0, !,
    CfC is (CfA + CfB - CfA*CfB).
cf_calc(CfA,CfB,CfC) :-
    CfA >= 0, CfB < 0, !,
    P is -CfB,
    minValue(CfA,P,M),
    CfC is (CfA + CfB) / (1 - M).
cf_calc(CfA,CfB,CfC) :-
    CfA < 0,CfB >= 0, !,
    P is -CfA,
    minValue(P,CfB,M),
    CfC is (CfA + CfB) / (1 - M).
cf_calc(CfA,CfB,CfC) :-
    CfA < 0, CfB < 0, !,
    P is -CfA,
    CfC is (CfA + CfB + P*CfB).

above_threshold(CfA) :-
    threshold(T),
    (CfA > T
    ;
    CfA < -T).
threshold(0.2).
```

9.5 THE SYSTEM CONTROL LEVEL

In RP-FMAS, system level control is implemented by a blackboard (Englemore and Morgan 1988). For a detailed discussion of blackboard architectures see Chapter 2 Section 2.5.3.

The blackboard itself is a temporary fact base in working memory where the system stores information provided by the user, conclusions inferred during a consultation, goals for the forward chaining inference engine, and requests for particular modules to perform some task. An example of the state of the blackboard for RP-FMAS might be:

```
fact(species('Red Pine'),100,user,102414).
fact(age(32),100,user,102416).
fact(tpa(750),100,tpa_3,102418).
etc.
```

The system level control program is quite simple. A short start-up program written in PROLOG loads all components of the system, initializes the blackboard, and activates a blackboard scheduling module. The blackboard scheduler is implemented as a simple forward chaining system:

```
bb_run :-
    repeat,
    one(bb_module(_)),
    retract(request(bb_stop)).
```

The clause **one/1** calls up each domain control module in turn until one of them executes. The domain control module checks to see if there is a request on the blackboard for it to become active or if there is a situation described by the facts on the blackboard that require some action by it. If there is, it does and the blackboard supervisor program checks to see if there is a request on the blackboard for the supervisor to stop operations. If there is, it stops. If no stop request can be found, the **retract/1** clause fails and backtracking occurs until the **repeat/0** clause is encountered. The repeat clause causes execution to go forward again and each domain control module is tried again. This logic keeps working the domain control modules until one of them places a stop request on the blackboard which causes program termination. These few lines also illustrate the power of a well designed PROLOG clause. Four lines of code are sufficient to write the entire supervisor module!

Each domain control module contains initial tests to see whether certain facts and requests are on the blackboard. If the correct pattern of facts and requests are found, the domain control module executes the rest of its code. Otherwise the domain control module retires until the next cycle of the scheduler, giving other domain control modules an opportunity to execute. Each domain control module must be written with care to ensure that it executes when and only when it logically should during the operation of the FMAS.

A trivial way to handle this is to have each domain control module put a request on the blackboard identifying which domain control module should execute next. Each domain control module must also remove the request that activated it. Even this simple method has considerable power since a single domain control module might request action by any of a number of other domain control modules depending on the outcomes of its own actions. So even if domain control module activation is entirely request driven, complex patterns of branching are possible.

The blackboard architecture with its distributed control fits KBS development well for several reasons. Unlike a conventional KBS that assumes total control of a consultation, a multiple-knowledge source KBS often provides a variety of services to the user. The user has the option to pick and choose from among these services. Individual services can be provided by separate domain control modules. Other domain control modules can then assume the task of describing the services available and scheduling those that are requested. Further discussion of the theory of blackboard system operations can be found in Englemore and Morgan (1988).

Besides the simple scheduler described above, the blackboard component of DSSTOOLS includes utilities for adding facts to and removing facts from the blackboard, for maintaining a trace file in which blackboard transactions are recorded, and for writing and reading data files used to communicate with non-PROLOG components of a KBS. The blackboard component also includes utilities for saving and reloading the contents of the blackboard. These utilities are useful during debugging of a KBS.

9.6 SUMMARY

This chapter outlined the major features necessary to develop a KBS such as the Red Pine FMAS using modules from a PROLOG toolkit, DSSTOOLS. Composing meta-PROLOG interpreters is complex task that needs study and practice. The variety of meta- interpreters that can be built is infinite. Different needs require different meta-interpreters. Although meta-PROLOG programming is decidedly more complex than native PROLOG programming, the rewards are ample.

Readers are not expected to learn all they need to know about meta-PROLOG programming from this one chapter. We do, however, believe that we have presented enough conceptual structure and practical examples to help readers enter this field successfully. A fully functional example of meta-programming as implemented in DSSTOOLS, RP-FMAS is available to readers by referring to the Frontmatter. Studying this example in detail along with the explanatory material in this chapter should provide the reader with enough knowledge to be able to build their own meta-interpreters for PROLOG. Consulting the references cited will also help readers make the best possible use of the material presented in this chapter.

CHAPTER 10

System Evaluation and Delivery

Composers usually make corrections to their manuscript scores, and art-historians ... constantly discover the rejected first-thoughts of the artist, hidden under the visible layers of paint. Sometimes corrections are minimized ..., and sometimes they are impossible ... Even so, the artist evaluates the production, so as to do better next time.

M. A. Boden

A central question in system delivery revolves around how much the user surrenders his or her own judgment and accountability to the expert, human or computer, and how much the user retains. This is an extremely important philosophical question with profound social and legal implications. The issue has not been forced to the forefront before now because conventional computer programs have been used primarily to report objective results based on algorithms that can be validated for accuracy against the real world. Interpretation of factual results and any potential actions suggested by the results have usually been left up to the judgment of the user. We customarily hold humans accountable for their actions based on their judgments.

Enter knowledge-based systems, which produce judgment as their routine product. How should the non-expert regard these "machine experts"? Should their judgments be regarded as a substitute for the human expert or should they be regarded as little more than convenient information sources like books? It seems likely that considerable debate on this issue will continue for some time to come.

Without delving further into social and legal acceptance of KBSs we examine in this chapter some of the technical aspects of system evaluation and delivery. Prior to actual delivery of a KBS, some attempt must be made to determine the operational value of the system. This stage is referred to as system evaluation. Basically, the questions that need to be answered are: Does it perform in the way it was designed?, and Is the level of performance acceptable?, and Does it perform for the users in the way they would like? In some sense, this aspect of system development represents the transfer of the technology, or knowledge, to end-users and their organizations, along with some guarantee of functionality.

A total evaluation consists of *verification*, *validation*, and *usability* assessment. In the next several sections we investigate these topics in more detail individually. Because it makes little sense to validate a system that has not been verified, or assess usability of a system that has not been validated, we impose a definite sequential order for performing these three aspects of evaluation. Following those topics, we consider the comprehensive evaluation of a knowledge system in its entirety; this includes the above three topics as subcomponents. Finally we present other issues related to actual delivery of a system to an organization or user community.

10.1 SYSTEM EVALUATION

As in the system development stages presented in the previous chapters, system evaluation should occur in parallel with the other parts of system construction. Even back as far as the feasibility study prototype (if it exists), there should be some verification, validation, and utility assessment performed (Figure 10–1). If evaluation occurs throughout the entire process, a final evaluation, prior to deliver, should uncover very few problems of a major nature (Schmoldt 1989). This is particularly important with respect to verifying and validating a system, because errors with system correctness can be difficult to pinpoint and rectify very late in the development process. System testing and refinement can and should be accomplished little by little during development so that correctness can largely be assumed when the final pieces are put together.

Traditional computer programs, using algorithmic solutions to solve problems, are difficult to evaluate as correct, so that absolute correctness can never be guaranteed. The only true way to evaluate a program is to run it under all possible sets of input data. This is a combinatorial impossibility, however, so computer scientists have resorted to algorithm analysis approaches to evaluation. KBSs are even more difficult to evaluate because the best solution cannot be derived in a rigorous mathematical fashion. The best one can do in this case is to apply a number of techniques that have been found useful for identifying

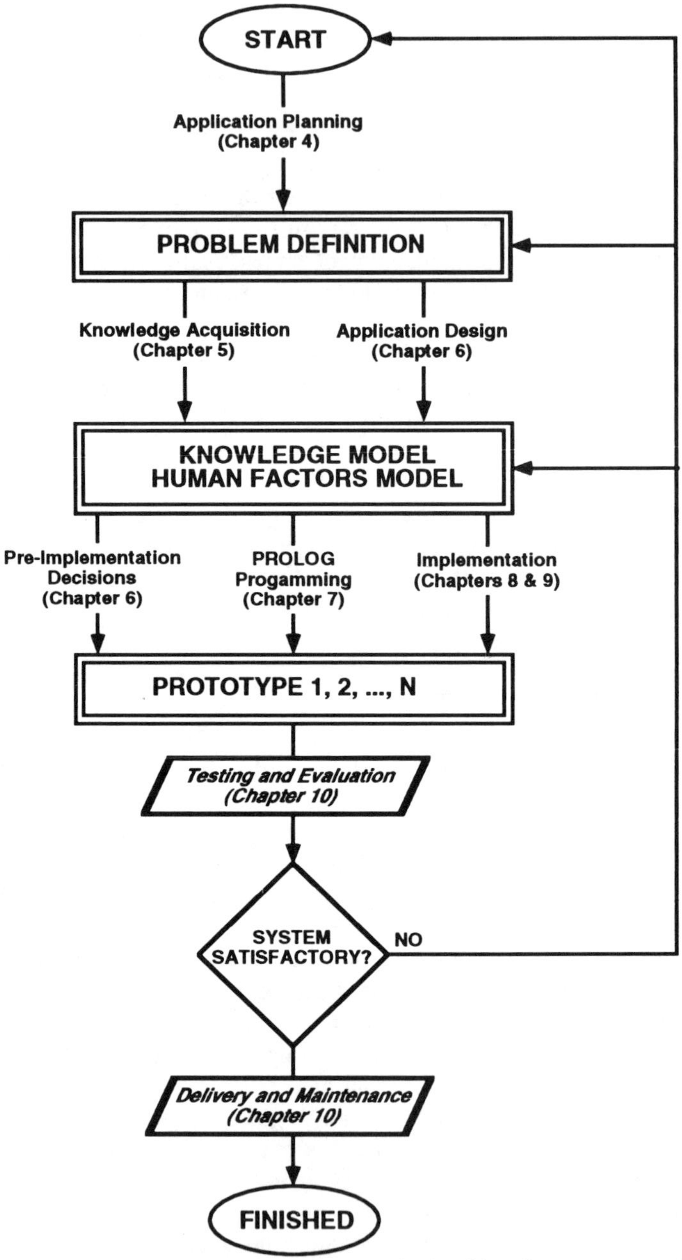

Figure 10–1. As each new prototype is created it should undergo some evaluation of performance and functionality. Any shortcomings that are detected may cause revision to be made at any of the previous development stages.

and correcting errors in knowledge-based systems. When applied carefully, these methods can help produce a system that works well over a range of domain problems.

10.1.1 Evaluation: Some General Issues

Sell (1985) identified five basic requirements for knowledge systems. These are:

1. *consistency* – is the advice free from contradiction?;
2. *completeness* – can it solve any problem in its domain?;
3. *soundness* – can it deliver the right answers?;
4. *precision* – is the strength of answers in line with the data and knowledge on hand?; and
5. *pragmatics* – can those for whom it was designed use it?

In addition, we would like to add two other stipulations, namely, that the system be:

6. *representative*, i.e. it should solve problems in the way the design intended and
7. *robust*, i.e. it should degrade gracefully as the quality of input data is reduced.

The items numbered (1), (2), and (6) together constitute **verification**. Verification evaluates whether a system is operating according to specification. Soundness (3) and precision (4) are aspects of **validation**. Validation assesses whether correct answers are being produced. **Usability** is the final evaluation step that assesses whether a system meets human factors considerations, operational environment requirements, and organizational needs; these include pragmatics (5) and robustness (7). Robustness helps ensure that a user is not left stumped by a knowledge system when the data are not "just right." As you might expect, usability is somewhat less objectively evaluated than the other two, but specific operational and organizational requirements along with user feedback can help us analyze usability issues.

10.1.1.1 Evaluating the Whole, or Parts. Another consideration for the evaluator is whether the entire system can, or should, be evaluated as a complete unit. For *usability assessment* the entire system needs to be the focus of evaluation because interfaces between system components are a very critical aspect of how well the system performs in its operating environment. When a system contains relatively autonomous expertise from different sources, however, it generally becomes more efficient and simpler to *verify* and *validate* individual aspects of the knowledge base separately. This is particularly true of blackboard systems, but may also be quite useful for applications in which a solution can be broken down into independent subgoals. In the later case, each subgoal could be verified and validated separately. Then, because only a small portion of the entire knowledge resident in a system has to be scrutinized and tested at once, the expert's time is minimized, test cases are much smaller, and errors are easier to locate.

No matter how modular a system is, however, a holistic evaluation needs to be undertaken to corroborate proper meshing of individual pieces. This is particularly important in blackboard systems, where significant processing may occur in the kernel that coordinates specialists' activities.

10.1.1.2 What Claim to Evaluate. The system evaluator must also be aware of which claim for system proficiency is being tested. Selection of a particular claim determines to some extent the verification and validation criteria. Table 10–1 lists two different claims and distinguishes differences in verification and validation associated with those claims. Depending on what claim system developers have settled on as their goal, the questions that need to be answered to verify and validate become relatively obvious. Note that usability is not affected because the issues it addresses are outside the internal operation of the system and address the system's place within organizational and user contexts.

When the goal for a system is "captures expertise", then only that system's knowledge model needs to be verified and the system should be validated against the expertise that the claim specifies it has captured, in this case, an expert. Then, our metric of comparison is the expert whose knowledge we have modeled in the system, i.e. the system expert. This idea may seem contrary to other authors (e.g., O'Keefe et al 1987), because we have made a distinction between different claims. In the second row of the table, when the claim is one of "credible solutions to real-world problems", then verification must address both the knowledge model and the system expert (either or both could be faulty). This is because the claim forces an evaluator to consider the competence of the expert (in relation to real-world problem solving) in addition to the

Table 10–1. Exactly how evaluation of a knowledge system proceeds is determined by which claim the developer attempts to prove. Two possible claims are that the system captures human expertise or that the system can reliably solve real-world problems.

Claim	Verification	Validation	Usability
Captures expertise	Verify knowledge model	Validate against system expert(s)	User
Credible answers to real-world problems	Verify knowledge model and system expert(s)	Validate against real-world test cases or independent expert(s)	User

program's successful portrayal of that competence. Validation in this case cannot use the system expert as a yardstick (at least not exclusively), rather system performance (and implicitly the system expert's performance) must be compared with observed results or with assessments by a panel of recognized experts.

Because most systems claim to perform respectably on real-world problems, the scenario represented in the second row of Table 10–1 usually applies. Successful evaluation in this latter case can also imply that the system expert's knowledge has been captured. In that sense, evaluation of the claim in row two subsumes the claim in row one. The reverse is not true, however; the claim in the first scenario, along with its successful evaluation, cannot imply superior performance due to the uncertainty that a system expert is truly expert.

10.1.1.3 Evaluation Errors. During all three phases of evaluation, there needs to be some awareness of the two types of errors that may occur. This situation is represented in Table 10–2. Whenever the actual truth of any state of affairs cannot be determined, but some decision must be made about the belief in that state (i.e. whether to accept it as true or not accept it as true), then the two-by-two scenario of Table 10–2 holds. Because of this uncertain state of the world, regardless of our beliefs there still exists the possibility of erroneous outcomes. In the table, the word "legitimate" is used to refer to any of, verification or validation or usability. When a system is illegitimate, but our action is to accept the system as legitimate, then the user of that system incurs some risk because the system may not always be correct (Type II error, O'Keefe et al. 1987). When a system is legitimate, but our action is not to accept the system, then the system developer incurs risk due to losses associated with system development efforts and any additional efforts to prove the system is legitimate (Type I error, O'Keefe et al. 1987). However, Type II errors may affect the system builder, also, through reprimands or job loss or loss of peer respect, and Type I errors may result in foregone performance enhancements for the user. So, either type of error can have serious consequences for both developers and users. Often, Type II errors are the ones which receive the most concern, because errors in judgment and their consequences are deemed more serious than are any unnecessary or unsuccessful efforts to avoid them.

In some sense we can take the position that a Type I error is the default condition when no system is, in fact, built. This is analogous to having a valid system (in this case a phantom), but not accepting it and not utilizing it. The only losses and risk for an unbuilt system are those incurred by users who are unable to use a nonexistent productivity tool. When an actual system is built, however, then the risk of Type I errors increases because of investment costs for development, and also, risk resulting from the added potential for Type II errors. This perceived increase in risk may provide some sort of analytical

Table 10–2. The decision to accept or not accept a KBS as a legitimate representation of knowledge carries with it certain risks for the developer and the end user.

		State of the Knowledge System	
		Legitimate System	Illegitimate System
Action Taken Toward Knowledge System	Accept as Legitimate	Acceptance is Consistent with Performance	Type II error
	Do Not Accept as Legitimate	Type I error	Non-Acceptance is Consistent with Performance

explanation for organizational hesitance to invest in systems of this sort, or any uncertain venture in general.

10.1.1.4 Test Cases. Oftentimes test cases for use in verification, validation, usability, or even knowledge acquisition, are in short supply. In many situations good, accurate records of past decisions or problems/results are not available in large quantities. A viable alternative to actual test cases is the use of *hypothetical* ones. Here, someone unconnected with the development effort creates pseudo test cases. Caution must be exercised in assuring that these hypothetical cases are also realistic, however. One advantage of this approach is that it may be possible to create good examples which might otherwise only be rarely encountered. Schmoldt (1987) reports the use of such hypothetical examples for testing and refinement of rule bases, and Senhen (1988) utilized hypothetical cases for knowledge acquisition. These types of test cases should probably not be used for system validation because a knowledge system's performance in real world scenarios will, in general, be more convincing.

When selecting and using any type of test case, one should ask whether a particular test case represents reality. That is, can the data of an example case be used by a rational and knowledgeable expert to arrive at the correct answer? For some test cases there may be insufficient information to arrive at a correct answer. When this describes the typical situation for real world cases that the system must eventually deal with, then system design should take into account incomplete data, and provide answers that reflect this ambiguity.

Test cases are often used to target specific parts of a system during evaluation. Some particular aspects of a system that might be tested are: (1) certain types of reasoning or certain sections of the logic, (2) particular functional capabilities (e.g. menus, reasoning traces, consistency checking), (3) performance on typical problems, (4) performance on difficult problems, (5) gradual system degradation as input data quality deteriorates, and (6) appropriate sensitivity to changes in data. Because there are many different aspects of a system that need to be tested, a suite of separate tests should be used. As modifications are made to a system during development, these tests should be applied again and again to ensure that the system does not regress in performance.

10.1.1.5 Expert vs. EXPERTS. As we verify and validate a system developed with the help of an expert, we may need to enlist the assistance of outside experts (EXPERTS), who were not involved in building the system. This raises several questions: (1) If expertise is rare, how can we find enough experts to participate in evaluation? (2) Is there adequate agreement on solutions to natural resource problems? and a related question, (3) Are there correct answers to natural resource problems? Answers to the first 2 questions depend somewhat on the third question. So, let's address that one first.

Looking back to chapter one, we noted there how interconnected natural resource problems are. Physical and biological systems are very tightly coupled, and each of those two super-systems contains many component subsystems. Scientists are just beginning to understand the complexity of these interactions. So, to say that there is an absolute correct answer is unrealistic. There may, however, be a "best" answer, but that best answer may be different today than it is tomorrow, when we might know more about the natural resource systems that we are dealing with.

Given our inability to produce an absolute correct answer in most cases and the temporary nature of our current knowledge, it would seem that there would be little agreement among so-called experts. But, in fact, this is not the case most often. Due to the scientific peer review process for publications, and due to the common experiences encountered by trained experts while practicing their specialty, most experts are in general agreement about the majority of specific natural resource problems (e.g., diagnostic situations). Of course, the larger, more poorly defined problems can lead to large disagreements among experts, e.g., global climate change. There will also be varying levels of expertise, often directly proportional to one's length of service. But, within a limited range of seniority, there is usually good agreement on solutions to natural resource problems. This consensus we have referred to previously as *community knowledge*, which must be present to some degree to make any KBS development effort worthwhile.

The answer to the remaining question is that the availability of expertise varies between disciplines and between geographic regions. But, it is usually possible to find qualified experts in most fields if one looks at federal, state, and local sources within any geographic area. A great many scientific gaps in our knowledge still exist, but we have a substantial base of empirical training for a growing group of professionals. These trained specialists are becoming more abundant and their cooperation should enable us to perform system evaluation with confidence.

10.1.2 Verification

As noted above, verification consists of three components: consistency, completeness, and representativeness. Completeness is generally the easiest to check. The developer must make sure that all possible solutions are represented and can be reached in some way. For pre-enumerated solution type problems, representing all solutions is often quite easy to accomplish. Several methods are discussed below that help ensure that each solution can be reached from the data describing a problem. There is no guarantee that all appropriate combinations of the data have been accounted for, but there are techniques to catalogue the ones that have been used to make inferences.

Although easiest, consistency can be the most time consuming aspect of verification. Nguyen et al. (1987) list redundancies, conflicts, subsumptions, unnecessary conditions, and circular reasoning as major consistency problems. Nguyen et al. (1987) focused on rule-based systems, but the same ideas carry over to other representational schemes. The techniques presented in this chapter address most of these consistency issues.

Representative reasoning is extremely important if a system is to reflect the knowledge it is designed to capture. This means that the system reasons from data to solutions using the strategies present in the knowledge model. This is very easy to check, because if a system reasons properly on several typical examples, it is likely that it will work similarly on all others. This is particularly easy to verify when the knowledge base has been organized in a logical and modular fashion. See for example the organization of rule sets in Schmoldt and Martin (1989), Figure 2–2. Unusual or exceptional cases may be the only area where this broad claim will not apply. Oftentimes exceptional cases are treated differently by a knowledge model and, hence, must be verified separately.

10.1.2.1 Transitive Verification. A useful strategy that may be used to verify a knowledge base is the idea of *transitive verification*. We refer to Figure 10–2 depicting three phases of knowledge system development and their verification links. Basically, transitive verification suggests a two stage methodology. In the first stage, the knowledge system developer examines the translation of the knowledge from its source into a knowledge model. The second stage focuses on the transmission of the knowledge model into an implementation (i.e. into a system). If these two steps can be verified separately, then there will be a certain level of confidence, using a transitive rationale, that the knowledge has been implemented correctly. While the idea of transitive verification does not provide a new or useful verification tool, it does suggest a strategic approach

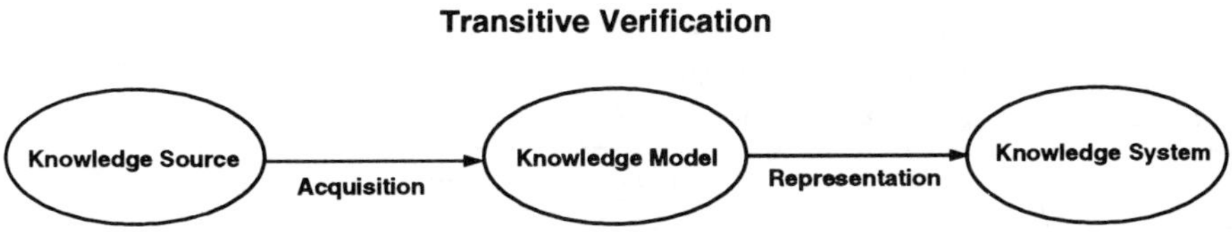

Figure 10–2. The task of knowledge system verification can be made more tractable by considering the two separate steps of verifying the knowledge acquired and the knowledge represented.

that allows a KBS developer to decompose the verification process into two smaller and, hopefully, more tractable tasks. Specific verification methods presented in the next several sections can then be applied to the two stages separately.

10.1.2.2 Execution Tracing. Recording various aspects of an executing KBS can be very informative. Depending on how much detail one records (see the list below), it is possible to follow every step in the reasoning process. To create a running log of knowledge system activities requires that a large number of example cases are available. Schmoldt (1987) used hypothetical test cases during this part of system evaluation. Walters and Nielsen (1988) suggest that hypothetical cases can be very efficient because complete examples with solutions are generally not necessary to examine specific areas of KBS reasoning. Consequently, many similar test cases, varying only in one or two data items, can be easily created.

Almost any aspect of KBS operation can be recorded in a trace log. Some of the most common items of information included are:

1. rules evaluated,
2. rules actually fired,
3. facts established,
4. class instances or objects created,
5. I/O performed, and
6. entry/exit of particular system components.

Many shells provide some facility for tracing the first four of these. The last two must often be hand coded by the programmer. The last information item (6) can be very critical when an application contains many highly interactive components. The contents of an execution trace can then be examined to determine if and where reasoning steps may have detoured from what were expected.

10.1.2.3 Knowledge Matrices and Tables. Most hypotheses are established by the interaction of several different factors. For example, a hypothesis,

Table 10–3. A knowledge matrix displays the relationships between three independent factors and their impact on a dependent factor, erodibility. The values, "low", "moderate", and "high" are possible erodibility values.

Erodibility	Percent Riparian Area Burned			Burn Pattern
	0 – 20	20 – 40	> 40	
0 – 20	low	low	low	
Slope 20 – 80	low	low	moderate	spotty
> 80	low	low	moderate	
0 – 20	low	low	low	
Slope 20 – 80	moderate	moderate	moderate	mosaic
> 80	moderate	high	high	
0 – 20	low	low	low	
Slope 20 – 80	moderate	moderate	high	extensive
> 80	high	high	high	

Y, may be inferred from the values of factors X_1, X_2, and X_3. Using the hypothesis and factors, a matrix can be constructed to ensure that the correct relationships have been specified and that all possible combinations of the factors have been accounted for (Braun 1989). The entries in the matrix represent the values of *Y*, the row and column headings correspond to the *X*'s. In contrast, for quantitative models a mathematical expression is often used to describe *Y* in terms of some function, *f*, of the *X*'s, e.g. $Y = f(X_1, X_2, X_3)$. This is a very compact and comprehensive representation. With the qualitative models inherent in knowledge-based systems, the values of factors are often ranges or categories or vague terms, so matrices represent the relationships more easily.

For a more concrete example, let's look at the matrix in Table 10–3. Here, the hypothesis we are interested in is *erodibility* of a tract of land following a wildfire. The factors identified as influencing erodibility are *percent slope*, *percent of riparian area burned*, and the *pattern of the burn* in the watershed. This matrix representation allows us to easily see what the relationships between all pertinent factors and *erodibility* are.

A matrix like this could also be part of a knowledge model specification because it succinctly displays valuable domain knowledge. It could also be used as part of a structured interview session in knowledge acquisition by focusing discussion on specific factors, their ranges, and the corresponding values of a dependent factor.

Consistency and completeness can readily be checked by examining this matrix and the resulting implementation of it in the program. All logically possible combinations of the factors should be represented and the associations between factors and hypothesis should be traceable to specific inferences in the implementation. This provides us then with an example of transitive verification. First, this matrix can be compared with the knowledge source, and then its implementation can be scrutinized to verify a reliable translation of the matrix.

Each hypothesis in the knowledge base can be specified in this way, whether the hypothesis is a final solution or an intermediate assertion. This approach has the nice side-benefit of forcing the system developer to be more structured in specifying the knowledge base. Because each table corresponds to only one hypothesis, each table is easy to construct and to understand. Other tables, in turn, will use those hypotheses in association with other hypotheses to infer still other hypotheses. If knowledge matrices (or their equivalent representation, tables) are used initially during knowledge acquisition, then many aspects of knowledge base verification become a trivial operation.

An equivalent, and possibly more useful, representation of the knowledge in Table 10–3 appears in Table 10–4. This knowledge table contains a column for each of the independent factors (*percent slope*, *percent riparian area burned*, and *burn pattern*) and one for the dependent factor (*erodibility*). There is one row for each possible combination of values for the independent factors. In this particular case, the total number of rows needed would be 3 X 3 X 3 = 27. While this may seem like a more verbose way to represent the same contents as Table 10–3, it is really not that onerous. For example, in Table 10–3 notice that when *slope* = 0-20%, *erodibility* is *low* regardless of the values of the other two factors. Table 10–4 represents this observation by using a "?" in the columns corresponding to *percent riparian vegetation burned* and *burn pattern* to indicate "we don't care what the values of those factors are, it's unimportant." Therefore, this one row contains knowledge that would otherwise be spread over nine rows (3 X 3). As a result of this and other shortcuts, a total of 9 rows in Table 10–4 covers all 27 cases. The real advantage of the representation in Table 10–4 is that if-then rules can immediately be written to correspond to each row of the table. For the example above, we could write the concise rule:

$$\text{IF } 0 \leq \text{slope} \leq 20 \text{ THEN erodibility} = \text{low.}$$

Again, as in Table 10–3, this new table serves as a knowledge model and also as a template for converting that model into a system representation.

10.1.2.4 Knowledge Diagramming

Hillman (1988) describes a graphical knowledge diagramming technique that can extend the specificity of the knowledge model presented in Chapter 6. Recall, that presentation contained a high level description of the knowledge in

Table 10–4. A knowledge table creates an alternative representation to the knowledge matrix. Each row of the table contains a set of values for each of the independent factors and a corresponding value for the dependent factor. This table format allows one to code rules directly from the entries of the table.

Percent Slope	Burn Pattern	Percent Riparian Area Burned	Erodibility
0 – 20	?	?	low
20 – 80	mosaic	?	moderate
> 80	extensive	?	high
?	spotty	0 – 40	low
20 – 80	spotty	> 40	moderate
> 80	mosaic	> 20	high
> 80	mosaic	0 – 20	moderate
20 – 80	extensive	0 – 40	moderate
20 – 80	extensive	> 40	high

terms of a sequence of events for red pine management. Each of these discrete activities in turn consisted of objects, e.g. density, growth, timber objectives, and possibly additional activities. Hillman expands on this knowledge modelling approach by graphing knowledge down to the level of primitive events and objects, i.e. events and objects that cannot be further reduced. Figure 10–3 contains an abbreviated example of Module 5 where more details of the knowledge model have been fleshed out. These objects and events can then be coded directly into the language or shell chosen for system implementation using rules, frames, objects, or whatever representation scheme has been chosen.

Again, because these diagrams are presented in a manner which is readily understood by anyone familiar with the domain and can also be quickly coded into some representation and inference methods, knowledge diagramming allows a developer to treat verification in a two-step fashion, *knowledge → model → implementation*. But more than this, knowledge diagramming and knowledge tables/matrices allow a knowledge model to extend both forward and backward as the translation link between a knowledge source and an implementation of that knowledge in the form of a system. Both of these techniques can be used for system design, but are just as important and useful for verification, and that is why we emphasize them here.

10.1.2.5 Dependency Tables. As a knowledge base grows in size, it also grows in complexity. New facts, new values for facts, new rules, new frames, new objects, and new relationships must be tracked to ensure that additions and changes maintain consistency and move toward completeness. One helpful aid for accomplishing this is the use of dependency tables (Nguyen et al. 1987).

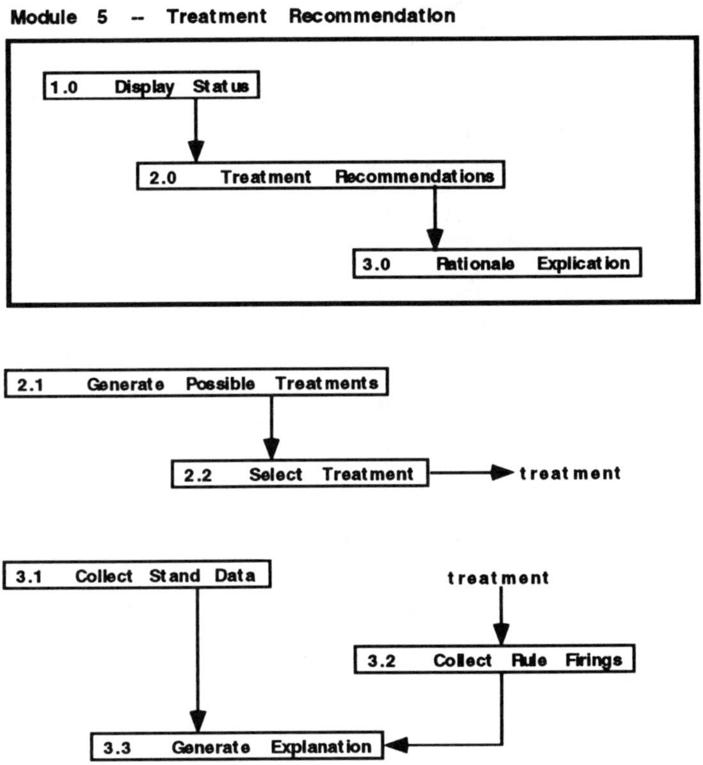


```
2.1.1
        IF product objective is pulpwood
           AND age < 60
           AND 90 < BA < 120
        THEN Recommend "Do nothing at this time"

2.1.2
        IF product objective is sawtimber
           AND age > 20
           AND stem density > 900
        THEN Recommend "Thin to 600 trees per acre"
```

Figure 10–3. Knowledge diagramming as presented by Hillman (1988) dissects events, objects, and rules into modular entities that are organized into a hierarchical structure. A portion of module 5 from the RP-FMAS knowledge model of Chapter 6 is depicted here.

R1: IF erodibility = moderate or erodibility = high
and establishment likelihood = moderate
THEN seed ryegrass at 8 lbs./acre

R2: IF soil repellency = high
and slope = steep
and burn pattern = extensive
THEN erodibility = high

R3: IF riparian area burned = high
THEN burn pattern = extensive

R4: IF burn pattern = extensive
THEN riparian area burned = high

R5: IF fire intensity = high
THEN soil repellency = high

R6: IF brush remaining = low
THEN fire intensity = low

R7: IF years since fire > 15
THEN brush remaining = high

R8: IF fire intensity = high
THEN years since fire > 15

	R1	R2	R3	R4	R5	R6	R7	R8
R1		X						
R2			X		X			
R3				X				
R4			X					
R5						X		
R6							X	
R7								X
R8						X		

Figure 10–4. Dependency tables summarize the interconnections between knowledge base rules.

An example of such a table and the KB rules from which it was constructed appear in Figure 10–4. Each row heading and each column heading of the table corresponds to one of the rules. Whenever a rule, e.g. R1, has a fact in its premise whose value can possibly be determined by another rule, say R2 (i.e. the fact appears in the conclusion of R2), then the corresponding row and column, i.e. (R1, R2), is marked in the table. If one limits rules to a single conclusion clause and minimizes the number of non-atomic facts (intermediate hypotheses) in rule premises, then it becomes very easy to explicitly pinpoint the rules that conclude any particular fact or clause. One merely searches the rules marked in any row corresponding to a rule that has the fact of interest as a premise condition.

A very nice feature of dependency tables is the ability to find circular reasoning. Circular reasoning occurs when somewhere in a particular chain of reasoning, a fact that one is trying to prove appears in the premise of another rule being used. R3 and R4 in our example are two such rules. In trying to prove that the *burn pattern* = *extensive*, we could try to prove that the *percent of riparian area burned* = *high*. But to prove the latter, we would refer again to *burn pattern*, hence, an infinite loop. Many systems do not detect and handle these situations, even still, loops should be avoided from a logical standpoint.

When two rules such as R3 and R4 are immediately adjacent within a KB, it will usually be readily apparent that circular reasoning exists. On the other hand, it is often the case that two rules involved in circular reasoning may be greatly separated within the rule structure of a system. A dependency table then has great benefit for discerning such problems. This potential can be better appreciated by examining a possible sequence of steps used to fire R1, that is, we try R2, R5, etc. or R1 → R2, R2 → R5, R5 → R6, R6 → R7, R7 → R8, and R8 → R6. Because the same rule, R6, appears twice in this sequence, we know that there exists the possibility of an infinite loop at run time (i.e. execution of the knowledge base). Whether an infinite loop actually occurs at runtime depends upon the actual values assigned to the facts.

Some aspects of completeness may also be checked using this table. Each column, except for one (corresponding to the top level goal), should have at least one mark in some row. In this example, R1, which infers a seeding decision, is the goal to be proved. There should also be enough marks in any particular row, corresponding to a rule, to account for all rules that: (1) might infer any non-atomic facts in that rule or (2) might infer any possible values for those non-atomic facts. In this way it becomes relatively easy to check whether all non-atomic facts have rules to infer their value, and also whether all the possible values for any particular fact can be deduced. So, for example, in R2 there are three facts, two of which are non-atomic and each can only be inferred from a single rule, hence, there are two marks in the row for R2. For most similar examples, there would be several rules to infer possible values for each of *repellency*, and *burn pattern*. Dependency tables help a developer see the interactions of rules and the consequences of changing particular rules. As a greater amount of modularity is introduced into a knowledge base, there exists less opportunity for unforeseen interactions between rules and, hence, dependency tables can remain small and manageable.

10.1.2.6 Regression Testing. Because the system development process occurs iteratively, there will often be a need to test and re-test successive prototype versions. As many additions and changes are made to a series of prototypes, however, it becomes difficult to guarantee that newer prototypes operate properly on past test cases. Hence, it becomes important to maintain a battery of test cases that have been successfully applied to past versions of the system. Then, as changes are made to create new prototypes, these test cases may be reapplied to each new prototype. Traditional software engineering methodology refers to this application of test cases as *regression testing*. These tests help ensure that prior functionality and performance has not degraded with changes to the prototype. Implied in this testing is a continual updating of regression tests to include new system features and capabilities. So, the collection of test cases will grow in size and breadth along with the system.

10.1.3 Validation

Whereas KBS *verification* provides for some assurance that a KBS operates as intended, KBS *validation* assesses how correctly a system performs. A typical validation process involves running a set of test cases through a KBS and then comparing those answers with historical data or with answers provided by experts. Because of the nature of KBS applications (i.e. as expert systems), usually the most common comparison is with human experts. Comparisons between system and expert responses might be made using statistical tests directly, or comparisons might use an intermediate stage, where a third set of "experts" (Schmoldt and Martin 1989, Yu et al. 1979b) evaluates agreement or correctness of both sets of responses.

Because statistical validation involves testing a hypothesis, e.g. "system performance is indistinguishable from human experts" or "system performance exceeds that of human experts", there are several statistical methods that may be applied. The use of probability theory implies, however, that test cases should be randomly selected from the population of all possible real world cases. Such a requirement is difficult to guarantee in practice. To provide a completely reliable confidence in system performance over a wide range of problem types would also require a very large set of test cases. Sufficient time to run these cases through a KBS and to evaluate/compare results can prohibit the use of large numbers of cases, however (Schmoldt 1989).

Another issue that deserves consideration is concern over biasing and blinding in evaluations (Chandrasekaran, 1983). Evaluators may be consciously or unconsciously biased toward computers, especially when computer systems invade previously human domains. Yu et al. (1979a) found this to be the case when medical experts were asked to evaluate computer-generated solutions to test cases. In a subsequent evaluation (Yu et al. 1979b) medical experts compared solutions provided by both computer and other experts, evaluators were not informed about the origin of particular solutions. Schmoldt and Martin (1989) used a similar design when evaluating PREDICT. In addition to removing potential bias, introducing comparative experts provides a standard by which both system performance and expert performance can be compared in a meaningful and tangible way (Schmoldt 1987).

10.1.3.1 Solution Types. Because system solutions may assume different forms, different statistical tests need to be applied. There are basically three different solution types: (1) a single, best solution without any weighting score, (2) a single, best solution with a weighting score, and (3) multiple-response solutions with weighting scores. For example, in our red pine silvicultural system we might only list a single answer, such as "thin to ninety square feet of basal area per acre". In the second case, we might append a score to that solution, indicating a measure of likelihood, confidence, or belief in the solution, e.g. "thin to ninety square feet of basal area per acre" (75). Or as a third possibility, we might list several solutions, each with a weight attached to it.

thin to 90 square feet of basal area per acre (75),
thin to 110 square feet of basal area per acre (55),
do nothing at this time (50).

The latter situation provides the user with the maximum amount of information when no single solution can be unambiguously proposed. It is also the most flexible, allowing the user some latitude in selecting a particular solution as an answer to his or her problem.

10.1.3.2 Hypothesis Testing. Statistical hypothesis testing involves comparing a sample from some population of individuals to another population with assumed characteristics. A comparison is usually made using one of the *parameters* from the population and the corresponding *statistic* of the sample. If the parameter and the statistic are "sufficiently close", then the hypothesis being tested (usually the null hypothesis) is not rejected. We are only able to say "not rejected" here because empirical data can *never* prove a hypothesis, they can only support it or else refute it (disprove it). "Sufficiently close" is determined by: the number of observations in the sample, the dispersion of observed values relative to their average value, the level of test error the experimenter is willing to accept, and the probability distribution assumed for the population. A typical pair of hypotheses for testing are presented below:

H_0: There exists no distinguishable difference between system and expert.

H_1: Some evidence exists for differential ability between system and expert in this task.

10.1.3.3 Statistics. When we deal with a system that produces a single final solution with no weighting score, then one of several tests may be appropriate. First, let's look at the situation in which we are comparing system performance directly to an expert. Here, we would have a single number indicating the number of agreements between the system and the expert. Because evaluation of each case results in success/failure, i.e. agreement/disagreement, the underlying probability distribution is the *binomial.* So, if empirical results gave us agreement on k cases out of n, and we wished to test whether the system had an agreement rate of p, then we can create a probability distribution for n cases with a success probability p (most statistical software packages will do this automatically). We can then examine this distribution to see how likely it would be to observe k or less successes given a success probability p.

For example, suppose that our system agreed with an expert 17 times out of 20 cases and we wanted to test the hypothesis that the true agreement rate is 98% of all cases, we would construct the probability distribution presented in the second column of Table 10–5. From this table we note that the cumulative probability of observing 17 or less agreements equals .0071 when the probability of agreement is 0.98. This places the observed agreement rate (0.85) far out on the tail of the distribution when the true agreement rate is 0.98. We would interpret this to mean that it is very unlikely that the agreement rate is as high as 98% given our sample of 20 test cases. So, the real rate of agreement is most likely less than 98%. Performing a similar comparison for smaller and smaller values of p, we can find the largest success rate that our experimental results cannot refute subject to some acceptable testing error. If we are willing to accept as much as 5% probability of error in our hypothesis test, then we can select a p value of 0.95 in Table 10–5 and can say that the performance rate determined by our experiment (0.85) is not sufficiently low to allow us to reject the null hypothesis. The probability of 17 or less successes given a true success rate of 0.95 is 0.0755, which is above our allowed error of 0.05.

We can also calculate a 95% confidence interval surrounding our estimate of 0.85, the experimental rate of agreement. The normal distribution (a continuous function) can be used to approximate the binomial distribution and provide us with an approximate confidence interval. It turns out that when p is near 0.5, the normal distribution is a good approximation for the binomial even when there is a small number of trials n. For moderate values of n, the normal approximation is quite accurate as long as the mean value np is more than 3 standard deviations from 0 and n (Mosteller et al. 1961). The standard deviation of a binomial distribution is the square root of npq. In our example, the mean is 17 and 3 standard deviations equals 1.5, so the normal distribution will give a good approximation. The mean of a normal distribution (in this case 17) plus or minus 2 standard deviations (in this case 1.0) gives us a 95% confidence

Table 10–5. Cumulative probabilities for the binomial distribution with n = 20 observations and success rates of 0.98, 0.95, or 0.96.

k	$p = 0.98$ Prob(X ≤ k)	$p = 0.95$ Prob(X ≤ k)	$p = 0.96$ Prob(X ≤ k)
0	0.0000	0.0000	0.0000
1	0.0000	0.0000	0.0000
.			
.			
.			
16	0.0006	0.0159	0.0074
17	0.0071	0.0755	0.0439
18	0.0599	0.2642	0.1897
19	0.3324	0.6415	0.5580
20	1.0000	1.0000	1.0000

interval of (16, 18) for the mean, which translates into the interval (0.8, 0.9) for p. In general, given the above caveats concerning n and p for a normal approximation to the binomial we can create a standard normal statistic Z using the following formula:

$$\mathrm{P}(X \geq x) = \mathrm{P}\left(\frac{X - np}{\sqrt{npq}} > \frac{x - \frac{1}{2} - np}{\sqrt{npq}}\right) = \mathrm{P}\left(Z > \frac{x - \frac{1}{2} - np}{\sqrt{npq}}\right)$$

The value x is the number of successes in the binomial and the 1/2 is a correction for changing from the discrete binomial to the continuous normal.

When we are comparing system performance to expert performance as judged by an independent and impartial evaluator, there are two possible tests. Which one is used depends on the exact hypothesis that the experimenter wishes to test. If, as in the first test above, we are testing whether a system reaches a level of proficiency equal to an expert, then a similar analysis is appropriate where the p value used would be the percentage of correct responses of the expert. We could perform a binomial test to ask what is the probability of obtaining X or fewer successes given the probability of success p of the expert? On the other hand, if we wish to test whether a system and an expert could have equal performance, then we could use the normal approximation discussed above and perform a paired t-test. Readers can refer to any statistical text on how to carry out such a parametric test. Alternatively, a Chi-square test for goodness of fit could be utilized, given certain restrictions on cell frequencies (see Bhattacharyya and Johnson 1977).

For a Chi-square test a contingency table is constructed, where each row corresponds to a system's or an expert's performance and contains two cells.

Table 10–6. A contingency table for comparing the performance of a KBS with that of an expert. A Chi-square test may be applied to the frequencies in this table.

	Correct	Incorrect
System	14	6
Expert	17	3

One cell contains the number of correct solutions and the other contains the number of incorrect solutions. For example, if a system was correct in 14 out of 20 cases and an expert 17 out of 20, then we would have the 2 X 2 cells in Table 10–6. Here, we are comparing the counts 14 correct/6 incorrect with 17 correct/3 incorrect and asking whether they could be from the same distribution. The calculations necessary to obtain a Chi-square statistic are not described here, details can be found in any statistical text (e.g. Bhattacharyya and Johnson 1977) or a statistical software package may be used (e.g. MINITAB, Ryan et al. 1976). This statistic can then be compared to the appropriate value in a Chi-square table corresponding to the degrees of freedom and confidence level desired. A Chi-square table can also be used when a system is compared to multiple experts (Schmoldt and Martin 1989) or when the experimenter wishes to check multiple experts for consistency. It is important to remember that the Chi-square test is only an approximate test, and is very sensitive to small numbers of observations.

Many knowledge-based systems produce a measure of confidence, probability, or belief as part of any solution. When this occurs, it becomes possible to use a paired t-test to compare the difference between an expert's and a system's answers. If the solutions of each has, instead, been scored by an evaluator, then the difference in evaluator scores may be used. This is discussed by O'Keefe et al. (1987) and has been applied by Schmoldt and Martin (1989) and Yu et al. (1979b).

The most complex and difficult situation occurs when any particular answer consists of several possible solutions and their associated confidence values. There are several possible approaches to this problem. One, used by Schmoldt and Martin (1989) and Yu et al. (1979b), has evaluators score each answer in its entirety. This provides a single score for each test case. These can then be tested using a t-test in a similar fashion to the above single-solution scenario. This method avoids the difficult comparison of multiple solutions, with their relative scores and rankings. It is left to the subjective analysis of evaluators.

In an alternative approach, O'Keefe et al. (1987) suggests the use of Hotelling's one-sample T^2 test. Because the confidence values assigned to various solutions of an answer may be correlated, simultaneously applying paired t-tests may be inappropriate. For Hotelling's test, a vector of differences is constructed for each possible solution in an answer. The means of these difference vectors

Table 10–7. When an answer consists of multiple solutions, each with some measure of belief, it becomes very difficult to determine when two answers agree. Two answers with identically ranked solutions (scenario A) might be less in agreement than two answers where the solutions have the same relative magnitude (scenario B).

		Solution 1	Solution 2	Solution 3	Solution 4
Scenario A	Answer 1	85	75	30	20
	Answer 2	80	30	30	20
Scenario B	Answer 1	80	75	25	20
	Answer 2	75	78	30	25

are then simultaneously compared using Hotellings statistic. In some sense, this statistic captures agreement in both the solutions listed and their relative rankings.

A third statistic can be used to look at the relative rankings of various solutions in an answer. Spearman's rho statistic produces a measure of correlation between rankings. However, as Parsaye and Chignell (1988) point out, looking at ranking alone ignores a great deal of information. For example, in the two scenarios of Table 10–7, the two answers in scenario A agree in rank, but it is not clear whether they convey the same information. The two answers in scenario B do not agree in rank, but each seems to convey almost identical information, and, except for the reversal, agree quite closely. Schmoldt (1989) introduced an extension of Shannon's (1948) measure of information to distinguish between different test cases. Basically, that technique (called Noise Amplification) produces a single numeric value that indicates the amount of decisionmaking information present in a set of data that represent a particular problem to be solved. Noise Amplification was applied in a rule refinement procedure and consequently did not address reversals like those in case B. Also, it was not designed to compare answers like those below but rather to compare the informational content of input data; however, it could be modified to do so. Because these types of answers contain so much information (some of it ambiguous), it becomes difficult to identify a single metric that allows one to say that two answer are in agreement or to select one answer as clearly preferable to another.

While all the above tests tell us how accurate a system may be, it may also be important to end users to know how precise a system is. That is, even though a system may be wrong on occasion, if it is close to the correct answer in most circumstance, occasional errors may not be a serious problem. When we have some measure of proximity to correct answers, as occurs when we have confidence values or evaluator's scores, then we can estimate solution variability and compare a system's variability with that of human experts. Schmoldt (1987) made such comparisons and concluded that high consistency is a valuable trait for a KBS, as it is for any human expert. In that study, evaluators assigned numerical scores (on a 0-10 scale) to each answer. How variable these scores were over the set of test cases provided a measure of consistency. Precision, or repeatability, is extremely important when a system operates in an advisory and decision support role.

10.1.3.4 Error Analysis. Now that some tests have been conducted to assess the validity of a system, what do we do with the results? It is unlikely that a system will perform flawlessly, so some changes are probably indicated by the tests. Under a verification assumption, i.e. we have successfully translated the knowledge source into a system, it would appear that validation errors indicate a difference of opinion between the knowledge source and our performance metric (either other experts or "correct" answers). If this view is taken, then it may be necessary to uncover where these differences lie and attempt to correct them, i.e. reach consensus between experts. Resolving differences does not necessarily mean changes to the system. A faulty system should not always be assumed as the root of discrepancies. Use of system reasoning traces, explanation capabilities, and confusion matrices (see below) can be useful to identify and correct any differences.

On the other hand, one may wish to dismiss performance differences as reflective of poorly understood aspects of the subject area, ones that cannot be resolved given the current state of knowledge in the subject area. Then, the system developer may do nothing at all or, alternatively, may attempt to incorporate dissenting opinions into the system alongside previous reasoning paths. The resulting system would contain a number of bifurcations, or branch points, corresponding to differing opinions. When a branch point is reached during a consultation, the system could follow two separate paths of reasoning, reflecting the two differences of opinion. This bifurcated reasoning may or may not, however, actually result in different final answers in all cases. Oftentimes, there are several different lines of reasoning that might be used to reach the same answer, each may possess some validity.

We can also retract our previous verification assumption and, instead, treat validation differences as resulting from erroneous knowledge. That is, knowledge incorrectly translated from the source to the knowledge model or from the knowledge model to the implementation. Again, traces, explanation, and confusion matrices may be used to identify areas of discrepancy. Then, examination of the two stages of knowledge translation (Figure 10–2) can illuminate where system construction erred.

When anyone makes an error in selecting among possible alternatives, it is generally due to an inability to distinguish between two of the alternatives, the one selected and the one that should have been selected. That is, there exists some confusion between those alternatives. *Confusion matrices* have been suggested by Parsaye and Chignell (1988) as one way to present and analyze system errors. Each cell (i,j) in a confusion matrix contains the relative frequency with which alternative i has been confused with alternative j. For these numbers to have much significance, however, there must be quite a few test cases available for each alternative. A confusion *cost* matrix can also be defined that indicates how serious each (i,j) confusion is. By multiplying corresponding cells of these two matrices we can obtain a final confusion matrix that reflects the total performance of a system. Attention can then be focused on those areas that have the greatest importance for realizing adequate cost-benefit performance.

10.1.3.5 A Validation Case Study. This section presents the validation of a diagnostic knowledge-based system named PREDICT (Schmoldt 1987). PREDICT contains knowledge about 28 different insect and disease problems of red pine growing in Wisconsin USA. Given a set of observations about a red pine stand, PREDICT produces a diagnosis that consists of a list of potentially damaging agents along with a measure of confidence that each agent might be the actual cause of damage. Unless injury characteristics are strongly diagnostic for a single agent, several agents are often implicated by a PREDICT diagnosis. Such an approach to diagnosis was intended to reflect the incomplete, uncertain, and vague nature of typical field observations about red pine injury.

An experiment was designed to compare the performance of the PREDICT KBS to that of recognized experts and to less-trained field foresters. The reason for including field foresters in the experimental design was to demonstrate that there exists a performance gap in the population of individuals who might be called upon to make diagnoses (see Figure 1-4c). We intended to show that it could be possible to raise the level of competence for this population by using a knowledge-based system as a diagnostic aide.

Two separate versions of PREDICT were constructed using different KBS development software. Although both versions of PREDICT use essentially the same knowledge base, there are some differences in operation that could affect performance. Therefore, both versions, labeled hereafter, PREDICT1 and PREDICT2, were tested.

Five human subjects were selected for the comparison, a forest pathologist, two forest entomologists, and two foresters. All of the subjects have work experience with both insect and disease problems. The two foresters had previously attended training sessions in forest pest diagnosis, given by the Wisconsin Department of Natural Resources, but they did not have advanced training in entomology or pathology. None of these experimental subjects was involved in any way with the development of PREDICT.

The 20 test cases selected for system evaluation ranged from easy to very difficult for diagnosis. Also, the cases were selected to include many of the more common and economically important pests. None of the cases had been used during the refinement phase of the project. Each of the five human subjects was shown only the written damage report for each case, the same information that was input to the two versions of PREDICT. The subjects were instructed to work alone (consultations with other people were prohibited), assess each case with the aid of any reference materials they normally used, and identify all pests that could possibly be causing the stand damage. They were also asked to assign a numerical value (20-100%) to each pest indicating their confidence in that pest.

The diagnoses in this experiment could not be evaluated on the basis of a single, correct answer. Also, it is rare to find an individual skilled in both insect and disease diagnosis. So two evaluation teams, consisting of one entomologist and one pathologist each, were chosen to evaluate the diagnoses of the five human subjects and the two versions of PREDICT. None of the members of the evaluation teams was involved with the development of PREDICT.

The two evaluation teams worked independently and arrived at diagnoses for each of the test cases. Each team was then shown the diagnoses from the human subjects and the two versions of PREDICT, but the diagnoses were scrambled so the evaluation teams did not know their sources. Also, there was no way the teams could determine if a diagnosis came from one of the human subjects or one of the versions of PREDICT.

Each evaluation team was asked to provide a comprehensive score, on a scale from zero (worst) to ten (best), for each diagnosis (non-integer values were permitted, but rarely used). This score is a subjective rating of the complete diagnosis, including an assessment of the appropriateness of the agents listed and their associated confidence values. Also, each evaluation team determined, for each diagnosis, if the assessment of the single most likely cause of stand damage was in agreement with that of the evaluation team.

The results of the evaluation are summarized in Table 10–8. Disagreement between either version of PREDICT and the evaluation teams, as to the most likely cause of stand damage, occurred in a total of 6 out of the 20 test cases; and in four of these cases the evaluation teams disagreed with each other. Most notably, both versions of PREDICT failed to diagnose Armillaria root rot, and produced false positives for root collar weevil and bark beetles. Evaluation Team 2 disagreed with several of PREDICT's sawfly diagnoses, but Team 1 did not. Disagreements on particular test cases were later analyzed by another entomologist, who concluded that valid diagnoses were made by PREDICT in all but one instance.

Chi-square tests were performed to identify significant differences between the numbers of diagnoses in agreement. The results are presented in Table 10–9. The differences between the versions of PREDICT and the experts are not significant, but PREDICT1 performed significantly better than both foresters,

Table 10–8. Total number of diagnoses in agreement, average comprehensive scores, and standard deviations from the 20 test cases.

Subject	Evaluation Team 1			Evaluation Team 2		
	Total no. in agreement[a]	Comprehensive[b]		Total no. in agreement[a]	Comprehensive[b]	
		Ave.	Std. dev.		Ave.	Std. dev.
Pathologist	16	7.20	3.22	14	6.40	3.08
Entomologist 1	16	7.75	2.73	14	6.35	2.68
Entomologist 2	16	7.40	2.98	13	7.00	2.73
Forester 1	12	5.60	3.56	13	6.50	3.07
Forester 2	9	5.05	3.43	8	5.05	2.98
PREDICT1	18	7.45	2.50	15	7.05	2.56
PREDICT2	17	6.80	2.21	15	6.70	3.13

[a] The number of diagnoses in agreement is based on the subject's and evaluation team's assessment of the single most likely cause of stand damage in each case.

[b] The comprehensive score is a subjective rating (0, worst – 10, best) of the complete diagnosis, including an assessment of the appropriateness of the agents listed and their associated confidence values.

and PREDICT2 performed significantly better than Forester 2. There is no significant difference between the two versions of PREDICT.

With the exception of Team 2's evaluation of PREDICT2, the standard deviations of the comprehensive scores (Table 10–8) are lower for the versions of PREDICT than for any of the human subjects, and in many instances they are significantly lower statistically. This suggests that PREDICT performs at a more consistent level than the human subjects. Consistency was a development objective during the construction of PREDICT, and is one of the reasons for listing several potential pests when a single pest cannot be diagnosed with certainty.

Differences between mean comprehensive scores for each pair of subjects are presented in Table 10–10. No significant differences were found between the versions of PREDICT and the three experts. Both versions of PREDICT performed significantly better than Forester 2, but Forester 1's performance is not significantly different.

It is interesting to note that the mean difference in comprehensive scores between the two versions of PREDICT, although small (0.50), is statistically significant. The two versions of PREDICT are almost identical in their assessments of the single most likely cause of damage. However, PREDICT2 listed more potential pests (an average of 2.9 per case), and the evaluation teams tended to assign lower ratings for producing very long lists. While shorter lists seems desirable from the view of experienced pest specialists, subsequent

Table 10–9. Chi-square tests for significant differences between the numbers of diagnoses in agreement with the evaluation teams', for each pair of subjects.

	Pathologist	Entomologist 1	Entomologist 2	Forester 1	Forester 2	PREDICT1
Entomologist 1	0.00 NS					
Entomologist 2	0.06 NS	0.06 NS				
Forester 1	1.45 NS	1.45 NS	0.91 NS			
Forester 2	8.72 **	8.72 **	7.37 **	3.21 NS		
PREDICT1	0.67 NS	0.67 NS	1.15 NS	4.01 *	13.65 **	
PREDICT2	0.29 NS	0.29 NS	0.62 NS	2.99 NS	11.85 **	0.08 NS

Note: NS = not significant, * = significant at 5% level, ** = significant at 1% level.

discussions with Wisconsin DNR pest management personnel indicated that, in field applications, foresters would benefit more from the longer lists of potential agents. The two versions of PREDICT differ in their threshold values for eliminating pests from consideration. PREDICT2 is more conservative, and this results in longer lists. The elimination threshold can easily be changed to conform to the average list size of the human experts (1.7 pests per case).

10.1.4 Usability

How well an application system performs in actual use can be viewed in three different areas: end users, intended use of the system, and the organization. With relation to end users we can identify several categories that play a significant role in usability.

1. **Correct skill level**—can users use the application effectively without extensive training?
2. **Compatibility**—is the system compatible with current methods and standards for performing work, physically, technically, and culturally?
3. **Training**—is there an enhancement of users' understanding of the subject area as a result of the system?
4. **Productivity**—has the level of performance by the user group been improved by the system?

The requirements analysis written earlier in project development becomes very important at this stage. It can provide some guidelines as to what is expected of the final product. Characteristics not included in the requirements should not necessarily be ignored here, they should instead be classified as non-critical. Non-critical characteristics may, however, point to enhancements

Table 10–10. Mean difference in comprehensive scores between subjects (column subject minus row subject). The significance level of each mean difference, based on a paired t-test, is shown in parentheses. Comprehensive scores were combined from both evaluation teams.

	Pathologist	Entomologist 1	Entomologist 2	Forester 1	Forester 2	PREDICT1
Entomologist 1	−0.25 NS (.70)					
Entomologist 2	−0.40 NS (.52)	−0.15 NS (.72)				
Forester 1	0.75 NS (.23)	1.00 NS (.054)	1.15 * (.044)			
Forester 2	1.75 ** (.0088)	2.00 ** (.0001)	2.15 ** (.0004)	1.00 NS (.058)		
PREDICT1	−0.45 NS (.42)	−0.20 NS (.72)	−0.05 NS (.92)	−1.20 NS (.072)	−2.20 ** (.0005)	
PREDICT2	0.05 NS (.93)	0.30 NS (.61)	0.45 NS (.38)	−0.70 NS (.28)	−1.70 ** (.0065)	0.50 * (.018)

Note: NS = not significant, * = significant at 5% level, ** = significant at 1% level.

for future system revisions and, as such, are worth considering. Also, it may be the case that new benefits are realized that were not written into the original requirements and, hence, become bonus benefits.

One of the most important items in the requirements should be the intended use, or function, of the final system. Looking at natural resources, we might consider management, research, and teaching to be three broad functional categories. Within the area of natural resource management we might further delineate specific management activities at certain levels, such as the planning level or the operational level. Some systems may be useful in several functional categories, but these systems usually turn out to be large and complex. On the other hand, smaller systems, that are targeted for a specific functional niche, have a greater likelihood for success.

Usability must also be evaluated from the perspective of the organizational climate. While certain users may find a system useful, if the system is not portable within an organization or does not interface well with other users' needs or does not interface well with other information systems in the organization, the overall organization receives less benefit than any small group of users within the organization. For the greatest good of an organization, there are often many factors to consider that have more global impact than the needs of some small user group. So, the perspectives of an organization may not always be coincident with those of a particular user group.

Many of the eventual impacts upon an organization may not appear until a system has been operating for a period of time. One way to expose some of these effects in advance of full system deployment is to supply a preliminary final product, often referred to a beta-test version. A beta version allows users an opportunity to find some remaining programming bugs and glitches. It also gives the organization a feeling for how well the new software can be absorbed into current work habits. Following this pre-delivery release, a final summary evaluation can assess how well the system scores on a wide variety of functional capabilities.

10.1.5 Summary Evaluation

The three aspects of evaluation presented above, verification, validation, and usability, provide a sense for the ultimate value of a system. But each does so separately. We still need to aggregate these separate assessments in some way that will allow us to make a summary evaluation. We need to determine if the system can in some way increase the ability to do more, or to do things more effectively. Two possible global evaluation procedures are presented below.

10.1.5.1 Questionnaires. One way to integrate several different assessments is to create a questionnaire that covers a large number of topics. First, it will be necessary to solicit numerous respondents to work with the candidate system. These individuals should represent all the groups that have some investment in the system's success: users, experts, and management personnel. If the application covered by the system is highly technical, management personnel may need to work alongside someone more technically familiar with the application area. Second, after spending some minimal amount of time with the system, each should complete a questionnaire. And preferably, it should be administered soon after a respondent's extensive-use period.

A questionnaire may contain statements about many aspects of system operation and use. Topics, such as understandability, speed, performance, transparency in the management information system (MIS) environment, user interface, and problem-solving power, are often included (Shneiderman 1987). Each can be further broken down into detailed questions about various aspects of a topic. For example, in the questionnaire excerpt of Figure 10–5, the user-interface topic has been subdivided into subtopics such as screen display, function key usage, terminology, etc. Also, each of these has been further dissected into attributes, e.g. different screen attributes. Because a questionnaire covering many topics can easily become quite lengthy, it is most expedient to make the respondents' selection of answers very similar over all questions. Bipolar responses with an underlying numerical scale work quite well. In each case, antonymous terms anchor each end of the scale. Between five and nine numerical gradations are usually sufficient to capture most subjective assessments. A respondent then selects a number from the scale which most closely approximates his or her feeling about that attribute.

Item	Low end	High end	Scale	
Amount of use by respondent			1 2 3 4 5 6 7 8 9 10+	
Screen display is	inappropriate	appropriate	0 1 2 3 4 5 6 7 8 9 10	NA
Reverse video is	inappropriate	appropriate	0 1 2 3 4 5 6 7 8 9 10	NA
Color differences are	inappropriate	appropriate	0 1 2 3 4 5 6 7 8 9 10	NA
Screen organization is	disorderly	orderly	0 1 2 3 4 5 6 7 8 9 10	NA
Volume of information is	cluttered	uncluttered	0 1 2 3 4 5 6 7 8 9 10	NA
Volume of information is	scant	adequate	0 1 2 3 4 5 6 7 8 9 10	NA
Sequence of displays is	confusing	clear	0 1 2 3 4 5 6 7 8 9 10	NA
Next screen in sequence is	unpredictable	predictable	0 1 2 3 4 5 6 7 8 9 10	NA
Maintain a sense of position	impossible	easy	0 1 2 3 4 5 6 7 8 9 10	NA
Return to previous display	impossible	easy	0 1 2 3 4 5 6 7 8 9 10	NA
Function keys are	ineffective	useful	0 1 2 3 4 5 6 7 8 9 10	NA
Meanings are	confusing	clear	0 1 2 3 4 5 6 7 8 9 10	NA
Next appropriate key is	confusing	clear	0 1 2 3 4 5 6 7 8 9 10	NA
Consistent position across displays	never	always	0 1 2 3 4 5 6 7 8 9 10	NA
Terminology is	inconsistent	consistent	0 1 2 3 4 5 6 7 8 9 10	NA
Pest management terms are	inconsistent	consistent	0 1 2 3 4 5 6 7 8 9 10	NA
Computer terms are	inconsistent	consistent	0 1 2 3 4 5 6 7 8 9 10	NA
Computer-related terms are used	too frequently	appropriately	0 1 2 3 4 5 6 7 8 9 10	NA
Relates to pest management	distantly	closely	0 1 2 3 4 5 6 7 8 9 10	NA
Definitions are	incomprehensible	understandable	0 1 2 3 4 5 6 7 8 9 10	NA

Figure 10–5. Questionnaires can provide a great deal of information about users' perceptions about a system. This information can then be used in statistical hypothesis testing.

Scores for each attribute across all respondents may then be ranked separately. For example, the values over all respondents for the attribute "function key meanings" would be ranked from 1 to *n*, where *n* is the number of respondents Similar rankings would be performed for each of the other attributes. Spearman's rank correlation statistic or Kendall's tau statistic may then be used to test hypotheses about associations between various attributes. In particular, any attributes that correlate with the length of time each respondent worked with the system are usually quite informative. Also, when examining the original scoring data, consistently low scores by most respondents across a number of attributes may reflect some basic flaw in the way the system operates.

10.1.5.2 Evaluation Using the Analytic Hierarchy Process. As Liebowitz (1986) notes, evaluation of a knowledge system necessitates the use of both qualitative and quantitative factors. The Analytic Hierarchy Process (AHP) does this well by dismantling an evaluation problem into smaller constituent parts. These smaller parts may also be further broken down into each of their component parts. This sort of hierarchical decomposition can be seen in the example of Figure 10–6. Evaluation criteria listed at the top level of this example are: performance, user-interface, understandability, and extent of application. Their position in the top level signifies that these are very important, or summary, criteria. Lower levels contain more detailed criteria. A portion of the second level, relating to the performance criterion, might be: typical cases, hard cases, and economically important cases. More and more layers may be added to any depth that seems appropriate for the evaluation task. Such an analytical tree is constructed by an evaluation team consisting of development team members and perhaps other interested persons or those with technical input. As noted in the caption of Figure 10–6, this approach can be used to evaluate different systems and different experts.

At each level of the hierarchy pairwise comparisons are made between all criteria subordinate to a particular node. An eigenvector from the matrix of paired comparisons represents the priority values for those criteria. Details for the calculations may be examined in Liebowitz (1986) or Saaty (1980). Any systems(s) or expert(s) being evaluated are then scored on each criteria present at the leaf nodes of the hierarchy tree. Those scores are multiplied by the priority values of intervening nodes as one walks up the tree to the root. The sum of all the weighted scores results in a final evaluation score for each system or expert. Final evaluation scores for competing systems or for human experts may be compared to those of the candidate system.

A powerful advantage of this approach is that it permits a systematic decomposition of *all* evaluation criteria. Pairwise comparisons are easy and straightforward, and the mathematics enables one to compute a final score that

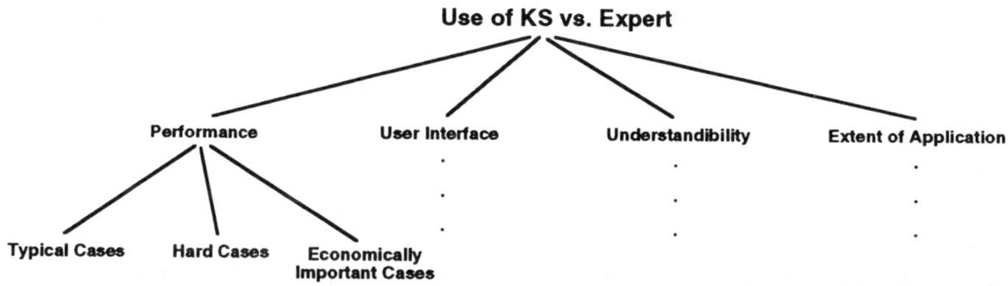

Figure 10–6. The Analytical Hierarchy Process permits the evaluation of all aspects of a knowledge-based system. Only a few of the higher level issues, important in an evaluation, appear in this diagram. Each system/expert being compared by this method is scored on each of the factors at the leaf nodes. Evaluation scores for each system/expert are then multiplied by the priority values contained in the intervening nodes between the leaf nodes and the root node. The final evaluation score computed at the root node incorporates the evaluation scores at the leaf nodes and the relative contributions of all factors at each level in the hierarchy.

reflects both the evaluation structure and the importance attributed to each evaluation factor. This approach is also quite flexible because questionnaires, discussed previously, can be used to provide some of the comparisons needed in the hierarchy.

10.2 DELIVERY

Final deployment of an operational system involves a number of additional activities. Even an "intelligent" KBS does not operate as an autonomous agent. Minimally, two support pieces should be included with the software, these are documentation support and maintenance responsibilities. These two items help make the software more useful now and help ensure that it will continue to be useful in the future.

10.2.1 Documentation

Operational use of any computer software requires two forms of documentation—*user* documentation and *system* documentation (Walters and Nielsen 1988). User documentation should contain informational material that thoroughly circumscribes the problem addressed by the software—what types of problems are addressed in the subject area and how the software fits into the MIS structure that currently exists. It should also carefully define the uses and limitations of the program. In some cases it may also be possible to suggest certain helpful hints that may allow the user to do more work or work more efficiently with the system.

The other aspect of documenting a knowledge system provides the capability for others to maintain and enhance the system, i.e. techical documentation. Evolution of a knowledge base, in most real situations, is unending. Thus, maintenance personnel must be given a well-documented basis for keeping the knowledge base accurate and current. There are some people who feel that the structure of KBSs makes them practically self-documenting because they are more transparent and more easily understandable than traditional software. While there may be some truth to this statement, KBSs do deal with knowledge and, hence, tend to be more complex on a conceptual level than information-oriented software. Therefore, it becomes critically important to carefully describe internal system structure, reasoning mechanisms used, search strategies employed, and alternative models and implementation methods *not* used (Walters and Nielsen 1988).

10.2.2 Maintenance

Maintenance and support are both extremely important in the life cycle of traditional software. Combined, they often represent a large revenue segment of the computer industry (Newquist 1988). Because systems become larger over time there must be some early planning for maintenance. As Newquist (1988) suggests, knowledge engineers must also become "knowledge janitors". The preceding paragraph discussed documentation for maintenance activities, now lets look at some of the tasks included in ongoing maintenance activities.

First of all, there should be an assignment of responsibility for performing maintenance tasks. In many organizations someone in the MIS department will be charged with these duties. This need not be the case, however. Considering that a system deals with knowledge structures rather than information structures, it may be more appropriate to assign maintenance work to someone more familiar with the KBS's particular subject area.

Next, some records of use and performance can provide important information, especially during the early days of delivery. Because it is very difficult to thoroughly validate system performance in testing stages, ongoing performance scrutiny can identify additional weaknesses. If performance is not found to be lacking, then users, managers, and developers can feel confident in the capabilities of the system. Records of use can be helpful to indicate who is using the system and how often they are using it. Little-used systems might be functioning poorly or might be difficult to use with other software or might have poor interface features, or there might be psychological reasons for non-use. The extensive investment required to bring a system to this stage means that it should be working properly and people should be using it often.

Third, system size and complexity growth over time must be carefully monitored. Difficulties associated with updates and extensions to a system's knowledge increase rapidly with the size of the knowledge base and its complexity. One of the highly promoted advantages of KBSs is their ease of modification. This is not necessarily true in practice. Careful construction and organization of a knowledge base is essential to make this potential advantage a reality; a knowledge model description helps with this task. Jacob and Froscher (1990) suggest subdividing a knowledge base into separate modules, where each module contains a description that specifies: local facts and hypotheses of the module, hypotheses inferred in the module and used elsewhere, and hypotheses inferred in other modules and used in the current module. Developing systems that are modular can make modification less cumbersome because only small portions of the total knowledge base are affected by any change. A well-detailed knowledge model helps create this modularity. A blackboard architecture or a similarly modular implementation certainly becomes useful, as well.

Fourth, if a particular system begins as a very small system but then expands to become a much larger one, it may still be possible to dissect the knowledge into fairly distinct and independent units. One way in which to accomplish this is to build separate systems over time covering the different functions as they are needed. Then, when an overall design has been developed for the encompassing subject area, individual systems can be linked together by well-defined interfaces.

Fifth, actual maintenance activities come in two types—*update* and *extension* (Walters and Nielsen 1988). Update maintenance involves the augmentation and modification of current system knowledge. This may be accomplished by changing the knowledge or facts or reasoning or heuristics or explanations in the knowledge base. As what we know about the external world changes, the representation of that knowledge in a KBS should change also. Extension maintenance refers to the addition of new functionality and features not previously present in the system. These might include changing the user interface, expanding the range of problems solved by the system, or linking the system into other software packages. By planning for the three considerations mentioned in the preceding paragraphs (i.e. maintenance responsibilities, records of use, and system growth), updating and extending a knowledge system can be accomplished more easily and in a more timely fashion.

10.3 SUMMARY

System evaluation and system delivery are two aspects of knowledge system development that often receive less emphasis than knowledge acquisition, knowledge representation, and search procedures. When evaluating a system, it is important to determine whether a system is to be compared to human expert performance or to some metric of historical success (i.e. test cases). Potential implications of an invalid system can have substantial impact on the type and extent of evaluation that is performed. The three major components of evaluation include: verification, validation, and usability. These may be applied to parts of a system separately or to the entire system holistically. Several techniques for verifying a system allow us to determine if current knowledge has been successfully translated into a knowledge model and then if that model has been accurately implemented in a system. Validation methods apply objective statistical tests to assess system accuracy or performance. In the final analysis, a verified and validated system means little unless users find it helpful or an organization gains some benefit from it. Actual placement of a system into a working environment requires documentation and maintenance support to ensure proper use and to encourage continued use into the future.

APPENDIX A Roots of Artificial Intelligence

There once was an ardent young suitor
Who programmed a female computer
But he failed to connect
The affective effect,
So there wasn't a thing he could do to' er.

There once was a passionate dame
Who wanted some things made plain,
So she punched up some cards,
Filled tape by the yards,
But – somehow – it just wasn't the same!

George A. Kelly (1963)

A.1 A DEFINITION OF ARTIFICIAL INTELLIGENCE

Artificial Intelligence, in the broadest sense, seems to be an attempt on the part of man to replicate himself in a machine, or to create different, perhaps even superior, artifacts. These creations are expected to think, to originate, to store, and to manipulate knowledge, and perhaps to self-propagate. They are required to exhibit human-like common sense, to explore their environment and deal with it. Partners and helpers to man, artificial intelligences are supposed to ease

---

Authors' Note: We asked our friend and colleague, Don Latham, to contribute a chapter expressing his views of the AI phenomenon. If there is such a thing as "fundamentalist" AI, it would be that AI's goal is to create intelligent agents—period. This has been the impetus for AI throughout much of history. This appendix provides an intriguing and entertaining historical account of where many ideas about artificial intelligence originated. Regardless of which AI goal we accept, the basics—representation, search, and learning—are still primary ideas.

the human condition and help create the utopia so avidly sought throughout human history, the Big Rock Candy Mountain.

Perhaps they will help to create a kind of utopia, caring for mankind so well that existence becomes bland and boring—then again, perhaps not. Intelligent machinery with free will might not care about the fate of its creators, destroying and supplanting them, hunting down the last poor remnants of the race *homo sapiens*.

The need to construct articulate, intelligent artifacts, regardless of the consequence, has been with us for a long time. Is artificial intelligence a possible peak in man's achievement or an oxymoron? Is it salvation or snake oil? As with many human endeavors, probably a little of each.

A.2 DREAMS OF INTELLIGENT ARTIFACTS

Because the term "Artificial Intelligence" first appeared in the 1950's, most people assume that the origins of those ideas appeared about the same time—or at least in the twentieth century. On the contrary, the notions of "intelligent servants", formal reasoning, and computation have existed for some time.

A.2.1 Ancient Dreams

The earliest reference I have been able to find pertaining to the construction of "artificially intelligent servants" is in the (translated) writings of Hermes Trismegistus. These were generated by several Greek authors between the second and fourth centuries AD. The manuscripts were attributed to one Hermes, who, it was claimed, had lived 3000 years before Christ. This attribution gave the pronouncements of Hermes the stamp of antiquity. The appearance of antiquity was necessary in those early centuries to market the ideas in any document, much like the appearance of newness in the twentieth.

A loose translation of the nature of the servants and instructions for making them are found in Hermes' writings (Yates 1964): "They are animated statues full of sensus and spiritus who can accomplish many things, foretelling the future, giving ills to men and curing them." ... "Our first ancestors invented the art of making gods. They mingled a virtue, drawn from material nature, to the substance of the statues, and 'since they could not actually create souls, after having evoked the souls of demons or angels, they introduced those into their idols by holy and divine rites, so that the idols had the power of doing good and evil.'".

The 16th century seems to have been a good one for artificial intelligence. Theophrastus Bombastus von Hohenheim, who called himself Paracelsus (right, the source of the word bombast) gave us some rather bizarre instructions for creating a homunculus, or little man, which "... has intelligence, and its mysterious manner of birth gives it the ability to investigate and communicate to us the secret of the most inscrutable mysteries." (according to Christian 1969). Roger Bacon, the first proponent of the experimental method, is supposed to have had a brazen head to answer his questions (Holyard 1957). Rabbi Loew of Prague, in the same century, is supposed to have created a homunculus by scratching the name of God on the forehead of a little clay man. This done, the creature ran amok and had to be stopped by erasing the name from its forehead (Scholem 1987).

Still later, in the 19th century, Mary Shelley, sitting around with the boys sipping sherry, decided that she could write as well as the best of them, and proved it true with *Frankenstein*, one of the all-time great AI stories. The Baron's monster (which has no name), when given life, went on a great quest to find out why it had been created, inadvertently doing a lot of damage in the process. It was, in fact, not particularly malevolent, merely ignorant, frightened, and a little clumsy. The beginning of the twentieth century gives us Cepek's play RUR (Russum's Universal Robots) and the term robot. Unfortunately, the robots Russum made were sexed, and two of them fell in love, leading the rest out on strike—so much for selfless servants.

A.2.2 Modern Nightmares

Since RUR motion pictures and the rise of science fiction literature have given us many examples of AI. Most of these, from the Forbin Project (Colossus) to the Terminator, have portrayed artificial intelligences as relentlessly inimical. When robots or AI are not malevolent, they are likely as not to be seen as comic characters such as C3PO in Star Wars, non-threatening to humans, even requiring looking after. There are of course benign exceptions, such as Gort in "The Day The Earth Stood Still." In written literature, the robots of Isaac Asimov and Clifford Simak are almost alone in their humanity coupled with rationally determined behavior (although it might be argued that the Terminator is quite logically constructed, considering its purpose, and did, in fact, become a "good guy", even unto self-sacrifice in its second film incarnation.). The reader can supply additional examples of modern dreams from direct experience.

A.3 REALIZATION

Humanity has finally arrived at the threshold of literal realization of his quest to be Godlike in creation. Computer systems, merged with motors, gears, hydraulics, light and strong plastics and metallic alloys, are becoming more commonplace daily in the workplace and even in our amusements (Disney World is not all done with mirrors). Sensing systems aping those of people exist and more are under development. And, in learning how to build the systems, we are learning more about *homo sapien* as well. We are learning to construct Hermes' statues (hardware) and the requisite incantations (computer programs) to give them sensus and spiritus, as promised.

A.3.1 Realize What?

The intelligence of these creations remains to be seen. Are we trying to create artificial personality or some kind of "pure" intelligence? The common view toward very intelligent people is one of coldness, relentless precision, lack of passion—what might be termed, almost, inhumanity. They are perceived as a threat. Indeed, such people are often described as machinelike.

A symbol for this kind of intelligence is chess. Of all the trappings of intelligent behavior, chess was thought to be the most complex. Ironically, it was one of the first real success stories of AI. Chess programs now play successfully at the grand master level, and small computers are readily available that can challenge and beat the casual player with frustrating regularity.

As the reader knows, this description of intelligence is far from the truth. In fact, intelligence in people is more often than not accompanied by a very caring personality. Some of the great human thinkers have also been exceptional humanitarians. If we humans do build intelligent machinery and we do project humanness into them, then they are more likely to be benign than malignant and cannot escape having a human personality. We are going to let our desire to create override our feeling of possible disaster from the results. What tools do we presently have to accomplish the task? What statues? Sensus? Spiritus?

A.3.2 Mind as a Tool

The primary tool is, of course, our own mind. Not only do we use our skills as artificers for construction of the statue, we use our analytic facility to study our minds as well as our environment and apply the results to our creation. All these studies are applied to the AI system (in no particular order): cognitive psychology, mechanical engineering, electrical engineering, anthropology, computer science, linguistics, behavioral psychology, ergonomics, mathematics, philosophy, physics, chemistry, metallurgy, biology, personality analysis, psychodynamics, ... In fact, it should be no surprise that the totality of human knowledge will be required to construct a complete artificial human system, and there is some doubt as to the sufficiency of that knowledge. Are we smart enough to pull it off?

The problems of constructing the statue will now be set aside. That is, we will not discuss power supplies, motors, input and output devices, and so forth.

At present, we need to concentrate on the mind of our creation, even though its mind and its body will eventually be one system, just as ours is, for our creation cannot help but reflect ourselves. To develop the mind, or at least the rational part of it, we need to consider the acquisition, representation, storage, manipulation, creation, and application of knowledge.

This will be a good trick. We don't really know very much about how we do these things with knowledge, and even less about why. In fact, the question of why has, except for a very few attempts, been left out of artificial intelligence. We will leave it out too, at this point.

A.4 KNOWLEDGE

A suitable definition of knowledge would seem to be simple. After all, we all know what knowledge is, don't we? It is, in fact, very difficult to define. Knowledge is one of those tough concepts that everybody has a feeling for, and that there seems to be a consensus about, but that is extremely difficult to express. We can sort of duck the philosophical question with an operational definition: that stuff that is acquired, stored, manipulated, and created by the mind and the body of *homo sapiens* is knowledge. Contrast this definition to that provided in Chapter 1.

A.4.1 Group Knowledge

A starting point with knowledge might be to see how we store and manipulate our "group" knowledge, i.e. the collective knowledge of some group of people. For most of our history, such knowledge was stored in the minds of the users of the knowledge, slowly added to, and then passed on through direct teaching and memorization. This is called oral storage, or the *oral tradition.* Written language was used later, first to keep accounts, then for history and boasting on the part of rulers, then for storage of knowledge. Even so, there were only a few great libraries, access was difficult, and a formal written language had to be mastered.

Gutenberg changed much of that with the printing press. Within 50 years of its introduction, the number of books went from a few hundred thousand to over 10 million, most of the new ones in the vernacular, hence accessible to more people, and many of them containing knowledge useful in more than the spiritual life. We now have truly monster libraries, with more knowledge than one person can ever hope to even scan, let alone master.

A.4.2 Hierarchical Structure

Consequently, we had to invent methods for storing and accessing this vast morass of knowledge, and came up with the hierarchical structure displayed, for example, in the Dewey Decimal system. It should be no surprise that this system resembles the same idea in the classification of bugs, animals, and plants known as the Linnaeian system. Thus, one natural storage system for knowledge seems to be a hierarchy of classes, with each child class including all the attributes of the parent class above it, and adding some of its own, until sufficient descriptive uniqueness has been reached to solve the problem at hand. This is one example of the AI concept of semantic networks, as well as one of the basic problems/subjects of AI, namely search, in this case through a tree structure.

At the end of a classification search, we find some particular object, e.g., a book (or a flower, for that matter). Many of these books, like this one, still follow the orderly hierarchical structure. Most, like novels, do not. Suddenly we are confronted with other units of knowledge. Even though the chapter, section, paragraph structure can be hierarchical, more often it is not. The units are more like clumps, or chunks. There's a good word for use in AI. Let's call them *chunks*.

A.4.3 Chunks

Can chunks be treelike in structure? Of course, but they can have other structures as well. A chemical formula is a chunk of knowledge. A road map, a computer program, how to wash one's hands are also chunks. Each of these is different in its composition, each has a sort of natural representation: a two-dimensional array, a list of simple sequential instructions, a behavioral set, respectively. Note that, for example, the same map chunk can be found by several differing hierarchies, such as scale, or political affiliation, or road quality.

All these chunks are differing examples of *knowledge representation*, one of the central problem areas of artificial intelligence. Of course, the techniques for manipulating chunks of knowledge are themselves chunks of knowledge, and have representations as well. Then there are techniques for using the techniques, and so on.

One of the problems one has with chunks is that there does not seem to be any hierarchy for them, or, for that matter, any other single convenient classification other than the simple list. This is itself, of course, a chunk of knowledge, how to make a list, and when to use it, or is the latter another chunk, separate from the first and loosely connected, or what? It is easy to see that too much contemplation of just how to do chunking might cause one to hit the road with a motorcycle (Pirsig 1974). Let's just make some lists, secure in the knowledge that they will be incomplete for purposes of knowledge representation and manipulation, and see if we can find some examples or analogs to help understand them.

A.4.4 Representation of Chunks

In no particular order, then, some representations are: semantic networks, which we have seen, frames, sets of rules, land maps, state spaces, neural networks, simple lists, any kind of database ever used, Martha Graham's dance notation, written music, a set of photographs of joy, anger, and the like, the Ten Commandments, ... Got the idea?

A.4.4.1 Frames and Semantic Networks. Many of the contemporary useful artificial intelligence efforts use frames, rules, neural networks, and state spaces. These are simple concepts, although implementation in artificial constructs is tricky. Minsky's frames (1975) represent chunks by first naming them. Then, the bulk, if you will, of the chunk is broken down into sub-chunks called slots. Each slot is a named thing, which represents an attribute, having a value. Also associated with the slot can be actions that are done on evoking the slot, or instructions for obtaining its value, or other things. It is easy to see that a semantic network can be subsumed in a set of nested frames. Frames can be much broader, however; note that parts of one frame can invoke parts of another frame, or even, although dangerous, parts of itself.

The easiest example usually given for a frame is person classification. One can build a frame called "person". The slots are then easily seen to be items like name, address, height, weight, eye color, and so on. If there were a slot called "employed by", that slot could contain an evocation of the frame for the employer, and recover that data as well. The weight slot, if empty, could have instructions for weighing, or call for a question about weight to be asked. In addition, some slots in the frame might inherit characteristics from a parent frame. Frames thus provide a very powerful knowledge representation, or data structure.

A.4.4.2 State Space. Another knowledge representation scheme is the state space. This scheme looks, as its name implies, as a set of objects, each having a state it can be in. No, not necessarily Nebraska or Idaho (although those are possibilities for, say, the state space of a fleet of trucks), but on or off, such as a set of switches, or positions in an array. Game boards are prime examples for this concept. A chess game consists of a sequence of states on the space of the board, locations of the pieces. Each piece in chess, for example, can be represented as a frame, with slots like point value, present value (may be blocked or threatened), name, position, color, and so forth. The state space is then the aggregate of the position slots. The dance notation referred to above might be mapped onto a state space representation, each state being a body position. A dance would then be a succession of states, and (to quote one of my favorite authors) so on. Can a state space be open-ended? You decide.

A.4.4.3 Artificial Neural Networks. An artificial neural network is an arrangement of processing elements called neurons which are, in different models, more or less like simple biological neurons. These have several inputs and one output, and a transfer function connecting them. If the sum of the signals at the inputs exceeds a threshold value, the output changes state. The output of one neuron serves as one input for another. When these neurons are connected in layers, such that the output of each neuron in one layer is connected as an input to every neuron in the next layer, the result is a crude approximation of some parts of the biological neural network called a brain. Artificial neural networks find, at present, their best use in pattern recognition and kindred applications. A simple rule base and its control mechanism can also be represented as a neural network. For an imagination stretching experience, consider a massively parallel processor, in which large numbers of computers are hooked together in a kind of neural net.

A.4.4.4 Rules. Another representation scheme to explore here is representation by rules. This, as can be seen in the remainder of the book, is a very handy representation indeed for many kinds of problems, especially diagnostic systems and logic constructions. A simple rule consists of an antecedent part and a consequent part. In the antecedent are all the conditions that must be met if the rule is to prove the consequent true, or established as fact for the purpose of the system. A simple example: IF the engine is idling, AND the idle is rough, (antecedents), THEN turn the idle mixture screw counterclockwise (consequent). Obviously, rules can be far more complex than this elementary example. There can, for example, be multiple consequents; also, there might be OR statements or NOT statements in the antecedents.

Real-world and/or real-time conditions may be antecedents. A temperature sensor outside the house is read by the computer, and a statement is invoked such as, IF the outside temperature is 70 degrees, THEN turn off the furnace system. The rule may be a general one, so that it can be used to test many values. And, essential to the construction of rule based systems, the antecedents of rules can be consequents of other rules.

Rules can be used in combination with the other constructs we have discussed. A rule or set of rules may, for example, be a slot in a frame, or a set of instructions connecting frames or frame slots. Further, rules may be invoked to handle transitions in state spaces, or as parts of chunks occupying positions in state spaces, or to inference over a semantic network. Hierarchies can be stated in terms of rules, or rules occupy nodes in hierarchies. Some rule sets can be expressed as a neural network. In fact, rules, frames, and state spaces all sort of come together in many AI systems. And all of them only begin to approximate the way a human intelligence stores its stuff.

A.4.4.5 Threaded Chunks. Remember the way a well-written paragraph is supposed to be composed? In particular, there should be a sentence or idea tying the paragraph to the ones just prior and just following. The paragraphs thus are supposed to be strung together like beads—a good idea. Let's call that one *threaded chunks*; one chunk tied to the next by some kind of formal system of tags. If you think this is nice and orderly, try unthreading something like Joyce's *Ulysses* or maybe a FORTH computer program. Nobody said the beads have to be "next to" each other, just next on the string! Still, this method is used in many knowledge manipulation schemes. In fact, Alan Turing (a British mathematician who developed many of the early ideas about computing machines) proved that all computation can be done this way using a string of binary bit chunks and a few simple operations. The string can of course be taken out of the chunks, and a new one generated using the same chunks. This is one of the techniques used in a recent computer knowledge organization called hypertext.

A shiny new idea? Well, no. In fact, when knowledge chunks were already old hat, orators and others who remembered knowledge were taught to string chunks together by imagining a walk through a familiar place, tying a chunk to each location or item along the walk. Thus, imagine coming in your front door. Tie a chunk to the door. Close the door, walk to your favorite chair. Tie another chunk to the chair, and so on to the kitchen, etc. Now, not only is this threaded, it is also indirectly addressed, since the thread does not connect the chunks, but the tags for the chunks. The chunks are elsewhere. No wonder Edward Feigenbaum, one of the leading lights of AI, has said "...it [the AI field] tends to reward individuals for reinventing and renaming concepts and methods that are well explored (Feigenbaum 1980)." Here, a pre-Aristotelian trick for remembering and processing chunks is resurrected with a new name and new originators.

A.5 GOALS

Now that we have some way of representing the knowledge our construct requires, what do we do with it, and how do we do it? What to do means that we are talking about goals. In the simple examples discussed, we wanted to tune a carburetor, record (or even invent) a dance, keep track of, or route, our trucks, or play a game of chess. Perhaps we want something constructed from parts, or to navigate a vessel. In the end, we will want the construct to set its own goals, and will need to invent reasons for the machine to use in selecting the goals. This is very vexing, because mankind has only the barest glimmer as to why it selects the goals that it does. Let us back off a little with our machine, then, and select its goals for it. We won't select something like "know and want everything" or "work for a vacation in Hawaii" but rather something simpler.

How simple do the goals have to be? For constructs, things like playing chess seem to be less difficult than, say riding a tricycle. For humans, the latter is proved by simple observation easier than the former. After all, we can give a four or five year old child a tricycle and within a very short time, the child can ride the vehicle down the sidewalk; not only that, but also converse with an imaginary playmate and hum a song while doing it. It almost seems as if machinery can easily carry out far more complex operations than a human, and stumbles badly over things that are child's play to a human. In short, machines cannot exhibit "common sense", whatever that is. That should be no surprise, machines are doing what we built them to do, number crunching and logical constructs.

Presented with a reasonable goal, however (that means one we know how to program), machines do just fine, provided we can supply the knowledge to the machine and create a way for the machine to use it. The set of chunks of knowledge used in the construct is called the domain of the knowledge. Note that as of now, programs have not been written that can extend a domain without the seeds of the extension being in place. That is, the knowledge to be gained is accessible to the machine within the program or construct of the machine, or is deduced from supplied data with supplied procedures.

In all systems to date, the goal or goals have driven both the knowledge representation and the methods used to reach the goal. In addition, the entire construct of the program is thoroughly understood by the author or authors of the program or construct. That is, the systems or constructs are reflections of the understanding of the makers. They are made, not born.

A.6 REACHING THE GOAL

With some goal or goals in mind for the program, then, what is the "how" of present AI? The programs that I am aware of, by no means exhaustive, have used such means as goal reduction, search through organized data structures or problem spaces, logical connections (called chaining) in rule bases, problem-solving logics, and generate-and-test. More than one of these techniques can be invoked, of course, such as is done in SOAR (Laird *et al.* 1986) or conceptual dependency programs (Schank and Reisbeck 1981).

A.6.1 Generate-and-Test

Generate-and-test seems on the surface to be so simple that it appears almost trivial. In this technique, a program is devised that will create or extend patterns or events. These patterns/events are partial or potential solutions to the problem being studied. Each pattern thus generated is then tested for its "fitness" as a solution. To demonstrate how powerful this technique really is, one need only examine the process of evolution—perhaps the most important and extensive example of generate and test. New genotypes are continually being generated, the resulting phenotypes are tested against the organism's environment. Those that prosper, pass the test; those that don't, pass into obscurity. The trouble with generate-and-test is that new patterns often cannot be accommodated, because specialized mechanisms must be built in for any particular set of patterns. Generate-and-test may be one of the major modes of functioning for the human mind, maybe one of the oldest modes. We all have lots of these generators built in, and perhaps some of them could be called common sense.

A.6.2 Goal Reduction

In goal reduction, an ultimate goal is broken into a sequence of sub-goals, which may themselves be further broken down, and so on. The sub-goals may or may not need to be done sequentially; that depends very much on the nature of the problem. Is this technique new? Hardly. Goal reduction, although not called that, is one of the fundamental methods used to teach things to people, and has been since teaching was invented. Suppose, for example, one needed to teach hand-washing. First, go into the bathroom. Second, locate the following items: soap, washbasin, faucet(s), towel. Third, turn on faucets. Fourth, test water for temperature—and so on. In this procedure, note that the person is being programmed, much like a machine might be programmed, the brain being plastic (not made of, but capable of learning). Of course, the machine program would be different. Roll into the cleaning room, locate the cleaning spray, open inspection panel 2, spray ...

In teaching this way, there are two approaches to the goal structure. One can start at the beginning, that is, going into the bathroom, and go forward step-by-step, or alternatively, begin at the final goal, that is, clean hands, teaching backwards by means of questions like "What do you have to do to get clean hands?". In fact, the teaching of elementary tasks for the developmentally disabled is done in just this way, either forward or backward, in a highly formalized fashion, and has been for a long time. In AI these methods are called forward and backward chaining in a rule base.

A.6.3 State-Space Search

A problem faced early on in AI research was the search problem. This is the problem generated in the famous "eight" puzzle, or the "missionaries and cannibals" puzzle, or checkers, or chess. That is, given a system in one state, how do you optimally search possible sequences of future states to reach a desired state? Several methods were devised. Specific ones can be found in the literature. The main feature of these methods is a measure of the "distance" from the current state to the desired state (usually called a heuristic function). Search can also be done in a problem space as in SOAR.

A.6.4 Genetic Algorithms and Emergent Behavior

The latest in "how to get there" is through emergent behavior and genetic algorithms. Genetic algorithms are self-modifying computer programs that evolve themselves into solutions to posed problems. These programs are kin to generate-and-test methods, keeping the best solutions, modifying them, and testing them against the problem. After several generations, the program settles down (hopefully) to a solution to the posed problem. The solution then can be kept as an "answer" (Koza 1992).

Emergent behavior is the term used to describe the gross behavior of a system that consists of many simple entities, operating locally, using very trivial rules of interaction with neighboring entities. Group behavior resulting from the simples rules of individuals can, of course, be extremely complex. Most examples of this technique have concentrated on replicating natural phenomena like flocking in birds or food-seeking in ants. A more complex form of emergent behavior is applied to robots, in that portions of the robot are autonomous, reporting to higher levels in simple ways. Does this remind you of the human body? It should.

A.7 INTERFACING, OR CONVERSING WITH THE STATUE

Having stated our goals, and the machinery for achieving them, how do we find out what the program has done? How do we interact with our statue? This calls for some kind of interface with the outside world. We can have the machine draw pictures, or make movies for us, or carry out observable actions. For many machines, this is preferred. For the most part, we'd like the machine to discourse rationally with us. The technology for either written or spoken discourse is now well advanced. Indeed, the programming of output, usually meaningful English or graphics (for our culture), is no particular problem in terms of comprehensibility, or even some semblance of style. Examples will be given in chapters of this text. The big sticking point is in devising something to understand the spoken or written word delivered as input and translate it into "machinese". This is going to be difficult, because we don't really know how people understand each other or, for that matter, themselves.

Solutions seem simple. For written input in a human language, just diagram the sentence, look up the agreed upon meanings of the words in a big dictionary that converts words to machine structures, and interpret the structures in terms the AI program can understand. Sounds easy, doesn't it? Try to apply this technique to the sentence "Sam gave Jane a baby." Several other linguistic input schemes have also stalled on sentences like this. In fact, application of linguistics to AI is one of the major problems at present. The fields of linguistics and AI cannot be separated, for intelligence and language are strongly connected. For now, we will have to adapt our input to the machine, that is, the human user will have to adapt himself or herself to the machine and suffer the syntax error messages until we can get something more meaningful going.

A.8 TESTING AND CONCLUSION

Finally, let us suppose a computer program, the sensus and spiritus of our statue, is completed. Can we determine how our statue stacks up against a Person? There is, in fact, a simple test (although it has been complicated by later improvement) stated by Alan Turing (of Enigma fame). Suppose we engage the suspect being in conversation using a simple terminal (no body language allowed). If we cannot tell, using any means possible under this restriction, whether the other end of the terminal connection is a Hermetic statue or a person, and it is a statue, then it can be said to be intelligent. In other words, if it quacks like a duck, then it's a duck. (My duckness test would be "can you eat it?" Reread the epigraph.)

It is here that reality rears its ugly head, for at present there are no computer programs that can pass Turing's simple test with regularity, or even the stricter yet seemingly simpler test of learning how to ride a tricycle in the real world. We have yet to put the construct into the machinery, animate our statue, get our golem going. Although we dream in animated Technicolor with surround sound now, we still program our computers one line or function or object at a time; that's the reality. They're still stupid, folks, but we won't give up. We'll have our statues yet.

The rest of this book provides some insight into one very useful application of AI called the knowledge-based system. With knowledge-based techniques, using rules and control mechanisms and knowledge bases, together with explanation facilities and graphics, you can transform knowledge of a carefully defined domain into organized constructs for teaching, diagnosis, and decision aids. Are these artificial intelligences? No, but think kindly of these systems, because they are the baby steps for the Hermitic Statue, can be very useful, and will stretch your mind besides. I hope you enjoy making these systems, because in the last analysis, the search for AI is simply lots of fun. In fact, as it is with many other things, it just may be that the search is more important than the goal.

A.9 SUGGESTED READINGS

Following are suggested readings for serious and not-so-serious use (the attentive reader will recognize that each of these chunks will contain threads to other chunks of equal interest): McCorduck 1979, Michie and Johnston 1985, Johnson 1986, Barr and Feigenbaum 1982, Winston 1984, Sowa 1984, Shapiro 1987, Morone and Hilbush 1987, Laird et al. 1986, Schank and Riesbeck 1981, Haugeland 1981, Lakoff 1987, and Simon 1988.

Bibliography

Amble, T. 1987. *Logic programming and knowledge engineering*. Reading MA: Addison-Wesley.

Anderson, J. R. 1983. A spreading activation theory of memory. *Journal of Verbal Learning and Verbal Behavior* 22: 261–295.

Arity Corporation. 1988. *Arity PROLOG version 5.1*. Concord, MA.

Asimov, I. 1951. *Foundation*. New York: Avon Books.

Barr, A., P. R. Cohen, and E. A. Feigenbaum. 1989. *The handbook of artificial intelligence*. Volume 4. Los Altos, CA: William Kaufmann, Inc.

Barr, A., and E. A. Feigenbaum, eds. 1981. *The handbook of artificial intelligence*. Volume 1. Los Altos, CA: William Kaufmann, Inc.

Barr, A., and E. A. Feigenbaum, eds. 1982. *The handbook of artificial intelligence*. Volume 2. Los Altos, CA: William Kaufmann, Inc.

Bassett, J. 1984. Red pine plantation management in the Lake States: a review. University of Michigan, School of Natural Resources, Ann Arbor, MI. IFSIM publication no. 3.

Bell, E. F. 1971. Simulating a silvicultural system. Ph.D. Dissertation, SUNY, Syracuse, New York. 161 pp..

Bellman, R., R. Kabala, and L. A. Zadeh. 1966. Abstraction and pattern classification. *Journal of Mathematical Analysis and Applications* 13: 1–7.

Benfur, R. A., and L. Furbee. 1989. Knowledge acquisition in the Peruvian Andes. *AI Expert* 4(11): 22–29.

Benzie, J. 1977. *Manager's handbook for red pine in the north central states*. USDA Forest Service, North Central Forest Experiment Station, St. Paul, MN., General Technical Report NC–33. 22 pp.

Berkshire Software Company. 1987. *Turbo shell*. 44 Madison Street, Lynbrook, NY 11563.

Bhattacharyya, G. K., and R. A. Johnson. 1977. *Statistical concept and methods*. New York: John Wiley & Sons.

Black, D., and J. Manley. 1987. A logic-based architecture for knowledge management. In: *Proceedings, 10th International Joint Conference on AI*, 1987 August 23–28, Milan, Italy. pp. 87–90.

Boose, J. H. 1985. A knowledge acquisition program for expert systems based on personal construct psychology. *International Journal of Man-Machine Studies* 23: 495–525.

Boose, J. H. 1986. *Expertise transfer for expert system design*. New York: Elsevier.

Borland International. 1988. *Turbo PROLOG version 2.0*. Scotts Valley CA.

Bowen, Kenneth A. and T. Weinberg. 1985. A meta-level extension of PROLOG. IEEE Computer Society, Symposium on Logic Programming, July 15–18, 1985, Boston, MA, Computer Society Press, pp. 48–53.

Bowerman, R. G., and D. E. Glover. 1988. *Putting expert systems into practice*. New York: Van Nostrand Reinhold.

Boy, G. A. 1991. Intelligent assistant systems. London: Academic Press.

Branscomb, L. M. 1979. Information: the ultimate frontier. *Science* 203:143–147.

Bratko, I. 1986. *PROLOG programming for artificial intelligence*. Reading MA: Addison-Wesley.

Braun, R. 1989. Expert system tools for knowledge analysis. *AI Expert* 4(10): 22–29.

Brokenshaw, D. M. Warren, and O. Werner, eds. 1980. *Indigenous knowledge systems and development*. New York: University Press of America.

Bronowski, J. 1973. *The ascent of man*. Boston: Little, Brown and Company.

Brown, G. C., W. Marek, V. Yeh, S. Meyhs, N. Freedman, and E. Bowen. 1985. Expert systems in agricultural pest management: an implementation example. *NCCI Software Journal* 1(2): 1–82.

Buchanan, B. G., and R. O. Duda. 1983. Principles of rule-based expert systems. In *Advances in computers*. Vol. 2, edited by M. C. Yovits. New York: Academic Press.

Buchanan, B. G., and E. A. Feigenbaum. 1978. DENDRAL and Meta-DENDRAL: Their applications dimension. *Artificial Intelligence* 11: 5–24.

Buchanan, B. G., D. Barstow, R. Bechtal, J. Bennet, W. J. Clancey, C. Kulikowski, T. Mitchell, and D. A. Waterman. 1983. Constructing expert systems. In *Building expert systems*. Edited by F. Hayes-Roth, D. A. Waterman, and D. B. Lenat. Reading MA: Addison-Wesley.

Buchman, R. G., S. P. Pederson, and N. R. Walters. 1983. A tree survival model with application to species of the Great Lakes region. *Canadian Journal of Forest Research* 13: 601–608.

Buckman, R. 1961. Development and use of three stand volume equations for Minnesota. *Journal of Forestry* 59: 573–575.

Buckman, R. 1962. *Growth and yield of red pine in Minnesota*. USDA Forest Service, Washington, DC USDA Technical Bulletin 1272.

Buckman, R. E., and R. F. Wambach. 1966. Physical responses and economic implications of thinning methods in red pine. In: *Proceedings, Society of American Forester Annual Meeting*. 1965 October 24–28. Detroit, MI. pp. 185–189.

Buech, Richard R., C. J. Martin, and H. M. Rauscher. 1990. An application of expert systems for wildlife: identifying forest stand prescriptions for deer. *AI Applications* 4: 3–18.

Bundy, A., and R. Welham. 1981. Using meta-level inference for selective application of multiple rewrite rules in algebraic manipulation. *Artificial Intelligence* 16: 189–212.

Buol, S. W., F. D. Hole, and R. J. McCracken. 1973. *Soil genesis and classification.* Ames, IA: Iowa State University Press.

Burke, J. 1978. *Connections.* Boston: Little, Brown and Company.

Carrico, M. A., J. E. Girard, and J. P. Jones. 1989. *Building knowledge systems: developing and managing rule-based applications.* New York: McGraw-Hill.

Casey, T. 1989. Picking the right expert system application. *AI Expert* 9: 44–47.

Chandrasekaran, B. 1983. On evaluating AI systems for medical diagnosis. *AI Magazine* 4(2): 34–37.

Chandrasekaran, B. 1986. Generic tasks in knowledge-based reasoning: high-level building blocks for expert system design. *IEEE Expert* 1: 23–30.

Chandrasekaran, B. and S. Mittal. 1983. On deep versus compiled approaches to diagnostic problem solving. *Int'l J. Man-Machine Studies* 19(5): 425–436.

Chandrasekaran, B., M. C. Tanner, and J. R. Josephson. 1989. Explaining control strategies in problem solving. *IEEE Expert* 4(1): 9–24.

Cherubini, M., L. Fanti, P. Torrigiani, and M. Zalloco. 1989. An integrated expert-system builder. *IEEE Software* 3: 44–52.

Chikayama, T. 1983. ESP – extended self-contained PROLOG as a preliminary kernel language of fifth generation computers. *New Generation Computing* 1: 4–13.

Cholawsky, E. M. 1988. Beating the prototype blues. *AI Expert* 3(12): 42–49.

Chomicki, J., and M. Minsky. 1985. Towards a programming environment for large PROLOG programs. In: *Proceedings, Symposium on Logic Programming*, IEEE Computer Society. 1985 July 15–18, Boston, MA. pp. 230–241.

Christian, P. 1969. *The history and practice of magic*, J. Kirkup and J. Shaw (Trans.). New York: The Citadel Press.

Clancy, W. J. 1983. The epistemology of a rule-based expert system – a framework for explanation. *Artificial Intelligence* 20: 215–251.

Clark, K., and F. McCabe. 1982. PROLOG: a language for implementing expert systems. In: Michie, D. (ed.), *Machine Intelligence 10*. Chichester, England: Ellis Horwood. pp. 455–470.

Cleaves, D. A., E. L. Smith, and William G. Bradshaw. 1987. Subjective assessments in the design and use of forest protection models. In: P. E. Dress and R. C. Field (eds.), *The 1985 Symposium on Systems Analysis in Forest Resources*, Proc. of the Society of American Foresters, Athens, Georgia, Dec. 9–11, 1985, pp. 386–401.

Clocksin, W. F., and C. S. Mellish. 1981. *Programming in PROLOG.* New York: Springer-Verlag.

Clocksin, W. F., and C. S. Mellish. 1987. *Programming in PROLOG.* 3rd. edition. New York: Springer-Verlag.

Coelho, H., and J. C. Cotta. 1988. *PROLOG by example: How to learn, teach, and use it.* Berlin: Springer-Verlag.

Cooke, N. M., and J. E. McDonald. 1987. The application of psychological scaling techniques to knowledge elicitation for knowledge-based systems. *International Journal of Man-Machine Studies* 26: 533–330.

Cooney, B. A. 1986. Representing knowledge through expert systems. *The Compiler* 3(4): 5

Coulson, R. N. and M. C. Saunders. 1987. Computer-assisted decision-making as applied to entomology. *Ann. Rev. Entomol.* 32:415–437.

Covington, M, D. Nute, and A. Vellino. 1995. *PROLOG programming in depth.* 2nd ed. Glenview IL: Scott, Foresman and Company.

Crawford, C. C., and J. W. Demidovich. 1981. Think tank technology for systems management. *Journal of Systems Management*, November 1981, pp. 22–25.

Cuadrado, J. L., and C. Y. Cuadrado. 1986. AI in computer vision. *BYTE* 11: 237–258.

Cuadrado and Cuadrado, 1988, chapter 8, page 15

Cullen, J., and A Bryman. 1988. The knowledge acquisition bottleneck: time for reassessment. *Expert Systems* 5(3): 216–225.

Dalkey, N. C., and O. Helmer. 1963. An experimental application of the Delphi method to the use of experts. *Management Science* 9(3): 458–467

D'Ambrosio, B. 1989. Extending the mathematics in qualitative process theory. *International Journal of Intelligent Systems* 4(1): 55–80.

Davies, M., and S. Hakiel. 1988. Knowledge harvesting: a practical guide to interviewing. *Expert Systems* 5(1): 42–50.

Davis, E. 1990. *Representations of commonsense knowledge.* Los Altos CA: Morgan Kaufmann.

Davis, J. R., P. Whignam, and I. W. Grant. 1988. Representing and applying knowledge about spatial processes in environmental management. *AI Applications in Natural Resource Management* 2(4): 17–25.

Davis, J. R., and J. L. Clark. 1989. A selective bibliography of expert systems in natural resource management. *AI Applications* 3: 1–18.

Davis, J. R. and P. Laut. 1989. An expert system to estimate trafficability in a remote region of Australia. *AI Applications* 3(1): 17–26.

Davis, R. 1980. Meta-rules: reasoning about control. *Artificial Intelligence* 15: 179–222.

Davis, R., and D. B. Lenat. 1982. *Knowledge-based systems in artificial intelligence.* New York: McGraw-Hill.

de Kleer, J. 1986. An assumption based truth maintenance system. *Artificial Intelligence* 28(2):127–162.

Dennett, D. C. 1985. *Elbow room: the varieties of free will worth wanting.* Cambridge MA: MIT Press.

Dennett, D. C. 1986. *Brainstorms: philosophical essays on mind and psychology.* Cambridge MA: MIT Press.

Doukidis, G. I. and R. J. Paul. 1992. A survey of the application of artificial intelligence techniques within the OR Society. In: Georgios I. Doukidis and Ray J. Paul (eds.) *Artificial Intelligence in Operational Research*, London: MacMillan Press, pp. 9–21.

Doyle, J. 1979. A truth maintenance system. *Artificial Intelligence* 12:231–272.

Duda R. O., and J. G. Gaschnig. 1981. Knowledge-based expert systems come of age. *Byte*, March, 238–249.

Duda, R. O., P. E. Hart, and N. Nilsson. 1976. Subjective Bayesian methods for rule-based inference systems. *Proc. AFIPS* 47: 1075–1082.

Durkin, John. 1993. *Expert systems catalog of applications.* Akron, OH: Intelligent Computer Systems Inc.

Englemore, R., and T. Morgan, eds. 1988. *Blackboard systems.* Wokingham, England: Addison-Wesley.

Erman, L. D., and V. R. Lesser. 1975. A multi-level organization for problem solving using many diverse, cooperating sources of knowledge. In: *Proceeding of the 4th International Joint Conference on Artificial Intelligence.* pp. 483–490.

Feigenbaum, E. A. 1979. Themes and case studies of knowledge engineering. In: Michie, D. (ed.) *Expert systems in the micro-electronic age.* Edinburgh: Edinburgh University Press. pp. 3–25.

Feigenbaum, E. A., P. McCordick, and H. P. Nii. 1988. *The rise of the expert company.* New York: Times Books.

Fischer, W. C. 1987. The fire effects information system. Pages 128–132 *in* Proceedings of the Symposium on Wildland Fire 2000, USDA Forest Service Gen. Tech. Rep. PSW–101, Pacific Southwest Forest and Range Experiment Station, Berkeley, CA.

Fischler, M. A., and O. Firschein. 1987. Intelligence: The eye, the brain, and the computer. Reading MA: Addison-Wesley Publishing Co.

Floyd, C. 1983. A systematic look at prototyping. In *Approaches to prototyping.* Edited by R. Budde. London: Springer-Verlag.

Forscher, B. K. 1963. Chaos in the brickyard. *Science* 142: 339.

Friedl, M. A., J. E. Estes, and J. L. Star. 1988. Advanced information-extraction tools in remote sensing for earth science applications: AI and GIS. *AI Application in Natural Resource Management* 2(2,3): 17–30.

Galitz, W. O. 1985. Handbook of screen format design. Wellesly Hills, MA: QED Information Sciences, Inc.

Genesereth, M. R., and M. Ginsberg. 1985. Logic programming. *Communications of the ACM* 28: 933–941.

Gevorkiantz, S. R. 1957. *Site index curves for red pine in the Lake States.* Lake States Forest Experiment Station, Technical Note 484.

Giarratano, J., and G. Riley. 1994. *Expert systems: principles and programming.* 2nd. ed. Boston MA: PWS-Kent Publishing Company.

Giere, R. N. 1984. *Understanding scientific reasoning.* 2nd edition. New York: Holt Rinehart and Winston.

Goldberg, A., and D. Robson. 1983. Smalltalk-80: the language and its implementation. Reading MA: Addison-Wesley.

Goldberg, D. E. 1989. *Genetic algorithms in search, optimization, and machine learning.* Reading MA: Addison-Wesley Publishing Company, Inc.

Gordon, S. E. 1989. Theory and methods for knowledge acquisition. *AI Applications in Natural Resource Management* 3(3): 19–30.

Gordon, S. E., and R. T. Gill. 1989. Question probes: a structured method for eliciting declarative knowledge. *AI Applications in Natural Resources Management* 3(2): 13–20.

Graesser, A. C., and L. F. Clark. 1985. *Structures and procedures of implicit knowledge.* Norwood NJ: Ablex Publishing.

Graham, I. and P. L. Jones. 1988. *Expert systems knowledge, uncertainty, and decision.* London: Chapman and Hall.

Grover, M. D. 1983. A pragmatic knowledge acquisition methodology. In: *International Joint Conference on Artificial Intelligence 83.*

Gudes, E., A. Shafrir, and S. Gerlitz. 1986. Software and knowledge engineering methods in building expert systems. In *Second International Expert Systems Conference.* Edited by *Expert Systems.* Oxford: Learned Information Ltd.

Hahn, G. J. 1985. More intelligent statistical expert systems: future directions. *American Statistician* 39(1): 1–8.

Halpern, Diane F. 1989. *Though and knowledge: an introduction to critical thinking.* Hillsdale NJ: Lawrence Erlbaum Associates.

Hammond, P. 1984. Micro-PROLOG for expert systems. In: Clark, K. L. and J. E. McCabe (eds.) *Programming in logic.* Englewood Cliffs NJ: Prentice-Hall.

Harmon, P. 1985. Small expert system building tools. *Expert System Strategies* 1(1).

Harmon, P., and D. King. 1985. *Expert systems: artificial intelligence in business.* New York: Wiley.

Harmon, P., R. Maus, and W. Morrissey. 1988. *Expert systems: tools and applications.* New York: Wiley.

Hart, A. 1986. *Knowledge acquisition for expert systems.* London: Kogan Page Ltd.

Hasling, D. W., W. J. Clancy, and G. Rennels. 1984. Strategic explanation for a diagnostic consultation system. *Int'l J. Man-Machine Studies.* 20(1): 3–19.

Haugeland, J., ed. 1981. Mind design. Cambridge MA: MIT Press.

Hebb, D. O. 1949. *Organization of behavior.* New York: Science Editions.

Hein, H., G. M. Kelermann, C. G. Thomas. 1987. X-Aid: a shell for building highly interactive and adaptive user interfaces. *IJCAI 87, Proc. of the Tenth Int'l Joint Conf. on Artificial Intelligence,* Aug. 23–28, 1987, Politecnico di Milano, Italy, pp. 97–99.

Helm, A., K. Marriott, and G. Lassez. 1985. PROLOG for expert systems: an evaluation. In: Kama, K. N., editor. *Proceedings, Expert Systems in Government Symposium.* IEEE Computer Society, pp. 294–304.

Hill, W. C. 1989. The mind at AI: horseless carriage to clock. *AI Magazine* 10(2): 28–42.

Hillman, D. 1988. Bridging acquisition and representation. *AI Expert* 3(11): 38–46.

Hoffman, R. R. 1987. The problem of extracting the knowledge of experts from the perspective of experimental psychology. *AI Magazine* 8(2): 53–67.

Hofstadter, D. R. 1979. *G—del, Escher, Bach: An eternal golden braid.* New York: Vintage Books.

Hogger, C. J. 1984. *Introduction to logic programming.* New York: Academic Press.

Holyard, J. 1957. *Alchemy.* Baltimore: Penguin Books, Inc.

Horowitz, E., and S. Sahni. 1978. *Fundamentals of Computer Algorithms*. Potomac MD: Computer Science Press, Inc.

Jackson, P. 1986. *Introduction to expert systems*. Reading, MA: Addison-Wesley.

Jackson, P., H. Reichgelt, and F. van Harmelen. 1989. *Logic-based knowledge representation*. Cambridge MA: MIT Press.

Jacob, R. J. K., and J. N. Froscher. 1990. A software engineering methodology for rule-based systems. *IEEE Transactions on Knowledge and Data Engineering* 2(2): 173–189.

Jeffers, J. N. R. 1991. Rule induction methods in forestry research. *AI Applications* 5: 37–44.

Johnson, G. 1986. *Machinery of the mind*. Redmond WA: Microsoft Press.

Johnson, S. C. 1967. Hierarchical clustering schemes. *Psychometrika* 32: 241–254.

Kahn, K. 1982. Intermission-actors in PROLOG. In: Clark, K. L. and S. A. Tarnlund (eds.) *Logic programming*. New York: Academic Press, pp. 213–228.

Kallio, E., and J. W. Benzie. 1980. *Red pine – an American wood*. USDA Forest Service, Washington, D. C. Leaflet FS–255, 4 pp.

Kelly, G. A. 1955. *The psychology of personal constructs*. New York: Norton.

Kline, P. J., and S. B. Dolins. 1985. Choosing architectures for expert systems. Final Tech. Report RADC-TR-85-192. Rome Air Development Center, Air Force Systems Command, Griffiss AFB, NY 13441–5700.

Knaus, R., and H. Blecker. 1990. Domain specific shells for experts in PROLOG. *AI Expert* 5: 40–47.

Koza, J. R. 1992. Genetic programming. Cambridge: The MIT Press.

Kowalski, R. 1979. *Logic for problem solving*. Holland NY: Elsevier North.

Ladkin, P. 1986. Time representation: a taxonomy of internal relations. In *Fifth National Conference on Artificial Intelligence*. Los Altos CA: Morgan Kaufmann Publishers, Inc.

Laird, J., P. Rosenbloom, and A. Newell. 1986. *Universal subgoaling and chunking*. Boston: Kluwer Academic Publishers.

Lakoff, G. 1987. *Women, fire, and dangerous things*. Chicago: University of Chicago Press.

Langley, P., and J. G. Carbonell. 1984. Approaches to machine learning. *Journal of the American Society for Information Science* 35: 306–316.

Langley, P., H. A. Simon, G. L. Bradshaw, and J. M. Zytkow. 1987. *Scientific Discovery: Computational Explorations of the Creative Process*. Cambridge MA: The MIT Press.

Laufmann, S. C., D. M. DeVaney, and M. A. Whiting. 1990. A methodology for evaluating potential KBS applications. *IEEE Expert* 12: 43–60.

Lazarev, G. L. 1989. *Why PROLOG?* Englewood Cliffs NJ: Prentice Hall.

Lecot, K., and D. S. Parker. 1986. Control over inexact reasoning. *AI Expert* Premiere Issue. pp. 32–43.

Lee, N. 1986. Programming with P-shell. *IEEE Expert* 1: 50–63.

Lemmon, H. 1986. COMAX: An expert system for cotton crop management. *Science* 233(4759): 29–33.

Lenat, D. B. 1980. AML An artificial intelligence approach to discovery in mathematics as heuristic search. In Davis R. and Lenat D. B. (Eds.), *Knowledge-Based Systems in Artificial Intelligence*. New York: McGraw-Hill.

Leonardi, L., P. Mello, and A. Natali. 1989. Prototypes in PROLOG. *Journal of Object Oriented Programming* 2: 20–28.

Leonardo. L., L. Saitta, and P. Torasso. 1985. Evidence combination in expert systems. *Int. J. Man-Machine Studies* 22: 307–326.

Leung, K. S., W. S. Felix Wong, and W. Lam. 1989. Applications of a novel fuzzy expert system shell. *Expert Systems* 6: 2–10.

Leung, K. S., and M. H. Wong. 1990. An expert system shell using structured knowledge. *IEEE Computer* 23: 38–47.

Liebowitz, J. 1986. Useful approach for evaluating expert systems. *Expert Systems* 3(2): 86–96.

Lloyd, J. 1984. *Foundations of logic programming*. Berlin, Germany: Springer-Verlag.

Logic Programming Associates. 1994. *WIN-PROLOG*. London: Logic Programming Associates, Ltd.

Lucas, R. 1988. *Database applications using PROLOG*. Chichester, England: Ellis Horwood, Ltd.

Luger, G. F., and W. D. Stubblefield. 1993. Artificial intelligence: structures and strategies for complex problem solving. Redwood City CA: Benjamin/Cummings Publishing Company.

Lundgren, A. L. 1981. *The effects of initial number of trees per acre and thinning densities on timber yields from red pine plantations in the Lake States*. USDA Forest Service, North Central Forest Experiment Station, St. Paul, MN. Research Paper NC-193.

Lundgren, A. L., and W. A. Dolid. 1970. *Biological growth functions describe published site index curves for Lake States timber species*. USDA Forest Service, North Central Forest Experiment Station, St. Paul, MN. Research Paper NC-36.

Lundgren, A. L., and D. M. Belcher. 1982. *A user's guide for REDPINE – a computer program for simulating growth and yield of red pine stands in the Lake States*. USDA Forest Service, North Central Forest Experiment Station, St. Paul, MN.

Mackintosh, A. R. 1988. Dr. Atanasoff's computer. *Scientific American* 259: 90–96.

Malpas, J. 1987. *PROLOG: a relational language and its application*. Englewood Cliffs NJ: Prentice Hall.

Marcus, C. 1986. *PROLOG programming*. Reading MA: Addison-Wesley.

Martin, G. L. 1980. Designing forest management decision support systems for small computers. Midwest Forest Mensurationists' Meeting, September 9–11, 1980, at Mackinac Island, MI. (Unpublished), 15 p.

McCarthy, J. 1980. Circumscription—a form of non-monotonic reasoning. *Artificial Intelligence* 13: 27–39.

McClelland, J. L., and D. E. Rumelhart. 1986. *Parallel distributed processing: explorations in the microstructure of cognition, vol. 2.* Cambridge MA: The MIT Press.

McCorduck P. 1979. *Machines who think.* New York: W. H. Freeman.

McDermott, J. 1982. R1: A rule-based configurer of computer systems. *Artificial Intelligence* 19.

McKeown, K. R., M. Wish, and K. Matthews. 1985. Tailoring explanations for the user. Pages 794–798 in *Proc. Ninth IJCAI,* Los Altos, CA: Morgan Kaufmann.

McRoberts, R. E., D. L. Schmoldt, and H. M. Rauscher. 1991. Enhancing the scientific process with artificial intelligence: forest science applications. AI Applications 5(2): 5–25.

Merritt, D. 1989. *Building expert systems in PROLOG.* New York: Springer-Verlag.

Michalski, R. S., J. G. Carbonell, and T. M. Mitchell, eds. 1983. *Machine learning: an artificial intelligence approach,* Palo Alto CA: Tioga Publishing.

Michalski, R. S., J. G. Carbonell, and T. M. Mitchell, eds. 1986. *Machine learning: an artificial intelligence approach, volume 2.* Los Altos CA: Morgan Kaufmann Publishers.

Michie, D., and R. Johnston. 1985. *The knowledge machine: artificial intelligence and the future of man.* New York: William Morrow and Company, Inc.

Millet, C., and M. Gilloux. 1989. A study of the knowledge required for explanation in expert systems. Pages 84–90 in *Proc. Fifth Conf. on Artificial Intelligence Applications,* March 6–10, 1989, Miami, Fla., IEEE Computer Society Press.

Mills, W. L. 1987. Expert systems: Applications in the forest products industry. *Forest Products Journal,* 37(9): 41–46.

Minsky, M. L., 1975. A framework for representing knowledge. In: P. Winston (ed.), *The psychology of computer vision.* New York: McGraw Hill. pp. 217–277.

Minsky, M. L., and S. Papert. 1969. *Perceptrons.* Cambridge MA: The MIT Press.

Mitchell, A. A. 1987. The use of alternative knowledge-acquisition procedures in the development of a knowledge-based media planning system. *International Journal of Man-Machine Studies* 26: 399–411.

Mittal, S., and C. L. Dym. 1985. Knowledge acquisition from multiple experts. *AI Magazine* 6(2): 32–36.

Morone, J. J., and M. R. Hilbush. 1987. *Experiencing artificial intelligence.* Blue Ridge Summit PA: TAB Books Co.

Moravec, H. 1988. *Mind children: the future of robot and human intelligence.* Cambridge MA: Harvard University Press.

Mosteller, F., R. E. K. Rourke, and G. B. Thomas, Jr. 1961. *Probability with statistical applications.* Reading MA: Addison-Wesley Publishing Company.

Nakashima, H. 1982. PROLOG/KR – language features. In: *Proceedings, 1st International Logic Programming Conference,* ADDP, September 1982, Marseilles, France. pp. 96–106.

Nau, D. 1983. Expert computer systems. *Computer* 16: 63–85.

Neches, R., W. R. Swartout, and J. D. Moore. 1985. Enhanced maintenance and explanation of expert systems through explicit model of their development. *IEEE Trans. Software Engineering* SE11(11): 1337–1351.

Newell, A. 1973. Artificial intelligence and the concept of mind. In: *Computer Models of Thought and Language*. R. C. Schank and K. M. Colby (Eds.), San Francisco: W. H. Freeman and Company. pp. 1–60.

Newell, A., and H. A. Simon. 1973. *Human problem solving*. Englewood Cliffs NJ: Prentice-Hall.

Newquist III, H. P. 1988. Struggling to maintain. *AI Expert* 3(8): 69–71.

Nguyen, T. A., W. A. Perkins, T. J. Laffey, and D. Pecora. 1987. Knowledge base verification. *AI Magazine* 8(2): 69–75.

Nute, D. E., H. M. Rauscher, D. A. Perala, G. Zhu, Y. Chang, and G. E. Host. 1995. A toolkit approach to developing forest management advisory systems in PROLOG. *AI Applications* 9(2): in press.

ÒKeefe, R. M. 1985. Expert systems and operational research—mutual benefits. *Journal of Operations Research* 36(2): 125–129.

ÒKeefe, Robert. 1986. Simulation and expert systems – a taxonomy and some examples. *Simulation* 46(1): 10–16.

ÒKeefe, R. M., O. Balci, and E. P. Smith. 1987. Validating expert system performance. *IEEE Expert* 2(4): 81–90.

Oliveira, E. 1984. Developing expert system builders in logic programming. *New Generation Computing* 2: 187–194.

Oliver, A. E. M. 1986. Control of rapid prototyping in expert systems development. In *Second International Expert Systems Conference*. Edited by *Expert Systems*. Oxford: Learned Information Ltd.

Olson, J. R., and H. H. Rueter. 1987. Extracting expertise from experts: methods for knowledge acquisition. *Expert Systems* 4(3): 152–168.

Olson, R. L., and J. D. Hanson. 1987. A knowledge-based deduction system for curve-shape analysis. *AI Applications in Natural Resource Management* 1(2): 3–12.

Olson, R. L., J. L. Willers, and T. L. Wagner. 1990. A framework for modeling uncertain reasoning in ecosystem management II: Bayesian belief networks. *AI Applications* 4(4): 11–24.

Paris, C. L. 1987. Combining discourse strategies to generate descriptions to users along a naive/expert spectrum. *Proc. 10th IJCAI*, Los Altos CA: Morgan Kaufmann

Parsaye, K., and M. Chignell. 1988. *Expert systems for experts.* New York: Wiley.

Parsaye, K., M. Chignell, S. Khoshafian, and H. Wong. 1989. *Intelligent databases*. New York: John Wiley & Sons, Inc.

Paulos, J. A. 1988. *Innumeracy: mathematical illiteracy and its consequences*. New York: Hill and Wang.

Pearl, J. 1986. On evidential reasoning in a hierarchy of hypotheses. *Artificial Intelligence* 28(1): 9–15.

Penzias, A. 1989. *Ideas and information: managing in a high-tech world.* New York: Simon & Schuster.

Pereira, L. 1982. Logic control with logic. In: Campbell, J. A. (ed.) *Implementations of PROLOG*. Chichester, England: Ellis Horwood. pp. 177–193.

Pirsig, R. 1974. *Zen and the art of motorcycle maintenance*. New York: William Morrow and Company.

Plant, R. E., and R. S. Loomis. 1991. Model-based reasoning for agricultural expert systems. *AI Applications* 5(4): 17–28.

Pople, H. E., Jr. 1982. Heuristic methods for imposing structure on ill-structured problems: The structuring of medical diagnostics. In: *Artificial Intelligence in Medicine.* P. Szolovits (Ed.), Boulder: Westview Press. pp. 149–154.

Porto, A. 1982. Epilog: a language for extended programming in logic. In: *Proceedings, 1st International Logic Programming Conference*, ADDP, Marseille, France. September 1982. pp. 45–63.

Post, E. 1943. Formal reductions of the general combinatorial problem. *American Journal of Mathematics* 65: 197–268.

Prerau, D. S. 1985. Selection of an appropriate domain for an expert system. *AI Magazine* 6(2): 26–30.

Pressman, R. S. 1982. *Software engineering: a practitioner's approach.* New York: McGraw-Hill Book Company.

Puccia, C. J., and R. Levins. 1985. *Qualitative modeling of complex systems.* Cambridge MA: Harvard University Press.

Pylyshyn, Z. W. 1986. *Computation and cognition: toward a foundation for cognitive sciences.* Cambridge MA: MIT Press.

Quinlan, J. R. 1983a. INFERNO: a cautious approach to uncertain inference. *Computer Journal* 26(3): 255–269.

Quinlan, J. R. 1983b. Induction of decision trees. In: *Machine learning: an artificial intelligence approach.* Edited by R. S. Michalski, J. G. Carbonell, and T. M. Mitchell. Palo Alto CA: Tioga.

Quinlan, J. R. 1986. Induction of decision trees. Pages 81–106 in *Machine Learning 1.*

Quinlan, J. R. 1990. Decision trees and decisionmaking. *IEEE Transactions on Systems, Man, and Cybernetics* 20: 339–346.

Ramm, Carl W. 1990. RPAL: a growth and yield simulation program for Red Pine plantations in the Lake States. *Northern Journal of Applied Forestry* 7: 96–97.

Rauscher, H. M. 1987. Expert systems for natural resource management. *FORS Compiler* 5: 19–27.

Rauscher, H. M. 1987. Increasing Scientific Productivity Through Better Knowledge Management. *AI Applications in Natural Resource Management* 1(2): 21–31.

Rauscher, H. M., and Robert Hacker. 1989. Overview of artificial intelligence applications in natural resource management. *The Journal of Knowledge Engineering* 2: 30–42.

Rauscher, H. M., J. W. Benzie, and A. A. Alm. 1990. A Red Pine forest management advisory system: knowledge model and implementation. *AI Applications in Natural Resource Management* 4: 27–43.

Rauscher, H. Michael and George E. Host. 1990. Hypertext and AI: A complementary combination for knowledge management. AI Applications 4: 58–61.

Rauscher, H. M, P. Janiga, K. Reynolds, and M. J. Twery (eds.). 1995. Decision support systems for natural resource management: theory and applications. *AI Applications* 9: (in press).

Reinhardt, E. 1987. An expert system for designing fire prescriptions. In: J. B. Davis and R. E. Martin (Eds.) *Proceedings of the Symposium on Wildland Fire 2000*. Gen. Tech. Rep. PSW-101, Berkeley CA. Pacific Southwest Research Station, Forest Service, U.S. Dept. of Agric. pp 223–225.

Reiter, R. 1980. A logic for default reasoning. *Artificial Intelligence* 13: 81–132.

Retz-Schmidt, G. 1988. Various views on spatial propositions. *AI Magazine* 9(2): 95–105.

Rewerts, C. C., B. A. Engel, J. B. Rogers, and D. D. Jones. 1990. An end-user interface for agricultural software. *AI Applications* 4:57–65.

Reynolds, K., M. Saunders, R. Olson, D. Schmoldt, M. Foster, D. Latham, B. Miller, J. Steffenson, L. Bednar, and P. Cunningham. 1995. Knowledge-based information management in decision support for ecosystem management in the Pacific Northwest U.S. Paper # 288 in 1995 ESRI User Conference Proceedings, Environmental Systems Research Institute, Inc., Redlands CA.

Rice, D. A., S. J. Nix, and A. J. R. Gillespie. 1989. A simple rule-based system for species selection in artificial regeneration. *AI Applications in Natural Resources Management* 3: 38–43.

Riorden, D. and K. J. Carden. 1990. Enhancing explanation in ecological expert systems. *AI Applications* 4: 11–18.

Ripple, W. J., and V. S. Ulshoefer. 1987. Expert systems and spatial data models for efficient geographic data handling. *Photogrammetric Engineering and Remote Sensing* 53(10): 1431–1433.

Roach, J. W., R. S. Virkar, M. J. Weaver, C. R. Drake. 1985. POMME: a computer-based consultation system for apple orchard management using Prolog. *Expert Systems* 2: 56–69.

Robinson, J. A. 1965. A machine-oriented logic based on the resolution principle. *Journal of the Association for Computing Machinery* 12(1): 223–41.

Ross, P. 1989. *Advanced PROLOG: techniques and examples*. Reading MA: Addison-Wesley.

Rossi, G. 1986. Uses of PROLOG in implementation of expert systems. *New Generation Computing* 4: 321–329.

Rowe, Neil C. 1988. *Artificial intelligence through PROLOG*. Englewood Cliffs NJ: Prentice Hall.

Rubinoff, R. 1985. Explaining concepts in expert systems: the clear system. Pages 416–421 in *Proc. Second Conf. Artificial Intelligence Applications*. IEEE Computer Society Press, Los Alamitos, CA.

Rumelhart, D E., and J. L. McClelland. 1986. *Parallel distributed processing: explorations in the microstructure of cognition, vol. 1*. Cambridge MA: The MIT Press.

Ryan, T. A. Jr., B. L. Joiner, and B. F. Ryan. 1976. *MINITAB student handbook*. Boston: Duxbury Press.

Saaty, T. L. 1980. *The analytic hierarchy process*. New York: McGraw Hill.

Sacerdoti, E. D. 1977. *A structure for plans and behavior*. New York: American Elsevier.

Safra, S. and E. Shapiro. 1986. Meta interpreters for real. In: H.-J. Kugler (ed). Proc. of the IFIP 10th World Computer Congress, Dublin, Ireland, Sept. 1–5, 1986, Elsevier Science Publishers B.V. (North-Holland), pp. 271–277.

Sage, A. P. 1991. *Decision support systems engineering*. New York: Wiley.

Salton, G.. 1991. Developments in automatic text retrieval. *Science* 253: 974–980.

Samuel, A. L. 1963. Some studies in machine learning using the game of checkers. In Feigenbaum, E. A. and Feldman, J. (Eds.), *Computers and Thought*. New York: McGraw-Hill. pp. 71–105.

Saveland, J. M., M. Stock, and D. A. Cleaves. 1988. Decision-making bias in wilderness fire management: Implications for expert system development. *AI Applications* 2(1): 17–29.

Schank, R. C., and R. P. Abelson. 1977. *Script, plans, goals, and understanding*. Hillsdale NJ: Lawrence Erlbaum.

Schank, R. C., and C. K. Reisbeck. 1981. *Inside computer understanding*. Hillsdale NJ: Lawrence Erlbaum Associates, Inc.

Schank, R. C. 1986. *Explanation patterns: understanding mechanically and creatively*. Hillsdale NJ: Lawrence Erlbaum Associates, Inc.

Schank, R. C. 1987. What is AI, anyway? *AI Magazine* 8(4): 57–66.

Schmoldt, D. L. 1987. *Evaluation of an expert system approach to forest pest management of red pine (Pinus resinosa)*. Ph.D. dissertation. University Microfilms International. 225 pp.

Schmoldt, D. L. 1988. PREDICT: an expert system for pest diagnosis. *The Compiler* 5(5): 29–41.

Schmoldt, D. L. 1989. Refining rule bases for classification knowledge-based systems. *AI Applications in Natural Resource Management* 3(3): 31–41.

Schmoldt, D. L. 1989. Knowledge management: an application to wildfire prevention planning. *Fire Technology* 25(2): 150–164.

Schmoldt, D. L. 1991. Simulation of plant physiological processes using fuzzy variables. *AI Applications in Natural Resource Management* 5(4): 3–16.

Schmoldt, D. L., and W. G. Bradshaw. 1988. A cumulative Delphi approach to knowledge acquisition. In *Proceedings of the Third Annual Rocky Mountain Conference on Artificial Intelligence*. Edited by Y. Wilks.

Schmoldt, D. L., and W. G. Bradshaw. 1991. A computer-based tutorial structure for learning and applying a complex process. *Computers and Electronics in Agriculture* 5: 265–283.

Schmoldt, D. L., and G. L. Martin. 1986. Expert systems in forestry: utilizing information and expertise for decision making. *Computers and Electronics in Agriculture* 1(3): 233–250.

Schmoldt, D. L., and G. L. Martin. 1989. Construction and evaluation of an expert system for pest diagnosis of red pine in Wisconsin. *Forest Science* 35(2): 364–387.

Schmoldt, D. L., and D. L. Peterson. 1991. Applying knowledge-based methods to the design and implementation of an air quality workshop. *Environmental Management* 15(5): 623–634.

Schmoldt, D. L., and H. M. Rauscher. 1994. A knowledge management imperative and six supporting technologies. Computers and Electronics in Agriculture 10(1): 11–30.

Schnupp, P., and L. W. Bernhard. 1987. *Productive PROLOG programming*. London: Prentice Hall.

Schnupp, P., C. T. Nguyen Huu, and L. W. Bernhard. 1989. *Expert systems lab course*. Berlin, Germany: Springer-Verlag.

Scholem, G. 1987. *Kabbalah*. New York: Dorset Press.

Schweickert, R., A. M. Rurton, N. K. Taylor, E. N. Corlett, N. R. Shadbold, and A. P. Hedgecock. 1987. Comparing knowledge elicitation techniques: a case study. *Artificial Intelligence Review* 1: 245–253.

Seal, R. H. 1966. Forestry as a system. PhD Dissertation, SUNY, Syracuse, New York, 180 pp.

Sell, P. S. 1985. *Expert systems: a practical introduction*. New York: John Wiley & Sons, Inc.

Selvidge, J. 1975. A three-step procedure for assigning probabilities to rare events. Pages 199–216 *in* D. Wendt and C. A. J. Viek (eds.), *Utility, probability, and human decision making*.

Senjen, R. 1988. Knowledge acquisition by experiment: Developing test cases for an expert system. *AI Applications* 2(2–3): 52–55.

Shafer, D. 1987. *Advanced turbo PROLOG programming*. Indianapolis: Howard W. Sams & Company.

Shafer, G. 1976. *A mathematical theory of evidence*. Princeton NJ: Princeton University Press.

Shannon, C. E. 1948. A mathematical theory of information. *Bell System Technical Journal* 27: 379–423, 623–656.

Shannon, C. E. 1950. Programming a computer for playing chess. *Philosophical Magazine* 7(41): 256–275.

Shapiro, E. 1983. Logic programs with uncertainties: a tool for implementing rule-based systems. In: *Proceedings, 8th International Joint Conference on Artificial Intelligence*, Karlsruhe, Germany. pp. 529–532.

Shapiro, S. C., ed. 1987. *Encyclopedia of artificial intelligence*. New York: John Wiley and Sons.

Shepard, R. N. 1962. Analysis of proximities: multidimensional scaling with an unknown distance function. I. *Psychometrika* 27: 125–140.

Shlaer, S., and S. J. Mellor. 1988. *Object-oriented systems analysis: modeling the world with data*. Englewood Cliffs NJ: Yourdon Press.

Shneiderman, B. 1987. *Designing the user interface: strategies for effective human-computer interaction*. Reading MA: Addison-Wesley Publishing Company.

Shoham, Y.. 1994. *Artificial intelligence techniques in PROLOG*. San Francisco: Morgan Kaufman.

Shortliffe, E. H., and B. G. Buchanan. 1975. A model of inexact reasoning in medicine. *Mathematical Biosciences* 23: 351–379.

Shortliffe, E. H. 1976. *Computer-based medical consultations: MYCIN*. New York: American Elsevier.

Simon, H. A. 1969. *Sciences of the artificial*. Cambridge MA: MIT Press.

Simon, H. A. 1979. *Models of thought*. New Haven CN: Yale University Press.

Simon, H. A. 1981. *The sciences of the artificial*, 2nd. ed. Cambridge, MA: MIT Press.

Simpson, H.. 1986. *Programming the IBM PC user interface*. New York: McGraw-Hill.

Slagle, J., and M. Wick. 1988. A method for evaluating candidate expert system applications. *AI Magazine* 9(4): 44–53.

Smith, P. 1988. *Expert system development in PROLOG and TURBO-PROLOG*. New York: Halsted Press.

Sowa, J. F. 1984. *Conceptual structures*. Reading MA: Addison-Wesley.

Spetzler, C. S., and C. A. Sta'l von Holstein. 1975. Probability encoding in decision analysis. *Management Science* 22: 340–358.

Starfield, A. M., and A. L. Bleloch. 1983. Expert systems: an approach to problems in ecological management that are difficult to quantify. *Journal of Environmental Management* 16: 261–268.

Starfield, A. M., and A. L. Bleloch. 1986. *Building models for conservation and wildlife management*. MacMillan Publishing Company.

Stefik, M,. and D. Bobrow. 1986. Object-oriented programming: themes and variations. *AI Magazine* 6: 40–62.

Sterling, L. 1984. Logical levels of problem solving. *Journal of Logic Programming* 2:151–163.

Sterling, L., and R. Beer. 1989. Meta-interpreters for expert system construction. *Journal of Logic Programming* 7: 163–178.

Sterling, L., and E. Shapiro. 1986. *The art of PROLOG*. Cambridge MA: MIT Press.

Sterling, L., and A. Lakhotia. 1988. *Composing PROLOG meta-interpreters*. Case Western Reserve University, Center for Automation and Intelligent Systems Research, Technical Report TR 88–125.

Sterling, L., and R. Beer. 1989. Meta-interpreters for expert system construction. *Journal of Logic Programming* 7: 163–178.

Stillings, N. A., M. H. Feinstein, J. L. Garfield, E. L. Rissland, D. A. Rosenbaum, S. E. Weisler, and L. Baker-Ward. 1987. *Cognitive science—an introduction*. Cambridge, MA: MIT Press.

Surko, P. 1989. Tips for knowledge acquisition. *PC AI* 3(3): 14–18.

Suzuki. N. 1988. Application-specific shell. *New Generation Computing* 5: 317–318.

Swartout, W. R. 1983. XPLAIN: a system for creating and explaining expert consulting programs. *Artificial Intelligence* 21: 285–325.

Takeuchi, A., and K. Furukawa. 1986. Partial evaluation of PROLOG programs and its application to meta-programming. In: Kugler, H. J. (ed.) *Proceedings, IFIP, 10th World Computer Congress,* 1986 September 1–5, Dublin, Ireland. North Holland: Elsevier Science Publishers BV.

Tanimoto, S. L. 1987. *The elements of artificial intelligence—an introduction using LISP*. Rockville MD: Computer Science Press.

Thomson, A. J., and C. M. A. Taylor. 1990. An expert system for diagnosis and treatment of nutrient deficiencies of Sitka Spruce in Great Britain. *AI Applications* 4(1): 44–52.

Tong, X., Z. He, and R. Yu. 1987. A tool for building second generation expert systems. In: *Proceedings, 10th International Joint Conference on Artificial Intelligence*. 1987 August 23–28, Milan, Italy. pp. 91–96.

Torasso, P., and L. Console. 1989. *Diagnostic problem solving*. New York: Van Nostrand Reinhold.

Townsend, Carl. 1987a. *Advanced techniques in turbo PROLOG*. Alameda CA: Sybex, Inc.

Townsend, C. 1987b. *Mastering expert systems with turbo PROLOG*. Indianapolis IN: Howard Sams & Company.

Turban, E. 1993. Decision support and expert systems: Management support systems. New York: MacMillan Publishing Co.

Tsang, E. P. K. 1987. The consistent labeling problem in temporal reasoning. In *Sixth National Conference on Artificial Intelligence*. Los Altos CA: Morgan Kaufmann Publishers, Inc.

Tversky, A., and D. Kahneman. 1974. Judgment under uncertainty: heuristics and biases. *Science* 185: 1124–1131.

Usery, E. L., and P. Altheide. 1988. Knowledge-based GIS techniques applied to geological engineering. *Photogrammetric Engineering and Remote Sensing* 54(11): 1623–1628.

Van de Ven, A., and A. Delbecq. 1971. Nominal vs. interacting group processes for committee decision making effectiveness. *Academy of Management Journal* 14: 203–212.

Vilain, M., and H. Kautz. 1986. Constraint propagation algorithms for temporal reasoning. In *Fifth National Conference on Artificial Intelligence*. Los Altos CA: Morgan Kaufmann Publishers, Inc.

von Winterfeldt, D., and W. Edwards. 1986. *Decision analysis and behavioral research*. Cambridge: Cambridge University Press.

Vrba, J., and J. A. Herrera. 1989. Expert system tools: the next generation. *IEEE Expert* 4(1): 75–76.

Waldrop, M. M. 1984. The necessity of knowledge. *Science* 223: 1279–1282.

Walters, J. R., and N. R. Nielsen. 1988. *Crafting knowledge-based systems: expert systems make realistic.* New York: Wiley & Sons.

Wambach, R. B. 1967. *A silvicultural and economic appraisal of initial spacing in red pine*. Ph.D. dissertation, Univ. of Minnesota, St. Paul.

Watanabe, M., T. Yamanouchi, M. Iwamoto, and Y. Ushioda. 1989. CL: a flexible and efficient tool for constructing knowledge-based expert systems. *IEEE Expert* 4: 41–50.

Waterman, D. A., and F. Hayes-Roth. 1978. An overview of pattern-directed inference systems. In *Pattern-Directed Inferences Systems*. Edited by D. A. Waterman and F. Hayes-Roth. New York, NY: Academic.

Weiskamp, K., and T. Hengl. 1988. *Artificial intelligence programming with turbo PROLOG*. New York: John Wiley & Sons, Inc.

Weiss, S. M., and C. A. Kulikowski. 1984. *A practical guide to designing expert systems*. Totowa NJ: Rowland and Allanheld.

Weiss, S. M., and C. A. Kulikowski. 1991. *Computer Systems That Learn*. San Mateo CA: Morgan Kaufmann Publishers, Inc.

Whittaker, A. D., M. L. Wolfe, R. Godbole, and G. Van Alem. 1991. Object-oriented modeling of hydrologic processes. *AI Applications* 5(4): 49–58.

White, W. B., and B. W. Morse. 1987. ASPENEX: an expert system interface to a geographic information system for aspen management. *AI Applications* 1: 49–53.

Wick, M. R., and J. R. Slagle. 1989. An explanation facility for today's expert systems. *IEEE Expert* Spring '89: 26–36.

Winston, P. H. 1984. *Artificial intelligence*. 2nd ed. Reading MA: Addison-Wesley.

Winston, P. H. 1984. Perspective. In *The AI Business, The Commercial Uses of Artificial Intelligence*. Edited by Patrick H. Winston and Karen A. Pendergast. Cambridge MA: MIT Press.

Yalcinalp, L. and Leon S. Sterling. 1990. Uncertainty reasoning in PROLOG with layered meta-interpreters. Case Western Reserve University, Center for Automation and Intelligent Systems Research, Technical Report TR 90–110, 16 pp.

Yates, F. 1964. *The art of memory*. Chicago: The University of Chicago Press.

Yin, K. M. 1987. *Using Turbo PROLOG*. Indianapolis IN: Que Corp.

Young, R. J. 1989. *Practical PROLOG*. New York: Ran Nostrand Reinhold.

Yu, V. L., B. B. Buchanan, E. H. Shortliffe, S. M. Wraith, R. Davis, A. C. Scott, and S. N. Cohen. 1979a. Evaluating the performance of a computer-based consultant. *Computer Programs in Biomedicine* 9: 95–102.

Yu, V. L., L. M. Fagan, L. M. Wraith, W. J. Clancey, A. C. Scott, J. Hannigan, R. L. Blum, B. G. Buchanan, and S. N. Cohen. 1979b. Antimicrobial selection by a computer. *J. Am. Med. Assoc.* 242: 1279–1282.

Zadeh, L. A. 1965. Fuzzy sets. *Information and Control* 8:338–353.

Zadeh, L. A. 1978. Fuzzy sets as a basis for a theory of possibility. *Fuzzy Sets Sys*. 1: 3–28.

Zadeh, L. A. 1979. A theory of approximate reasoning. In *Machine intelligence 9*. Edited by J. E. Hayes, D. Michie, and L. I. Mikulich. New York, NY: Hatsted Press, 1979.

Zasada, Z. A., and J. W. Benzie. 1970. *Mechanized harvesting for thinning sawtimber and red pine*. University of Minnesota, Agricultural Experiment Station, Miscellaneous Report 99. Forestry Series 9.

GLOSSARY

Abduction. Development of hypotheses to explain observations. In a diagnostic situation, for example, the usual cause-effect inference is reversed because it is necessary to infer a cause (hypothesis) from effects (observations).

Accommodation. The process by which new information is molded into existing memory structures.

AI or **Artificial Intelligence**. See **artificial intelligence**.

Algorithm. A procedure or recipe, consisting of a finite number of steps, that manipulates some data for the solution of a specific problem.

Analogical Learning. A learning method that is base on analogy, where similarities between a new situation and other, familiar situations are used to respond intelligently to the new situation.

Antecedent Reasoning. See **forward chaining**.

Architecture. (1) The organizing framework imposed upon knowledge-based systems and problem-solving activities. (2) The knowledge-engineering principles that govern selection of appropriate frameworks for specific knowledge-based systems.

Arity. The number of elements in a set or structure. Specifically, the number of arguments in a PROLOG structure.

Artificial Intelligence (AI). A multidisciplinary subfield of computer science that blends cognitive science, psychology, philosophy, neurophysiology, engineering, robotics, and linguistics and applies them to the study of intelligent behavior. Much of the work in AI focuses on generating intelligent behavior in machines.

Assertion. A statement declared as having some degree of truth.

Assimilation. The acquisition of new information which is molded into an existing memory structure by accommodation.

Atomic. Indivisible facts that are able to describe the problem area in the most fundamental terms.

Backward Chaining. A reasoning control strategy in KBSs whereby an attempt is made to satisfy, or establish, a top-level goal by generating and solving progressively simpler goals. These simpler, or intermediate, goals are rule premises in the chain of logical deduction from atomic facts to top-level goal.

Bayes' Theorem (or Rule). A probability based method for inexact reasoning in which probability is equated with truth. Then, the truth of a hypothesis, given evidence for its truth, is determined by the *a priori* probability of that hypothesis along with the probability of the current evidence given that the hypothesis is in fact true.

Belief. (1) A hypothesis about some unobservable situation. (2) A measure of the believer's confidence in an uncertain proposition.

Blackboard. A framework for knowledge-based systems in which knowledge can be arranged so that it is distributed and yet shared among a number of cooperating processes. Typically, the blackboard is partitioned for representing hypotheses at different levels of abstraction and mediates the cooperative activities of multiple "subexperts" or specialists or agents.

Breadth-First Search. A fanning search of a hierarchical state space where all possible transitions from the current state are investigated prior to expanding search to the next level of the hierarchy.

Certainty. A measure of confidence in the validity of a proposition, hypothesis, or inference statement.

Certainty Factor. A quantification of certainty that is used in Shortliffe and Buchanan's model of inexact reasoning. Numerically, certainty factors are the difference between an individual's measure of belief in a proposition or rule and that individual's measure of disbelief in that same proposition or rule.

Classification. A useful model for describing the category of problems exemplified by diagnosis. Problem instances are viewed as members of clusters, where each cluster corresponds to a particular diagnosis. Problem instances are assigned to these clusters (classes) on the basis of some metric of proximity.

Chunk. A conceptual entity having both content and structure and composed of data, information, or knowledge. Ideally, a chunk has unity, coherence, and adequate development; it is similar to a paragraph of written composition.

Chunking. The process by which items are grouped together in memory by an underlying principle or pattern.

Closed-World Assumption. A default understanding in many KBSs that only those statements known to be true are actually true and that all other statements not known specifically to be true are false.

Cognitive Psychology. The study of human information processing strategies. These strategies are described by postulating a series of internal states and computational procedures which describe the functional organization of the nervous system.

Cognitive Science. The study of human thinking processes, which can be used to develop computer models of cognition. Contributes theories and understanding of human information processing (IP) strategies that provide a model or basis of comparison for machine IP strategies.

Common Sense Knowledge. The knowledge (facts, procedures, representations, heuristics) used intuitively and most frequently in everyday living. It includes specialized knowledge about objects moving in space and time, people moving politically, economically, scientifically, etc. Generally, if an average person takes more than a couple of seconds to answer a question, then the topic does NOT deal with common sense knowledge. A comparable philosophical term would be "implicit" or "tacit" knowledge.

Concept. (1) A descriptive schema for a class of things or (2) a particular instance of the schema with some of its general properties specialized to characterize a specific subclass or instance.

Confirmation by Exclusion. A KBS control strategy that tries to find a satisficing solution by eliminating all potential solutions that are less acceptable.

Conflict Resolution. A strategy and procedure, in a forward-chaining KBS, to select one rule to evaluate out of many available rules.

Consequent Reasoning. See **backward chaining**.

Constraint-Based Reasoning. Method of reasoning in which knowledge is formalized as a set of constraints. A solution for a problem formulated in this way is one in which all constraints are satisfied to some predetermined degree. Constraint-based reasoning is applied in design, planning, and truth maintenance.

Control (of knowledge-based systems). Any procedure, explicit or implicit, that determines the overall order of problem-solving activities

Controller/Interpreter. See **reasoning engine**.

Conventional Systems. Computer programs characterized by the use of procedural languages, iterative algorithmic control flow, and lack of any real emphasis on explicitly represented knowledge (e.g. a payroll program).

Data. Undigested observations of facts and relationships, unorganized and unevaluated.

Data Base. In the context of KBSs, see **working memory**.

Data Driven. See **forward chaining**.

Data Structure. Data types that can be constructed within a particular computer language and contain are used to represent information in the computer. They access methods to create, modify, and manipulate that information.

Decision Key. A data item that "keys" on a single fact and can be used to select one of several reasoning paths.

Decision Process. The sequence of decision steps (states), and the strategy leading to that sequence, which results in a desired problem solution(s).

Decision Rule. An association between facts about a particular domain problem and a solution of that problem.

Decision Space. The set of all possible decisions or solutions states and their relationship to one another.

Decision Tree. A generalized search procedure that views the decision process as a tree structure. Any search for a problem solution begins at a solitary root node; solution states are leaf nodes. In this representation, each node has only one parent node. The tree is traversed by applying a decision key at each step.

Declarative Knowledge. Definitions, facts, and theories.

Deduction. Reasoning based on established inference procedures that can be applied to particular situations. Modus ponens (if-then rules) is a general inference procedure that can produce specific consequent propositions from problem-specific antecedent conditions.

Deep Knowledge. Knowledge that is based on first principles or an abstract model; a detailed understanding of the processes at work in some area of knowledge. Contrast this with **surface knowledge**.

Demon (or **Daemon**). A procedure that monitors a situation and responds by performing an action when an activating condition appears.

Depth-First Search. A diving search of a hierarchical state space where only a single new state in each level is visited during the dive into the hierarchy. The dive ends when a solution state is reach or when a dead end is reach. When reaching a dead end the depth-first search backtracks to the previous level, visits the next available state at that level, and then dives again into the hierarchy.

Domain. The particular problem area which a knowledge-based system is designed to operate in.

Domain Expert. A specialist; a person recognized by peers as especially competent, who works with a knowledge engineer to provide domain expertise to build a knowledge base. Examples are: medical diagnostician, car mechanic, credit checking clerk. On-the-job experience after formal training differentiates experts from non-experts.

Domain Ontology. A theory of reality or knowledge in a particular subject area that results from the analysis of factor, structure, and strategy knowledge in a that domain.

Epistemology. In philosophy, the investigation of the nature, source, and validity of knowledge.

Expert. See **domain expert**.

Expertise. The set of capabilities that underlies the high performance of human experts, including extensive domain knowledge, intuition that simplifies and improves problem-solving, meta-knowledge and meta-cognition, and compiled forms of behavior that afford great economy in skilled performance.

Expert System. A computer program that performs non-trivial tasks at a level comparable to human experts. The label "expert system" usually implies that the knowledge was acquired from human experts. Domain-specific knowledge is represented symbolically and reasoning processes are explained.

Expert System Development Tool. The reasoning engine, knowledge base creation tools, and other support software used to build expert systems.

Explanation. Motivating, justifying, or rationalizing an action by presenting antecedent considerations such as goals, laws, or heuristic rules that affected or determined the desirability of the action.

Explanation Facility. The ability of a knowledge-based system to explain the reasoning behind recommendations, and justify its conclusions, when queried by the user.

Exploratory Programming. A software development strategy that uses successive prototypes as focal points for trial-and-error programming. This technique usually requires a programming environment that allows rapid re-coding and execution of alternative approaches.

Facet. Allows control information to be attached to a slot in a frame. A facet provides the mechanism to represent various types of slot value restrictions. Facets may determine: permitted values for the slot, maximum and minimum values permitted in the slot, text of a question to be asked, whether the user should be asked by the system to provide a slot value.

Fact. A proposition or datum whose validity is accepted.

Fifth Generation Effort. Computer systems for the 1990s which are designed to communicate in natural language, reason intelligently, and incorporate enormous amounts of stored knowledge.

First Generation KBS. Knowledge-based systems that focus mostly on reproducing expert performance, and that do so primarily by the use of surface level knowledge.

First-Order Predicate Calculus. A type of formal logic that uses functions and predicates to describe relations between individuals. The predicate calculus extends propositional calculus by including predicates and inference rules for quantifiers and by allowing variable quantification over individuals. Statements about a problem would be expressed by formulas in the calculus. See **logic-based**.

Forward Chaining. A data-driven control strategy whereby all possible rule-base deductions are performed based on the current working memory knowledge. Because these deductions may have increased the base of currently known facts, all the rules are applied again, until a solution is deduced.

Frame. A knowledge representation scheme that associates one or more features with an entity. These features are stored in various slots with particular slot-values. Similar to *property-list*, *schema*, *unit*, and *record* in various writings.

Frame-Based. A knowledge representation approach that uses frames organized into networks of entities and their features.

Front End. An interface program that operates between a user and a complex application program. Its operation usually makes the complexity of the application program less daunting and more understandable to the user.

Full Algorithm. An algorithm that terminates in either a solution, or with a message that the problem cannot be solved.

Fuzzy Logic. Also called "possibilistic logic," is an extension of Boolean logic whereby each possible value of any entity is represented by a membership function. This function captures the inherent vagueness that descriptive values often contain. For example, in the statement, "Bill is tall", "tall" would be a membership function that maps height values onto the real-number interval (0, 1). Then to determine the truth of "Bill is tall", one needs to insert Bill's actual height into the function to obtain a number indicating Bill's degree of "tallness".

Fuzzy Information. Data that is uncertain or without sharp edges. The level of truth is not exclusively either true or false, but may assume a subjective value between true and false.

General Problem Solver (GSP). An early AI system, significant as the first system to separate problem solving methods from knowledge about the problem. GPS's emphasis was to reproduce the "generality" believed at that time to be the key to intelligent problem-solving behavior.

Generate-and-Test. A very general problem-solving method in which one uses data about the problem to generate possible solutions and then applies some testing procedure that eliminates unlikely solutions.

Goal Driven. See **backward chaining**.

HEARSAY-II. An expert system for speech understanding developed in the 1970's at Carnegie-Mellon University. It is particularly significant because it introduced the blackboard architecture.

Heuristic. A rule of thumb, or private knowledge, that generates a relatively easy solution to a difficult or poorly understood problem. Unlike an algorithmic approach, a heuristic approach does not guarantee an optimal solution to a problem.

Heuristic Search. Searches that apply domain knowledge to limit the search space and to pick the path most likely to lead to a solution.

Hierarchy. A recursive tree structure that is composed of a single root node from which descendant nodes emanate. Each node forms a subtree.

Human-Factors Engineering. Designing and building a human factors model.

Human-Factors Model. A combination of hardware, software, and a user model that together define the interaction of machine and user.

Hybrid System. A system that combines several different knowledge representation methods and reasoning capabilities into a single integrated system.

Inexact Reasoning. Very few statements about real-world entities are known with certainty to be either true or false. A variety of techniques exist for representing and reasoning about uncertain propositions.

Inference Engine. See **reasoning engine**.

Inference Trace. A textual or graphical display of the inferences that were used to derive a conclusion.

Information. A collection of data that has at least some level of organization, but it remains unevaluated in terms of a justified belief in its truth.

Inheritance. The passing of knowledge from frame to frame down through a taxonomy from general to specific. For example, knowing that a car has wheels allows a Ford to automatically inherit wheels. The reuse of data and operations through specialization.

Instance. An example of a class or category that exhibits the class's characteristics.

Instantiation. Assigning specific values to variable elements in some knowledge structure, e.g. an if-then rule.

Intelligent Front End (IFE). See **front end**. A front end program that uses AI techniques to perform its task "intelligently".

Interpreter. See **reasoning engine**.

Introspection. The process of examining how one thinks or feels about things in the world, literally "to look inward".

Knowledge. The representation of real-world constructs, such as facts, rules, events, heuristics, and perception-based performance. The body of truth, information, and principles acquired through education, training, and experience. Knowledge means being aware of facts, and it also means understanding how to apply that factual knowledge.

Knowledge Acquisition. The process of collecting, organizing, and formalizing knowledge from some source, usually human experts, for use in a computer program—also, in a general sense, learning.

Knowledge Acquisition Program. A set of algorithms used to create a knowledge base. The program applies variably sophisticated techniques to elicit knowledge from a human expert and then structures that knowledge for use in a knowledge-based system.

Knowledge Base. A database that stores knowledge needed to solve problems in a specific domain.

Knowledge-Based Systems (KBS). Or Knowledge Systems. Computer programs that solve problems ordinarily requiring specialized knowledge. This knowledge need not necessarily be acquired from human experts. KBS are often characterized by their use of declarative languages, search, and knowledge representation.

Knowledge Elicitation. Knowledge acquisition applied to a human specialist. See **knowledge acquisition.**

Knowledge Engineer. The designer and builder of, in addition to KBSs, all systems that incorporate or emulate intelligence in their operation.

Knowledge Management. Knowledge management includes: assessing existing knowledge for gaps, filling those gaps with new knowledge, evaluating that new knowledge for consistency with existing knowledge, and delivering knowledge to its consumers (managers, scientists, policy makers, and the public at large). By managing knowledge in this way, we are able to treat the human knowledge base as a collection of commodities that can be examined, created, improved, and distributed.

Knowledge Model. An ontology for the knowledge that is particular to a knowledge-based system. It provides an unambiguous description of organized knowledge.

Knowledge Representation. Formalizing and storing knowledge into some structure (often using symbols) so that those symbols can be automatically processed by machine (computer).

Knowledge Source. Most generally, any repository of knowledge, e.g. books, people, software. More specifically, in the AI literature, a knowledge source is a software "specialist" that receives a segment of knowledge from the blackboard, and operates upon that knowledge independently of other knowledge sources. A control specialist sequences the access of knowledge sources to the main blackboard.

Language. A mode of structured communication that consists of some finite set of symbols (alphabet), a vocabulary of finite length strings of symbols (words), a syntax for organizing those vocabulary elements into formulae (sentences), and a semantics that attaches meaning to formulae of the language.

Learning. The cognitive activities involved in acquiring new knowledge and in applying existing knowledge to novel situations.

Line of Reasoning. A coherent sequence of facts and inferences that together explicate the steps followed in some particular thought process.

LISP. *Lis*t *P*rocessing, a symbol manipulation language that uses a small set of primitives to operate on lists, as the primary data structure.

Logic. Is a formal language that includes syntactic rules, a semantics, and an inference procedure.

Logic-Based. A knowledge representation scheme based on some form of propositional or predicate logic. Assertions and inferences are expressed as formulas in this logic. See **PROLOG**.

Logical Subsumption. A situation in which one set of clauses logically infers another. In knowledge bases where evidence is accumulated incrementally, subsumption must be recognized and avoided.

Machine Learning. An area of AI with the goal of building software that can learn from its own experience, by modifying its heuristics and knowledge base, and thereby reduce the narrow constraints or "brittleness" of knowledge bases. Machine learning research is intended to expedite the knowledge acquisition process.

Massively Parallel Computer. A computer characterized by an architecture that produces high-speed computations through a large number of individual processors operating relatively independently.

Means-End Analysis. Problem solving in which the differences between a current problem state and a goal state are all that is relevant to the process of solving the current problem (as opposed to scheme acquisition). The goal-state is the most critical feature of means-end problem-solving.

Membership Function. In fuzzy set theory, a function with the range (0, 1) that describes the degree to which a set of values are members of a fuzzy set.

Meta-cognition. The capability to think, or introspect, about one's own thought processes.

Meta-Knowledge. Self-knowledge about what one knows. Knowledge about knowledge.

Meta-rule. A rule that prescribes the manner in which other rules should be employed.

Model. An archetype or pattern that is used to abstractly represent something. One can then use the model as a focal point for reasoning rather than reasoning directly about the phenomena of interest.

Module. A unit of system construction, usually defined to implement a single portion of the system.

Modus Ponens. A common syllogism used for inference, it states that if A is true and A implies B, then B is also true.

Modus Tollens. Also, referred to as the contrapositive inference syllogism, it states that if not B is true and A implies B, then not A is also true.

Monotonic System. A system in which propositions, once proven, never become invalid.

MYCIN. Rule-based system, developed at Stanford University in the 1970's, that diagnoses bacterial infections and recommends treatment. It is important because it represents the prototypical first-generation expert system and because is introduced certainty factors to deal with inexact reasoning.

Natural Language. The conventional method for exchanging information between people, e.g. Natural languages have evolved over time, through use, not by means of any formal specifications.

Natural Resources. Any physical, biological, or cultural asset that occurs in nature and may possess some measure of value, real or perceived.

Network. In general, a set of interconnecting lines, whose connecting points are often called nodes. The spatial organization of those lines and nodes implies some logical relationships between the concepts represented by the nodes. See **semantic network**.

Nonmonotonic System. A system of reasoning in which propositions are subject to retraction or revision over time.

Node. In general, a node is a point of intersection or ending. In abstract problem-solving terms, a node is a state such that each state has some logical relationship to certain other states. This logical relationship enables a decision procedure to traverse the set of states in some orderly and systematic way.

Object. A conceptual computer programming entity composed of imbedded variables and methods. Instance variables describe the object, and methods indicate the type of functions performed by the object. Because instance variables and methods are hidden, inter-object communication is performed through message passing. Messages indicate what functions are to be performed by the object, but not how they should be performed.

Object-Oriented Programming. A program that associates all data and functions with objects in the application. Messages are sent to objects requesting a particular function or service. The object's response uses a method or function to provide the requested action. The requesting message provides the necessary data from data stored with that particular object, or other objects. Unlike function or subroutine calls, the object receiving the message defines the function.

Operations Research. The application of mathematical methods to find optimal solutions for complex systems. Critical variables of the system are related mathematically in an objective function, which is optimized subject to certain constraints.

Procedural Knowledge. Knowing how to use factual knowledge to solve problems. See **meta-knowledge**.

Procedural Program. Useful in a knowledge-based system to perform numerical calculations, to structure tasks, to handle initialization and input-output, and to handle portions of the reasoning process that are not heuristic.

Procedure-Driven. Program execution follows a procedural set of steps pre-defined by an algorithm.

Productions. Pattern-action pairs whereby the action is performed whenever the pattern is matched.

PROLOG. *Pro*gramming in *Log*ic, a symbol-manipulation language that is based on first order predicate logic and uses resolution as a deduction mechanism.

Proof of Concept. A small to medium-sized program designed to demonstrate that a selected application can be successfully addressed by knowledge-based system technology. Ideally, it will not take more than a few months to build, and will demonstrate that a narrow section of the application's knowledge can be represented in software, installed on a computer, and used to answer queries.

Prototype. The next stage of a coded KBS after proof of concept; it is an evolving piece of software that exhibits some correct and complete knowledge representation and use of heuristics, but has yet to undergo extensive verification, validation, and user acceptance testing in an operational setting.

Qualitative Reasoning/Model. A way to describe phenomena in terms of causal, compositional, or subtypical relationships among objects and events. Qualitative models are represented by networks based on prototype subsumption, state transition, and structural and procedural composition.

Rapid Prototyping. An approach to the development of KBSs that involves quickly developing small-scale prototype systems as a basis and focal point for further system development. It allows knowledge engineers to quickly ascertain whether they are on the right track, and to verify that appropriate KBS building tools and knowledge representations, are used.

Reasoning Engine. The component of a rule-based system that embodies the inference and control mechanisms. It selects rules to evaluate, interprets the rule's instructions, matches the rules premises to working memory elements, and carries out the action(s) specified in the rule. Also, referred to as an inference engine.

Recursion. A process whereby an operation is performed using a previous result based on the same operation. A recursive process requires (1) some terminating condition and (2) some way to specify the result of the current operation using the result of the previous instantiation of the operation.

Reductionism. The philosophy that all higher order phenomena can be explained by lower level laws.

Resolution. An automated technique for inferencing using formulae of predicate logic in Horn clause form.

Rule. In general, a law or regulation that governs behavior or operations. In KBSs, an antecedent-consequent pair, which can support inference.

Rule-Based. A knowledge representation scheme identified by the use of *if-then* rules to express domain knowledge.

Rules of Inference. A specification of how to derive conclusions from premises; a syllogism.

Satisfice. Achieve *any* solution that satisfies all imposing constraints. This is in contrast to optimization methods that find the single "best" solution.

Schema or Frame. A conceptual organization that represents a stereotyped situation, event, or object. Describes a present-state framework into which actions are organized, a structure into which new information can be incorporated or accommodated.

Script. A frame-like knowledge representation structure in which the entities being represented are sequences of events. See **frame**.

Search. The process of examining different solutions and different solution paths in a problem space.

Second Generation Knowledge-Based Systems. KBSs that use deep knowledge or models to reach a solution and explain reasoning.

Semantic Networks. A knowledge representation scheme where nodes, representing objects and other entities, are connected by directed labeled arcs representing the relationships between the nodes.

Semantics. Specifies the meaning of syntactic structures in a language.

Semi-Algorithm. An algorithm that does not always terminate in a solution or a message that the problem cannot be solved.

Sensitivity Analysis. An analytical technique in which a single parameter is incrementally increased and decreased, to examine the effect of the parameter on total system behavior.

Shell. The reasoning engine, knowledge representation, and support facilities of an knowledge-based system after all of the domain specific knowledge has been removed.

Solution Space. See **decision space**.

Slot. Location where knowledge about attributes of a frame can be stored. For example, if a frame represents an automobile, a slot for color could take on any of the possible values of "color".

State Space. Many problems can be described by states which represent one's position relative to a start state and the goal states. The collection of states forms a network and is referred to as the space of all possible states in the problem. For most significant problems, a state space is very large, e.g., possible board positions in chess.

Surface Knowledge. Anecdotal and experiential knowledge that does not rely on any deep understanding of the subject matter. See **heuristic**.

Symbol. A physical pattern representing information–usually occurring in a structure of related symbols. An abstract entity which may be a letter, digit, punctuation mark, or even a word when viewed as a single symbol by an automaton.

Symbolic Data. A string of characters that stands for some real-world concept.

Symbolic Manipulation. Performing operations on symbols and symbol structures with the belief that those operations represent the processing of the actual meanings of those symbols.

Syntax. Pertaining to the form or structure of a symbolic expression, as opposed to its meaning or significance. (Contrast **semantic**).

Temporal Logic. System of symbolic logic that accommodates the dynamic nature of objects and events through time.

Tracing. A facility provided by a reasoning system that keeps a record (trace) of all reasoning activities; it allows the programmer and user to determine how the system arrived at its conclusions.

Transition. An established procedure that allows movement from one state to another in a state space search.

Tree. See **hierarchy**.

Truth Maintenance System (TMS). A system of symbolic logic that preserves consistency in beliefs that change over time.

Truth Value. A value assigned to some statement indicating belief in its truth.

Turing Test. A hypothetical test, created by Alan Turing, in which a human must judge whether an unidentified entity that it is blindly communicating with is a human or computer. Failure or indecision by the human judge means that the computer has behaved as intelligently as the human competitor. Turing claimed that this test can be used as a definition of intelligence.

Usability. Evaluation of software to determine how well it fits the needs of the intended user group. This includes consideration of human factors issues, operational environment, and organizational needs.

User Interface. In our formulation, the user interface consists of the physical medium that the computer provides. This includes hardware and software.

Validation. Assurance that the software actually helps the user as intended in the original requirements and objectives and that it performs to an acceptable level of competence.

Verification. Checking the internal consistency and completeness of software to assure that it operates according to specifications.

World View. Dynamic model of a task or problem that reflects an individual's grand perspective of the subject area. This model may include conceptual abstractions, data, procedural knowledge, uncertainties, temporal relationships, strategies, and heuristics.

Working Memory. The component of a rule-based system that contains the set of facts or assumed propositions.

INDEX